THE
INTERNATIONAL SERIES
OF
MONOGRAPHS ON PHYSICS

International Series of Monographs on Physics

112. R. M. Mazo: *Brownian motion: fluctuations, dynamics, and applications*
111. H. Nishimori: *Statistical physics of spin glasses and information processing - an introduction*
110. N. B. Kopnin: *Theory of nonequilibrium superconductivity*
109. A. Aharoni: *Introduction to the theory of ferromagnetism: second edition*
108. R. Dobbs: *Helium three*
107. R. Wigmans: *Calorimetry*
106. J. Kübler: *Theory of itinerant electron magnetism*
105. Y. Kuramoto, Y. Kitaoka: *Dynamics of heavy electrons*
104. D. Bardin, G. Passarino: *The standard model in the making*
103. G. C. Branco, L. Lavoura, J. P. Silva: *CP violation*
102. T. C. Choy: *Effective medium theory*
101. H. Araki: *Mathematical theory of quantum fields*
100. L. M. Pismen: *Vortices in nonlinear fields*
99. L. Mestel: *Stellar magnetism*
98. K. H. Bennemann: *Nonlinear optics in metals*
97. D. Salzmann: *Atomic physics in hot plamas*
96. M. Brambilla: *Kinetic thoery of plasma waves*
95. M. Wakatani: *Stellarator and heliotron devices*
94. S. Chikazumi: *Physics of ferromagnetism*
92. J. Zinn-Justin: *Quantum field theory and critical phenomena*
91. R. A. Bertlmann: *Anomalies in quantum field theory*
90. P. K. Gosh: *Ion traps*
89. E. Simánek: *Inhomogeneous superconductors*
88. S. L. Adler: *Quaternionic quantum mechanics and quantum fields*
87. P. S. Joshi: *Global aspects in gravitation and cosmology*
86. E. R. Pike, S. Sarkar: *The quantum theory of radiation*
84. V. Z. Kresin, H. Morawitz, S. A. Wolf: *Mechanisms of conventional and high* T_c *superconductivity*
83. P. G. de Gennes, J. Prost: *The physics of liquid crystals*
82. B. H. Bransden, M. R. C. McDowell: *Charge exchange and the theory of ion–atom collision*
81. J. Jensen, A. R. Mackintosh: *Rare earth magnetism*
80. R. Gastmans, T. T. Wu: *The ubiquitous photon*
79. P. Luchini, H. Motz: *Undulators and free-electron lasers*
78. P. Weinberger: *Electron scattering theory*
76. H. Aoki, H. Kamimura: *The physics of interacting electrons in disordered systems*
75. J. D. Lawson: *The physics of charged particle beams*
73. M. Doi, S. F. Edwards: *The theory of polymer dynamics*
71. E. L. Wolf: *Principles of electron tunneling spectroscopy*
70. H. K. Henisch: *Semiconductor contacts*
69. S. Chandrasekhar: *The mathematical theory of black holes*
68. G. R. Satchler: *Direct nuclear reactions*
51. C. Møller: *The theory of relativity*
46. H. E. Stanley: *Introduction to phase transitions and critical phenomena*
32. A. Abragam: *Principles of nuclear magnetism*
27. P. A. M. Dirac: *Principles of quantum mechanics*
23. R. E. Peierls: *Quantum theory of solids*

Brownian Motion

Fluctuations, Dynamics, and Applications

Robert M. Mazo
University of Oregon

CLARENDONPRESS · OXFORD
2002

Great Clarendon Street, Oxford OX2 6DP
United Kingdom

Oxford University Press is a department of the University of Oxford. It furthers the University's objective of excellence in research, scholarship, and education by publishing worldwide. Oxford is a registered trade mark of Oxford University Press in the UK and in certain other countries

First published 2002
First published in paperback 2009
Reprinted 2014

Published in the United States of America by Oxford University Press
198 Madison Avenue, New York, NY 10016, United States of America

British Library Cataloguing in Publication Data
Data available

Library of Congress Cataloging in Publication Data
Data available

ISBN 978-0-19-955644-1

PREFACE

Brownian motion was discovered early in the nineteenth century and explained in the early years of the twentiethth century. Now, at the beginning of the twenty-first century, it is still a subject of considerable interest and current research. It has applications in chemistry, physics, mathematics, engineering, biology, finance, and perhaps elsewhere. This book is an attempt to present the theory of Brownian motion in a way understandable to readers with training in physics and chemistry. In particular, a background in statistical mechanics equivalent to that obtained from a graduate course is assumed.

I do not feel competent to discuss the ramifications and applications of Brownian motion theory to all the fields just mentioned. The discussion in this volume is limited, for the most part, to applications to chemistry and physics. Although a few passing remarks may be made about other fields, I have not attempted any substantive treatment of them.

There is a standard source on the theory of Brownian motion, a review article by S. Chandrasekhar (1943). This has long been the point of entry for new students of the subject and can still serve that purpose today. But it is, after all, 58 years old. Much new has been learned in the past half-century, and it is my hope that this book, which tries to present some of it, will be a worthy companion to Chandrasekhar's review.

Since this is a book on theory, and uses mathematics freely, a few remarks about Brownian motion in the field of mathematics is in order. Brownian motion in mathematics is a subfield of the theory of probability. It concerns the behavior of an important class of stochastic processes. Convergence of sequences, measure theory, existence theorems, etc. are its concerns. I do try to give the flavor of this subject without most of the technical details. I have been incomplete, but I have tried hard not to be incorrect, so that the reader who wishes to pursue the mathematical theory from other sources will not have to unlearn anything. This book is a book on physical science, not on mathematics. The level of mathematics used should not be beyond that of a graduate student in theoretical physics or theoretical chemistry. It is not meant for the mathematician who wishes to study random processes. For example, the concept of martingale, very important in the mathematical theory, does not appear in this book.

Even in the domain of the physical sciences, the book is restricted to Brownian motion that can be described by classical mechanics. There have been studies of Brownian motion in quantum mechanics, but to cover these properly with the requisite treatment of needed background material would have increased the size of the book considerably. Since most of the existing applications and most of the existing theory are for classical systems I decided to limit the book to this area.

The book begins with a historical account of the development of Brownian motion from its experimental discovery by Robert Brown to the quantitative experiments of the early 1900s. Chapter 2 is a treatment of the probability theory needed for the remainder of the book. It is a rapid run through, but I believe it contains the required background.

Chapter 3 generalizes probability from static to dynamic processes; it introduces the theory of stochastic, or random, processes. These are processes evolving in time in a way governed by probability, rather than causal, laws. As stated above, we do not do the rigorous theory full justice.

Chapter 4 begins the discussion of the physical theory. In it we present the theories of Albert Einstein and Marian von Smoluchowski, who were the first to show convincingly that Brownian motion is a manifestation of molecular fluctuations.

Chapters 5 and 6 return to mathematical subjects. First we investigate what is meant by a differential equation for a random process, and then discuss the relationship between two different answers. Then we study a class of integrals over Brownian paths introduced by Norbert Wiener in his atttempt to put Einstein's theory on a rigorous mathematical footing. We also discuss Feynman's integral over paths in quantum mechanics. Strictly speaking this has nothing to do with Brownian motion, but the analogy is so strong that I decided to include a brief discussion.

In Chapter 7, I attempt to put some flesh upon the bare bones of the foregoing chapters by treating a number of soluble examples with considerable physical interest.

The next chapter, Chapter 8, presents Smoluchowski's generalization of Einstein's theory to the case when there is an external force acting on the Brownian particle. A particular application, Hendrik Kramers' theory of the passage of particles over potential barriers, has become important in condensed phase chemical kinetics, and I had to make a choice of how much space to devote to it. I decided not to discuss applications in detail, nor to dwell on its many elaborations. This may disappoint some readers, but is consonant with the way other applications are treated here.

Chapter 9 introduces a topic in the theory of probability, the random walk. This is essentially the study of the properties of the sum of independent, identically distributed, random variables as functions of the number of terms in the sum. It is a useful model for the trajectory of a Brownian particle. Entire books exist on this subject, and I treat only those parts of it directly relevant to our theme.

The next three chapters, 10 through 12, bring in the tool of statistical mechanics. They treat attempts to derive the phenomenological theory, on which the previous part of the book was based, from a microscopic, molecular, theory. There has been much success in this endeavor, but there are still some open questions.

Chapters 13 through 16 are devoted to varius applications of the theory thus far developed. There are, of course, many more applications that could have been included; some choices had to be made to keep the book within a reasonable size. The choices represent a compromise between my interests, my estimate of the topics' interest to others, my estimate of their importance, and the limits of my knowlege. The reader will have to decide if I have chosen wisely.

Up to this point, the book has treated systems with a single Brownian particle in an ambient medium. In Chapter 17 we study the properties of Brownian motion when the system is concentrated enough that interactions between Brownian particles become important. We restrict ourselves, as does most work, to semi-dilute solutions, that is, the regime where only the lowest order (in concentration) corrections to single particle behavior are important. The subject of this chapter is an important area of current research.

Thus far, the book has been concerned with ensembles of particle paths, and averages over them. In the final Chapter 18 we study certain aspects of individual realizations of Brownian motions. The three topics treated here are Brownian dynamics, a numerical technique, the fractal nature of Brownian paths, and deterministic chaos in Brownian systems. Brief introductions to the requisite mathematics are included. These are areas in which I expect to see more progress in the near future.

I should like to thank many colleagues for discussions of many of the topics treated here over many years. In particular I should like to thank Richard Bearman, Arieh Ben Naim, Jeffrey Cina, Ewa Gudowska-Nowak, Marina Guenza, John Leahy, Andrew Marcus, and Christian Van den Broeck for reading and criticising an early draft of the manuscript. Ruth Karras criticized the first chapter with the eye of a historian. Of course, I am solely responsible for any errors, omissions or infelicities that remain. Richard Koch and Jan Blankenship helped by teaching me how to put the manuscript in LaTeX form. The reference librarians at the University of Oregon Science Library and Lucylla Rutkowska of the Jagellonian University were of great help in locating some of the more obscure references. Finally, I should like to thank my wife Joan for putting up with me while I was writing this book.

Those who consider a thing proved
simply because it is in print are fools.
Maimonides

Eugene, Oregon R. M. M.
July 2001

CONTENTS

1

HISTORICAL BACKGROUND

1.1 Robert Brown

In the year 1803, Napoleon sold France's North American colonies to that new nation, the United States, for the small sum of sixty million francs. This sale is known in American history books as *the Louisiana Purchase.* President Thomas Jefferson, wishing to find out exactly what he had bought, sent out an expedition of exploration under the leadership of Meriweather Lewis and William Clark. Among the contributions of the Lewis and Clark expedition, one of the least significant would have the greatest impact, albeit indirectly, on physics and chemistry.

The Lewis and Clark expedition left in 1804, reached the Pacific Ocean in November 1805, and returned in 1806. The story of the expedition, based on Lewis' and Clark's journals, makes a marvelous adventure story. In addition to geographical and ethnographical information, the expedition also brought back botanical specimens. A genus of plants from among these specimens was named after Captain William Clark. One member of this genus was *Clarkia pulchella,* a wildflower found in the Pacific Northwest of the United States.

Specimens of *Clarckia pulchella* were brought to England in 1826 by the Scots botanist David Douglas, who made an expedition to the Pacific Northwest in the years 1824–1828. It is he after whom the magnificent Douglas fir (*pseudotsuga menziesii*) is named.

By the year 1827, Robert Brown (1773–1858) was a renowned botanist. Darwin (much later) referred to him as '*facile princeps botanicorum.*'(Darwin 1896). Mrs. Darwin,quoted about a dinner party in 1839, said

> Mr. Brown, who Humbolt calls 'the glory of Great Britain' looks so shy, as if forced to shrink into himself, and disappear entirely.(Browne 1950)

As a young man, Brown studied medicine at Edinburgh, but never finished his studies nor took a degree. He enlisted in a newly raised Scottish regiment and was posted to Ireland, where he was appointed Surgeon's Mate, although he seems to have spent more time collecting botanical specimens than attending to patients.[1]

Brown acquired some reputation as a botanist, and had come to the attention of Sir Joseph Banks who was organizing an expedition to Australia, or, as it was then called, New Holland. Banks had need of a botanist for the expedition, and

[1] The main source here for Brown's life is Mabberly (1985). Much of the scientific background is from Brush (1968). See also Kerker (1974).

offered the position to Brown; Brown's medical experience no doubt weighed in his favor. Robert Brown accepted as soon as he could sever his connection with the army, and his formal career as a botanist was begun.

In addition to collecting and classifying, Brown made several important discoveries in botany. Perhaps the one most celebrated by biologists is his discovery that plant cells are eukaryotic, that they have a nucleus. Among physical scientists and mathematicians, however, he is known primarily for the eponymous motion associated with his name

In 1827, Brown was investigating the way in which pollen acted during impregnation. He wanted to use non-spherical grains, in order to be able to observe their orientation. The first plant he studied under the microscope was *Clarkia pulchella*, whose pollen contains granules varying from about five to six microns in linear dimension. It is these granules, not the whole pollen grains, upon which Brown made his observations. Concerning them, he wrote

> While examining the form of these particles immersed in water, I observed many of them very evidently in motion... These motions were such as to satisfy me, after frequently repeated observation, that they arose neither from currents in the Fluid, nor from its gradual evaporation, but belonged to the particle itself.(Brown 1828)

This inherent, incessant motion of small particles suspended in a fluid is nowadays called *Brownian motion* in honor of Robert Brown. Similar observations had, in fact, been made earlier by other workers. Brown, however, was the first to give them serious scientific study, and to show that, in fact, they were not due to the living origin of the moving particles. It can be said that Brown, the botanist, showed that the phenomenon was not one of biology, but one of physics.

For example, Brown had strongly illuminated the specimens under his microscope and hence had heated them. This caused evaporation of the ambient fluid, and Brown asked whether this evaporation might be causing the motion which he observed. To answer this experimentally, he made a mixture of water containing particles with an immiscible oil and shook the mixture; small drops of water were formed in the oil, some containing only a single particle. These were stable and did not evaporate for some time. He wrote

> But in all the drops thus formed and protected, the motion of the particles takes place with undiminished activity, while the principal causes assigned for that motion, namely, evaporation, and their mutual attraction and repulsion, are either materially reduced, or absolutely nil.

Brown had his results printed in a pamphlet, entitled *A brief account of microscopical observations made in the months of June, July, and August, 1827 on the particles contained in the pollen of plants; and on the general existence of active molecules in organic and inorganic bodies*. This work was originally intended for private circulation, but was reprinted in the archival literature shortly after its appearance (Brown 1828). Brown used the word *molecule* in the title in a sense different from its current one. It referred to earlier teaching of the Comte de Buffon (1707–1786) who introduced this word for the ultimate constituents of the bodies of living beings. This had nothing to do with the later development of

Dalton's atomic theory in which the word *molecule* took on its modern meaning.

Brown's pamphlet must have had a popular vogue. In George Eliot's novel *Middlemarch*, published in 1872 but set in the years 1829–1832, she has one character, Fairbrother, say to another, Lydgate,

'Do you want to make an exchange? ... I have some sea mice— fine specimens—in spirits, and I will throw in Robert Brown's new thing, *Microscopic Observations on the Pollen of Plants*— if you don't happen to have it already.(Eliot 1989)

Brown published a second paper on the motion (Brown 1829). In it he reported the experiments on the oil –water emulsion mentioned above, and discussed previous observations by others which could have been interpreted as prior to his. He claimed that these previous observations were of motile animalcules or organisms. In one case in which an author claimed to have observed animated inorganic particles, Brown states that the observation is suspect because the author was susceptible to 'optical illusions'.

There was, however, one predecessor whom Brown overlooked. In July of 1784, Jan Ingen-Housz published a short paper entitled *Remarks on the use of the microscope* (Ingen-Housz 1789, 1–5) that contains the following lines[2]

...one must agree that, as long as the droplet lasts, the entire liquid and consequently everything which is contained in it, is kept in continuous motion by the evaporation, and that this motion can give the impression that some of the corpuscles are living, even if they have not the slightest life in them. To see clearly how one can deceive one's mind on this point if one is not careful; one has only to place a drop of alcohol at the focal point of a microscope and introduce a little finely ground charcoal therein, and one will see these corpuscles in a confused continuous and violent motion as if they were animalcules which move violently around.'

On the basis of these few lines, Van der Pas (1971) has argued that Ingen-Housz should be accorded priority in the discovery of Brownian motion. However, although Ingen-Housz doubtless observed the motion, he did not follow up his observation with any investigation of it, as did Brown. His ascription of the motion to evaporation was, as Brown showed in a simple experiment, quite wrong. Robert Brown still deserves pride of place with respect to the discovery of the motion now named after him.

1.2 Between Brown and Einstein

In the period between 1829 and about 1900 not much progress was made in the understanding of Brownian motion. For the first thirty years after Brown's observations, general interest in the phenomenon almost disappeared. In the 1850s, developments in the theory of heat and kinetic theory stimulated new experiments and conjectures. It is striking, however, that the founders and main developers of kinetic theory, Maxwell, Boltzmann, and Clausius, never published anything on Brownian motion.

[2]Van der Pas' translation from the original French.

Let us give a brief summary of some of the ideas and conjectures of that period; some of the names mentioned may be familiar because of their other contributions to physics; others will doubtless be new to the reader.

In 1858, Jules Regnault suggested, on the basis of his experiments, that the motion was due to the absorption of light, locally heating the fluid and giving rise to microscopic streaming motions or currents. Christian Wiener argued in 1863 that the motion must be attributed to internal molecular motions in the fluid. This has led some to yield to him the honor of being the first to suggest the true nature of Brownian motion. However, it must be mentioned that Wiener's ideas of molecular motion were based on the concept of two kinds of atoms, material atoms and aether atoms. His concept of Brownian motion was intimately bound up with the properties of these aether atoms, not a point of view which would be accepted today.

In the period 1874–1880, three Belgian Jesuits, J. Delsaux, J. Thirion, and X. Carbonelle argued that molecular fluctuations lay at the root of the phenomenon. There is a distribution of molecular velocities which will give rise to fluctuations of the fluid density, and consequently the pressure, on a local, microscopic scale. These fluctuations average out on the macroscopic scale. These three workers had the right idea, but did not work out any quantitative consequences which could be compared to experiment. They also did not realize the implication of their ideas for the second law of thermodynamics; perhaps this is not surprising, since statistical mechanics was yet to be developed at the hands of Gibbs and Boltzmann.

In 1888, Léon Gouy also introduced the idea of fluctuations, and suggested that Brownian motion might be an exception to the second law. Attention was brought to Gouy's ideas by Henri Poincaré. It is interesting that many of the ideas of the modern quantitative theory of Brownian motion were worked out in the 1900 doctoral dissertation (in mathematics) of Louis Bachelier, upon whose examining committee Poincaré sat. The subject of the thesis was a stochastic theory of speculation on the stock market, hardly a topic likely to excite interest among physical scientists (or among mathematicians, either). Although it contained many innovative mathematical ideas, it was largely ignored.

From its discovery, Brownian motion has been the subject of experimental investigation as well as theoretical speculation. We have already mentioned Brown's experiments designed to test the influence of evaporation. In fact, a great many of the speculations on the cause of the motion were the results of experimental work by their authors: the effect of solvent viscosity, of temperature, of added electrolyte, etc. One other kind of experiment had far reaching implications although they were not apparent until much later: the attempt to measure the velocity of the moving particles.

Christian Wiener and Sigmund Exner tried to measure the Brownian particle velocity by measuring the displacement of the particle in a cross-hatched field of view. Many years later, Felix Exner, Sigmund's son, repeated his father's measurements using the same technique, both for sample preparation and for measurement. But this way of measuring the velocity (by measuring $\Delta x/\Delta t$) gave puzzling results; the ratio of displacement to time interval depended on the time interval (which is normal) even when the time interval became very short (which is not normal). The difference quotient did not seem to have a limit for small times. It was not until Einstein's work of 1905 that this behavior was properly understood and it was realized that the velocity is not a useful thing to measure.

The reason is that the path of a Brownian particle, on the length scales available from observation through a microscope, is an extraordinarily irregular curve. In modern language one would say that it is a fractal, statistically self similar on all scales. Such curves are differentiable almost nowhere. Consequently the particle whose trajectory they represent has no velocity, as usually defined. We shall discuss the fractal properties of Brownian paths in Chapter 18.

1.3 Albert Einstein

Albert Einstein (1879–1955) is certainly the best known physicist of the twentieth century, perhaps even of all time (though Isaac Newton might be a competitor for this latter distinction). His fame lies mainly in the special and general theories of relativity, which he published in 1905 and 1916 respectively. It is not widely recognized that, in his younger days, Einstein was greatly interested in molecular physics. He made important contributions to this area, and the theory of Brownian motion is among these.

In 1900 Einstein finished his studies at the ETH (*Eidgenossische Technische Hochschule*) in Zurich. Unable to obtain a position as assistant, which would have enabled him to continue his studies, he supported himself by taking temporary secondary school teaching positions. Eventually, through the help of Marcel Grossman, a friend and future collaborator, in 1902 he obtained a position in the Swiss Patent Office as Technical Expert, Third Class. He found the work interesting, and it gave him sufficient free time to pursue his own interests in physics.

In the year 1905, Einstein published four papers that at once raised him to the rank of a physicist of the highest caliber. These were: the photon hypothesis (for which he later received the Nobel Prize), the first paper on Brownian motion, and his first two papers on relativity theory. He did not get the Ph.D. degree until the following year, 1906; interestingly, his thesis was on none of the above problems, but concerned the determination of the dimensions of molecules. In this work he developed the celebrated formula relating the viscosity of a suspension of particles to their volume fraction.

The first paper on Brownian motion was entitled *On the motion of particles suspended in a resting fluid demanded by the molecular theory of heat* (Einstein 1905). The introduction of this paper contains the words

'It is possible that the movements to be discussed here are identical with the so called 'Brownian molecular motion'; however the information available to me regarding the latter is so lacking in precision that I can form no judgement in the matter.'

A second paper (Einstein 1906a), however, was entitled *On the theory of Brownian motion.* In the period between the two publications he had apparently been convinced of the relevance of his considerations for the understanding of the phenomenon associated with the name of Brown. Einstein published two additional short papers on Brownian motion (Einstein 1907, 1908), but these were of relatively minor interest.

The 1905 paper is a remarkable document. While its notation and nomenclature seem a bit archaic, the ideas and their development are surprisingly modern. Einstein first argues that particles suspended in a fluid must exert the same osmotic pressure as particles dissolved in the fluid, the concentrations being the same. Thus the molecular-kinetic theory of heat requires that suspended particles possess a motion of the same character as that of dissolved ones. He thus realizes that the theoretical problem is really one of probability theory.

Einstein assumes that the position of a suspended particle as a function of time, in a time-coarse-grained sense, is a Markov process. He then writes down the Chapman–Kolmogorov equation for the transition probability,[3] and expands, assuming that the spatial range of the transition probability is small in a short time interval. The final result is a diffusion equation for the probability density of the position of the suspended particle. Of course, the concepts of Markov process and Chapman– Kolmogorov equation did not yet exist in 1905; we have imposed modern names on the techniques which Einstein independently utilized. This only emphasizes the originality of his concepts. We shall treat Einstein's development in detail in Chapter 4.

The solution of the diffusion equation was well known, and Einstein could immediately deduce that the mean square displacement of the suspended particle grows linearly with the time. The proportionality coefficient is just twice (for the one-dimensional case considered) the self-diffusion coefficient of the particle. The last section of the paper is concerned with the possible use of the relations derived to determine molecular dimensions, a subject of interest to Einstein at that time, and to which he returned again later in the theory of light scattering.

The result that the mean (actually root-mean-square) displacement of a particle is proportional to the square root of the elapsed time explains why previous attempts to measure the velocity of a particle undergoing Brownian motion were unsuccessful. In fact, if such a relationship were to hold down to the shortest times, the path of the particle would be non differentiable, and the velocity of the particle would not exist. Einstein was fully aware of this, and discussed it in his second paper (Einstein 1906b). He recognized that his assumption that the displacements of the particles in non-overlapping time intervals were statistically independent breaks down for very short times. The generalization to short times was given some years later by Ornstein (1920) and Fürth (1920).

The 1906 paper of Einstein included some technical improvements on the methods of the earlier paper, a treatment of the rotational Brownian motion of

[3] We shall define these terms in Chapter 3.

a spherical particle, and considerations of the limits of validity of the assumptions which were made in the theory.

After his work on Brownian motion, Einstein's interest turned to other areas of physics: quantum theory and relativity theory. He published a few papers in molecular physics in later years, but these were less important than the work of 1905 and 1906, and the bulk of his later research.

The papers of Einstein and of Smoluchowski, of which more below, are usually considered to have given the impetus to the experimental work which corroborated their predictions in all details. However, it is worth pointing out that Theodor Svedberg (1906) published an experimental paper on this subject in 1906, too early to have been influenced by Einstein and Smoluchowski. It was one of the earliest papers of this future Nobel laureate. The definitive experimental work was done by J. Perrin and his school (Perrin 1908a,b); it was definitely influenced by Einstein and Smoluchowski's work. It will be described after the contributions of Smoluchowski are introduced.

1.4 Marian von Smoluchowski

Marian von Smoluchowski (1872–1917) was born in Vienna. During his lifetime, Poland was not an independent state; it was partitioned between Russia, Prussia, and Austria. Smoluchowski's father was a lawyer from Krakow, in the Austrian part of Poland, who became an important civil servant of the Chancellery of the Austro-Hungarian Empire in Vienna. Marian entered the University of Vienna, where, among other things, he attended the lectures of Boltzmann; in later life he was called *der geistige Nachfolger Boltzmanns*, the intellectual successor of Boltzmann.

On the completion of his doctorate, he did post-doctoral work abroad. The years 1895–96 were spent in Paris with Lippmann, 1896–97 in Glasgow under Lord Kelvin, and 1897 in Berlin with Warburg. In 1898 he returned to Vienna, and in 1900 became professor in Lemburg (now Lvov). There he remained until 1913, when he accepted the chair of experimental physics at the Jagellonian University in Krakow. He received the offer of a chair at the University of Vienna, but Krakow kept him by making him rector (president) of the university. He died in 1917 as a result of the diphtheria epidemic of that year, aggravated by wartime conditions.

In an obituary, Sommerfeld (1917) compares Smoluchowski to Max Planck. He says that Planck is in the depth of his heart, a thermodynamician. Even though his quantum reform demands statistics, it is not a fundamental part of his manner of thinking.

> 'But for Smoluchowski, statistics were the breath of life; the second law represented only an approximate law, ...'

It seems to have been this love for molecular statistics which led him to the consideration of Brownian motion about 1900, though he did not publish his results until 1906 (Smoluchowski 1906a), under the impetus of Einstein's paper.

Smoluchowski first argued that the objections raised to a molecular-kinetic explanation of Brownian motion on the grounds that individual molecular impacts were too feeble to impart the observed velocities to particles were unjustified. While individual collisions indeed had very small effects, Smoluchowski showed, on the basis of a Bernoulli trial model, that fluctuations in the collision rate could give rise to velocities of the observed order of magnitude. He then adopted a more refined model, based on the free path kinetic theory of gases with persistence of velocities. He had had experience with persistence effects in prior work on heat conduction in dilute gases. Detailed calculations based on this model led him to a result that differed from that of Einstein only by a numerical factor of $\sqrt{(27/64)} = 0.65$ for the root-mean-square displacement. The dependence on the physical parameters of the particle and the ambient fluid were the same in the two theories. The discrepancy arose because the two authors had had to make approximations of different natures in carrying out the details of their respective theories. In later years, Smoluchowski referred to Einstein's numerical factor as the correct one.

Smoluchowski later studied Brownian motion in configuration space for particles under the influence of an external force (Smoluchowski 1913, 1915). In particular, he studied a constant force (e.g. gravity) and a linear restoring force.

Einstein's and Smoluchowski's scientific paths crossed again. The subject this time was the theory of the scattering of light near the critical state of a fluid, the critical opalescence (Einstein 1910; Smoluchowski 1908). Einstein also wrote a sympathetic obituary of Smoluchowski (Einstein 1917) with special reference to Smoluchowski's emphasis on the importance of fluctuations, both in the Brownian motion and the critical opalescence problems.[4]

1.5 Molecular Reality

Since the time of John Dalton (1766–1844) the idea that matter was made up of 'elementary' particles called atoms, and their unions, now called molecules, took strong hold in the scientific community. Especially after the First International Chemistry Conference in Karlsruhe in 1860, where Stanislao Cannizarro showed how the ideas of Amadeo Avogadro could be used to construct a rational table of atomic weights, atoms became a central idea in chemistry and physics. Yet there were skeptics.

Perhaps the most prominent of these skeptics were the physical chemist Wilhelm Ostwald (1853-1932) and the physicist Ernst Mach (1838–1916). These men argued that there was no direct experimental proof of the existence of atoms. To be sure, the atomic theory was in concord with the macroscopic facts of experimental physics and chemistry. They agreed that it was a useful way of summarizing experience. Yet the lack of direct experimental verification led them to

[4]Other sources of information on Smoluchowski's life and work are Ulam (1957) and Teske (1977).

deny atomic theory the dignity of an explanation of the nature of things. They maintained this attitude with great vigor.

Most scientists took the opposite point of view. The facts of chemistry and the kinetic theory of gases on the physical side provided such strong indirect evidence that they were willing to accept atoms as real.

The Einstein–Smoluchowski theory of Brownian motion provided ammunition for the atomists. How can one explain the incessant movement of the particles, which seems to contradict the second law of macroscopic thermodynamics? By fluctuations, say Einstein and Smoluchowski. Well, then, what is fluctuating? This can only be explained on the basis of atoms. The next key question is, what is the experimental evidence that the Einstein–Smoluchowski theory is quantitatively correct?

F. Exner and Th. Svedberg made early quantitative studies of Brownian motion (Exner 1900; Svedberg 1906) They did not have the Einstein and Smoluchowski results available; the experiments were not suitable for a detailed verification of the theory. This had to wait for the experiments of Jean Perrin (1870–1942), a convinced atomist.

These experiments were difficult. In particular, in order to study the dependence of the mean-square displacement of the radius of the particles, it was necessary to prepare a monodisperse suspension, not a trivial task. Perrin's early experiments showed substantial agreement with the theory (Perrin 1908a, 1908b), and he and his students continued refining the work. Later that year, Chaudesaiges, a student of Perrin, reported experiments in which Einstein's law was completely verified if he took Avogadro's number to be 6.4×10^{23}, close to the currently accepted value (Chaudesaiges 1908). Perrin published a long paper on his and his students' researches (Perrin 1909) and became an energetic proselytizer for the reality of atoms. He received the Nobel Prize in 1926 'for his work on the discontinuous structure of matter...'

Perrin was apparently also the first to realize that the path of a particle undergoing Brownian motion must have elements in common with the graph of a function which is not differentiable. Such functions had previously been studied by mathematicians and regarded as pathological cases, whose only importance was in illustrating what was really encompassed within the general concept of function. But now, according to Perrin, these so-called pathological functions can be seen to have a physical realization. This idea was taken up by mathematicians, and forms the basis for a branch of the theory of stochastic, or random, processes, itself a subfield of the theory of probability.

Perrin's experimental verification of the Einstein–Smoluchowski theory, together with the work of J. J. Thompson on the electron, was rapidly successful in persuading most of the anti-atomists that atoms really did exist. Ostwald, one of the most prominent skeptics, recanted in a new edition of his influential textbook (Ostwald 1909), saying

> this evidence now justifies even the most cautious scientist in speaking of the *experimental* proof of the atomic nature of space filling matter. What has up to now been

called the atomistic hypothesis is thereby raised to the level of a well founded theory ....

Only Mach was not convinced, and continued to consider the existence of atoms as only a hypothesis.

1.6 The Scope of this Book

In this chapter, we have outlined the ancient (nineteenth century) and early modern (first decade of the twentyth century) history of Brownian motion as a subject of scientific study. Our treatment has been brief. The impressive work by S. Brush (1968), which was an invaluable source for the preparation of this chapter, provides a thorough historical study of the matters treated here.

In addition to its important historical role in clarifying the importance of fluctuations and the meaning of the second law of thermodynamics, and its role in making the reality of atoms manifest, Brownian motion has proven to have many uses in science. It is a useful model for noise in physical systems. It provided the impetus for much of the theory of stochastic processes in mathematics. It gives a useful molecular model for the dynamics of macromolecules. It provides the basis for a numerical algorithm for simulation of molecular processes (called Brownian dynamics). And it even describes some aspects of the movements of the stock market. We shall attempt to describe all of these applications in future chapters, (except for the stock market).

The first chapter has deliberately been kept non-technical. There is not a single equation or formula in the chapter. As befits a story with historical pretensions, I have drawn on the original publications, as far as I have been able to determine them. Neither of these characterizations fits the rest of the book. It will be as technical as needed to present its message, and will refer to the most accessible or most complete exposition of the point at issue, often but not necessarily the original publication. Nevertheless, I have tried to present maintain some historical perspective in the references.

So let us now continue with our study of a field of chemistry and physics which has a remarkable pedigree, which is fairly well understood at the present time, and which still shows signs of future promising applications.

2

PROBABILITY THEORY

2.1 Probability

Most physical systems accessible to experimental observation consist of many particles: atoms, molecules, etc. In trying to understand their properties, we cannot hope to follow the motion of all the constituent particles in full detail.[5] The outcome of an experiment is not completely predictable from the limited data available about the initial condition of the system, even when the physical laws governing the system are known. Normally the best that one can do is to predict what will happen most of the time, and estimate how frequently deviations from this expectation will occur and what their magnitudes might be. The mathematical tool which has been developed to make such predictions is the theory of probability.

Let me be clear that we are not concerned here with the problem of estimating the true value of some physical quantity from a set of experimental data that inevitably includes some scatter. This is an important problem, but it is not the problem studied in this book. The methods used in attacking it are also based on probability theory, but the specialized probabilistic tools needed for its study are not treated here. This chapter reviews the concepts of probability theory that will be useful in the rest of this book. There is much more to the subject, which I shall not attempt to cover.

The mathematical setting for probability theory is the following. There is a space, Ω, called the *sample space*. The points of the sample space, designated by ω, are called *elementary events*. Subsets of Ω containing more than one point are called *compound events*, or just events. The idea is that the ωs are the possible results of experiments and Ω is the set of all possible results of experiments. For example, if the experiment is the throwing of a die, Ω consists of six points, which may be taken as the integers 1 ...6. Ω has 64 subsets, counting the empty set ϕ and the entire space, Ω .

To each of these subsets, A , there is assigned a number, $P(A)$ called the probability of A. This is supposed to represent the likelihood that the event A will occur in a given experiment or trial. How are these numbers to be assigned? There is a large philosophical literature on this question. For example, do probabilities reflect the degree of certainty or belief that an event has occurred or will occur? Or do they represent the relative frequency with which events will occur if the

[5]With modern high speed computers even this is possible for systems of several hundred particles. This is called the *molecular dynamics* method. But even here probability considerations enter through the initial conditions.

experiment is repeated sufficiently many times? There are other interpretations which have been offered also. Most textbooks on probability theory devote some space to discussion of the interpretation problem before beginning the technical development of the theory; a good brief treatment can be found in Gnedenko (1978). Most physical scientists adopt the relative frequency interpretation, and we shall do so here without further discussion.

The mathematical theory of probability is concerned with the relations between the probabilities of various sets, and the properties of various functions associated with these probabilities. To describe these relations, we need some notation. Let A and B be sets. We denote by AB the set of points common to both A and B. [6] It represents the simultaneous occurrence of both A and B. The symbols $A+B$ denote the set of points belonging either to A or to B or to both.[7] This corresponds to the occurrence of either A or B or both. The set of points in A but not in B is denoted by $A-B$. This corresponds to the occurrence of A and non-occurrence of B. The complement of a set, A, denoted by $\overline{A}$, is the set of points not in A; $\overline{A}= \Omega - A$. This corresponds to the non-occurrence of A. If AB is the empty set, ϕ, we say that the events A and B are *mutually exclusive.*

With these preliminaries, we can state the axioms at the foundation of probability theory. Even these are stated somewhat informally, since for internal consistency we must leave open the possibility that there are some subsets for which the probability may be undefined.

A probability function is a function, P , defined on the subsets of a space Ω with the following properties:

(1) $P(A) \geq 0$ for every A in Ω (for which P is defined).

(2) $P(\Omega) = 1$.

(3) If $AB = \phi$, then $P(A+B) = P(A) + P(B)$. That is, if two events are mutually exclusive, the probability of their joint occurrence is the sum of their individual probabilities.

(4) Let $\{A_j\}$ be any sequence of mutually exclusive results, $A_j A_k = \phi$ for $j \neq k$. Then $P(A_1 + A_2 + A_3 + \ldots) = P(A_1) + P(A_2) + P(A_3) + \ldots$. For spaces Ω with a finite number of points, this is a theorem, provable from (1)–(3). However, for infinite sample spaces it must be assumed as a separate axiom.

The reader acquainted with measure theory will realize that P is a measure on Ω with finite total measure, (normed to unity), given certain technical requirements on the sets A_j ; they must form what is called a σ-field or σ-ring (Halmos 1950). We shall not enter into these measure theoretic details here. Suffice it to

[6]In set theory this is usually denoted by $A \cap B$. The product notation is more common in probability theory. It is called the *intersection* or set theoretic product of the two sets.

[7]In set theory this is usually denoted by $A \cup B$ and is called the *union* of the two sets.

say that they lie behind the statement made earlier that there may be sets for which the probability is not defined.[8]

From the axioms of probability theory given above, one may derive a number of useful auxiliary formulae. For example

$$P(\phi) = 0 \tag{2.1.1}$$

$$P(A) = P(A\overline{B}) + P(AB) \tag{2.1.2}$$

$$P(A + B) = P(A) + P(B) - P(AB) \tag{2.1.3}$$

With these and similar results in hand, we can solve many problems dealing with finite sample spaces; that is, when the space Ω contains only a finite number of points. Such problems gave the original impulse for the development of probability theory; they describe the outcome of games of chance.

'Since traveling was onerous (and expensive), and eating, hunting and wenching generally did not fill the 17th century gentleman's day, two possibilities remained to occupy the empty hours, praying and gambling. Many preferred the latter. Hence it is not surprising that proficiency in the estimation of gambling odds kept close to the forefront of thestate of the combinatorial art and indeed motivated much of it." (Montroll and Schlesinger 1984)

But finite sample spaces will not suffice for the development of the theory of Brownian motion. Thus we may not dwell on the delightful problems associated with finite sample spaces, but proceed to introduce new concepts useful for infinite sample spaces.

When the sample space is finite, the meaning of the term 'at random' is usually clear. It means that the set of elementary events concerned all have the same probability. When the sample space is continuous, one must exercise more care in characterizing randomness.

A classical example, known as *Bertrand's paradox* illustrates this. What is the probability that a chord drawn at random in a circle has a length exceeding the side of an inscribed equilateral triangle? We give three solutions, illustrated in Fig. 2.1.

a) Fix one end of the chord on the circle and draw the diameter through the fixed end. All chords lying within ± 30 degrees of that diameter will be longer than the given chord. Hence the required probability is $1/3$.

b) Draw a diameter through the midpoint of the given chord. From the geometry of the construction it is clear that chords intersecting the diameter between $1/4$ and $3/4$ of its length will have the required length. Thus the required probability is $1/2$.

c) The midpoint of the chord can be anywhere in the circle. The chord's length will satisfy the given condition if its midpoint lies in a

[8]For ordinary Lebesgue measure on the line, it is known that there must exist non-measurable sets. These sets are very esoteric, and will not be met in practice.

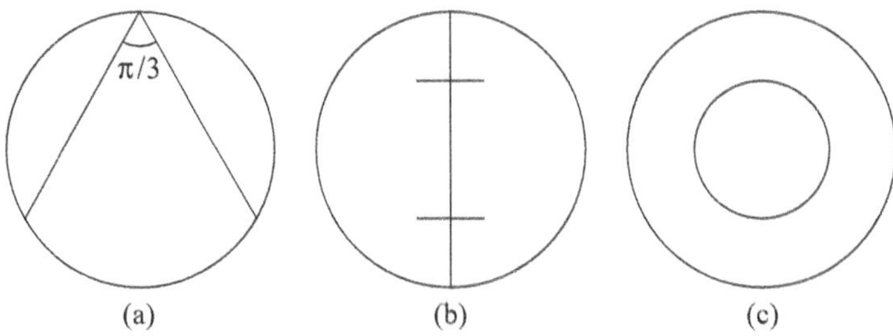

FIG. 2.1. Various solutions to Bertand's paradox. (a) End of chord picked randomly. (b) Midpoint of chord picked randomly along a diameter. (c) Midpoint of chord picked randomly in a circle.

> circle of radius 1/2. The ratio of the area of this circle to that of the original given circle is 1/4, which is the required probability.
>
> Three solutions: which, if any, is correct? The answer is that they are all correct. It is the statement of the problem that is at fault; the concept of choosing a chord 'at random' is not sufficiently clearly defined. Picking an angle at random, picking a point on a line at random, and picking a point in a two-dimensional area at random are not at all the same things. One must decide which meaning of the term 'at random' is intended. Only then can it be stated unambiguously which formulation of the problem is the one to be solved, and hence which answer is the correct one for the properly posed problem.

It is important to see more precisely why any of the three answers given to the Bertrand paradox can be correct, depending on the interpretation of the phrase 'at random'. The point to be emphasized is that the probability of a given event cannot depend on the variables used to describe that event. Let us see how this applies to interpretations (a) and (c). In interpretation (a), the probability that a chord's midpoint lies between r and $r+dr$ is $p_a(r)dr$. In interpretation (c), the probability is $p_c(r)rdr$ because for this interpretation, the midpoint must lie in an annular ring of width dr. Consequently, if the probabilities are to be equal

$$p_a(r) = rp_c(r)$$

Thus it is impossible that p_a and p_c are simultaneously constants.

On the other hand, picking a point randomly along as diameter means that p_a is constant. Picking a volume element randomly means that p_c is constant. Thus the two assumptions are inconsistent with each other, and it is not surprising from this point of view that they lead to different answers. The case of interpretation (b) can be dealt with in a similar manner.

2.2 Conditional Probability and Independence

So far we have been concerned with the probability of occurrence of certain events without regard to what other events may have occurred. A typical problem might be, what is the probability that a Brownian particle is found in the left-hand side of its container? But we may also be interested in the question, what is the probability that a Brownian particle is found in the left-hand side of its container if it is known that it was in that side a time T previously? This kind of probability is called *conditional probability.* To define it formally, we write

$$P(AB) = P(A \mid B)P(B) \tag{2.2.1}$$

$P(A \mid B)$ is the conditional probability that A occurs *given* that B has occurred. It is only defined if $P(B) \neq 0$.

A conditional probability is a genuine probability; it satisfies all of the axioms listed in Section 2.1. The sample space, however, is a different space than the space Ω of the unconditional probabilities. It is generally a subspace of the unconditional sample space.

As an elementary example, suppose you throw a fair die. The probability that the upper face bears more than three dots is 1/2. The probability that the face bears more than three dots if it is known that the number of dots is even is 2/3. In the first case the sample space is the integers $\{1, \ldots, 6\}$; in the second case the sample space is the points $\{2, 4, 6\}$.

We now turn to the concept of independent events. Qualitatively, we say that two events are independent if the occurrence of one has no influence on the occurrence of the other. Quantitatively, the events A and B are independent if

$$p(A \mid B) = P(A) \tag{2.2.2}$$

Alternatively

$$P(AB) = P(A)P(B) \tag{2.2.3}$$

It should be emphasized that the notion of independence is a property of the probability distribution being discussed, not of the set theoretical structure of the events.

The concept of conditional probability and independence can be extended to more than two events. We illustrate for the case of three events; the extension to more events is obvious. The conditional probability of C given A and B (that is, given AB) is defined by

$$P(ABC) = P(C \mid AB)P(AB) \tag{2.2.4}$$

The three events A, B, C are independent if

$$P(ABC) = P(A)P(B)P(C) \tag{2.2.5}$$

and, in addition, if the events A and B , A and C, and B and C are pairwise independent events.

Pairwise independence of the three events alone does not imply independence of the events. A simple example is the following. Suppose there is a circle divided into quadrants. One quadrant is colored red, one blue, and one green. The fourth has stripes of all three colors. One chooses a quadrant at random, say by throwing a tetrahedral die, or spinning a pointer.

Let R be the event 'red occurs in the quadrant chosen.' B and G are defined similarly for blue and green. S is the event 'the striped quadrant is chosen.'

Clearly $P(R) = P(B) = P(G) = 1/2$, since, for example, R occurs if either the red or the striped quadrants are chosen. Also $P(RB) = 1/4$, and similarly for the other pairs, while $P(S) = 1/4$. Now the event S and the event RBG are the same event. Consequently $P(RBG) \neq P(R)P(B)P(G)$, while $P(RB) = P(R)P(B)$, and similarly for the other pairs. Thus the events R , B , and G are not independent as a triple of events even though any two of them are independent.

2.3 Random Variables and Probability Distributions

In the elementary examples which we have used to illustrate concepts in the previous paragraphs, it was a simple matter to describe the events of the sample space in detail. In more complicated, and in particular more realistic, situations this is awkward, or just not possible. We do, however, often know the results of some numerical measurement carried out on the system. For example, suppose we have a box of electrical resistors, all nominally of the same resistance. Because of the vagaries of the manufacturing process, the actual resistances will differ from each other and from the nominal resistance. If a resistor is picked from the box at random, simultaneously a resistance value is necessarily picked at random from the set of resistance values of the resistors. The resistance is a real valued function defined on the set of resistors. Its value can be determined by experiment Thus we have a numerically valued function defined on the sample space.

Similar situations are very common; thus it is worthwhile to have a common vocabulary to describe them. A real or complex valued function defined on a sample space is called a *random variable.*[9] Random variables can also be vector valued, that is map a sample point onto an n-tuple of numbers.[10]

It is common practice in probability theory to denote random variables by bold face upper case letters, e.g. $\mathbf{X}$, and their numerical values by lower case medium letters, e.g. x. One could write $\mathbf{X}(\omega) = x$, where ω is a point in the sample space. Note the difference between probabilities and random variables. Both are functions on the sample space, but probabilities are *set functions*; their

[9]We have used the words *numerical valued* to allow for the possibility that we may be interested in complex valued as well as real valued functions.

[10]For the reader conversant with measure theory, a caveat is in order. Just as not every subset of a probability space need have a probability definable on it (some sets are not measurable), not every function defined on a sample space can be a random variable (some functions are not measurable). A random variable must be a measurable function in order to have a consistent theory. Those unacquainted with measure theory who wish to ignore this caveat may do so with impunity within the framework of this book.

arguments are sets of points in the space. Random variables are *point functions*; their arguments are single points in the space.

Since we often do not know very much about the detailed structure of the sample space,it is usually not possible to specify the form of a random variable in an analytic way. But there is a convenient way to characterize random variables analytically. We first discuss this for the case that the random variable is a real valued function. Let us define a function $F_{\mathbf{X}}(x)$ by

$$F_{\mathbf{X}}(x) = P(\mathbf{X} \leq x) \tag{2.3.1}$$

The notation $P(\mathbf{X} \leq x)$ means the probability that the random variable, $\mathbf{X}$, has a value less than or equal to the number, x. It is defined as the probability of the set of events, $\{\omega\}$, such that $\mathbf{X}(\omega) \leq x$. The subscript is placed on F to denote which random variable is being considered. It may be omitted if no confusion will result. Clearly F has the following properties:

a. $F(-\infty) = 0$
b. $F(\infty) = 1$
c. F is a non-decreasing function of x.

$F_{\mathbf{X}}$ is called by mathematicians the *distribution function* of the random variable $\mathbf{X}$. Physical scientists use the words *distribution function* to refer to a different, though related concept which we shall shortly define. Henceforth, to avoid confusion, whenever we use functions like F (rarely in this book) we shall refer to them as *cumulative* distribution functions.

The following theorem about the structure of cumulative distribution functions has been proved (Feller 1966): Every cumulative distribution function can be written as the sum of three parts: an absolutely continuous part $F_{a.c.}$, an atomic part F_a, and a singular but continuous part, F_s. For the definition of absolute continuity, see Titchmarch (1939); for our purposes it means that $F_{a.c.}$ can be written as an integral.

$$F_{a.c.}(x) = \int_0^x f(y)dy \tag{2.3.2}$$

where f is some integrable function. The atomic part, F_a, has jumps at a denumerable number of points $\{x_k\}$ and is constant between the jumps. The singular continuous part, F_s, is continuous everywhere but has its points of increase concentrated on sets of zero probability (measure zero). Such functions cannot be handled by ordinary analytical methods. If they do arise in a problem, the problem is reformulated to avoid their use. In this sense, they will not occur in practice. Feller (1966) gives examples where such singular functions arise but can be removed by a reformulation. Nevertheless, he insists that the singular distributions are conceptually important. In recent years, the concept of *multifractal*, a probability measure with fractal support has been introduced. This is a new example where singular continuous distributions may arise. Fractals will be discussed briefly in Chapter 18.

If we are willing to leave the domain of ordinary functions and introduce generalized functions (often also called distributions, a use of the word quite distinct from that of the preceding paragraphs), the atomic part can be written as a sum of Dirac delta functions

$$F_a(x) = \int_0^x \sum p_k \delta(y - x_k) dy \tag{2.3.3}$$

The function $f(x)$ of (2.3.2) is called the *density function* of $F_{a.c.}$ by mathematicians. Physical scientists sometimes use that name, but usually refer to $f(x) + \sum p_k \delta(x - x_k)$ as the distribution function of $\mathbf{X}$. This is the difference in terminology referred to earlier. We shall use the scientist's nomenclature rather than the mathematician's.

It is important to keep the distinction between probabilities and probability densities in mind. Probabilities are invariant to variable change. Probability densities are not. We saw this in our discussion of Betrand's paradox in Section 2.1.

The ideas introduced here can easily be extended to vector valued and complex valued random variables. For the vector valued case, we define

$$F_{\mathbf{X}}(x_1, x_2) = P(\mathbf{X}_1 \leq x_1; \mathbf{X}_2 \leq x_2) \tag{2.3.4}$$

and so on for more than two variables. For complex values random variables we need only consider a vector valued function with arguments the real and imaginary parts of the complex value.

Finally, it is useful to introduce the idea of independent random variables in analogy to the idea of independent events. Two random variables, $\mathbf{X}$ and $\mathbf{Y}$, are said to be *independent* if the events $(x_1 \leq \mathbf{X} \leq x_2, y_1 \leq \mathbf{Y} \leq y_2)$ are independent for all (x_1, x_2) and (y_1, y_2).

2.4 Expectations and Particular Distributions

The distribution function of a random variable contains a great deal of information about the variable and, in many cases, may be difficult to compute in sufficient detail. It would be convenient to have simple numerical (rather than functional) characterizations of the random variables that would give useful partial information about the distributions and yet be easier to compute. Some measure of the central tendency of the distribution, and of its spread are often all that are needed for specific problems. Of course, these measures do not always suffice. For example, if we are interested in the probability of occurrence of rare outliers, the central tendency of the distribution is not helpful. Such cases are uncommon.

The measure of central tendency that is commonly used is the *mathematical expectation* or *mean* of the variable $\mathbf{X}$. In the mathematical literature it is often denoted by the symbol $E(\mathbf{X})$ or by μ; in the physical literature $< \mathbf{X} >$ is more common. Another common notation is $\overline{\mathbf{X}}$ but this is hard to read when there are exponents present; we shall use it occasionally, but never for random variables with exponents. The definition is

$$< \mathbf{X} >= \int xp(x)dx \tag{2.4.1}$$

where $p(x)$ is the distribution function of $\mathbf{X}$. Every possible value of $\mathbf{X}$ is weighted by the probability of its occurrence, and the results are added. Thus x values of high probability are more important than those of low probability in determining the expectation. Consequently $< \mathbf{X} >$ usually gives an indication where the high probability values of x are likely to be, i.e. the central tendency.

Occasionally care is needed in interpreting the mean. Consider a simple game in which a player wins one unit or loses one unit on each play, each with probability 1/2. Let $\mathbf{X}$ denote the player's winnings after one play of the game, where $\mathbf{X} = 1$ for a win and -1 for a loss. Then $< \mathbf{X} >= 0$. But clearly zero is not a possible outcome; the only possible outcomes are ± 1. In this case the mean does not tell us very much. On the other hand, if the player plays the game many times, and $\mathbf{X}$ represents his cumulative gain instead of his gain in one turn, then $< \mathbf{X} >$ is a much more informative quantity. We shall return to this point below.

There are two other measures of central tendency that are sometimes used: the median and the mode. The median is that value of x for which $F_{\mathbf{X}}(x) \geq 1/2$, i.e. half of the cases will have values greater than x, half less. The mode is the value of x for which $p(x)$ is a maximum. There may be more than one mode for a given distribution; such distributions are called multimodal. These measures of central tendency are often used in descriptive statistics. They are harder to handle analytically than the mean, and do not often occur in probability theory.

A useful characterization of the width or spread of a distribution would be the mean of the deviation from the mean, $<| \mathbf{X} - < \mathbf{X} >|>$. The absolute value bars are essential, for if they were omitted, the result would be identically zero. Unfortunately, the mean of an absolute value is difficult to handle analytically, so it is customary to use a different measure of the width, called the *variance* σ. The variance is defined by

$$\sigma^2 =< (\mathbf{X} - < \mathbf{X} >)^2 >= \int x^2 p(x)dx - (\int xp(x)dx)^2 \tag{2.4.2}$$

As with the absolute value, the quantity being averaged is always positive and vanishes only when everything is concentrated at the mean. The *standard deviation* is defined as σ, the square root of the variance. Both measures of the spread offer qualitatively the same information; the standard deviation is easier to use. μ and σ^2 are but two members of a class of numbers characterizing the distribution function of a random variable, the *central moments*, μ_j, defined by

$$\mu_j =< (\mathbf{X} - < \mathbf{X} >)^j > \tag{2.4.3}$$

Moments of order higher than the second are often difficult to obtain, and consequently are not often met with in practice. But there is an interesting theoretical question: does a knowledge of *all* the moments determine the distribution function uniquely? The answer is no as can be seen from the following example.

> Example. Let $0 < \alpha < 1$. Then all of the distribution functions $\phi_\alpha(x) = (1/24)(1 - \alpha \sin x^{\frac{1}{4}}) \exp(-x^{\frac{1}{4}})$ have the same sequence of moments. The key step in verifying this is to show that $\int x^n \exp(-x^{\frac{1}{4}}) \sin x^{\frac{1}{4}} \, dx = 0$ for all integral n.

On the other hand, conditions on the moments are known which ensure that a sequence of moments satisfying these conditions uniquely defines a probability distribution. Roughly speaking, these conditions require that the moments do not increase too rapidly as j increases. We shall not delve further into this question here.

Moments can be defined about any point, a , not only about the mean. The definition is

$$\mu_j(a) = < (\mathbf{X} - a)^j > \tag{2.4.4}$$

It is easy to see that the choice $a = < \mathbf{X} >$ makes $\mu_2(a)$ a minimum over all possible choices of a. For $\mu_2(a) = < (\mathbf{X} - < \mathbf{X} > + < \mathbf{X} > - a)^2 > = \sigma^2 + (< \mathbf{X} > -a)^2$. Usually only the central moments ($a = < \mathbf{X} >$) and moments about the origin ($a = 0$) are used.

We now give two examples of probability distributions to illustrate many of the points made above:

> Example 1. *The Binomial Distribution.* This is a purely discrete distribution which occurs in many contexts. One can perform an experiment which has two possible outcomes. For the sake of definiteness we call these outcomes success (S) and failure (F). The probability of S in one experiment is p; the probability of F is $q = 1 - p$. The experiment is performed N times; each trial is independent of all of the others. What is the probability that there are exactly j Successes in the N experiments?
>
> The Ss and Fs can occur in any order. For each specific order of occurrence, the probability of occurrence of that ordered sequence is $p^j q^{N-j}$. But there are $N!/j!(N-j)!$ such sequences for a given j. If we denote the random variable 'number of Successes in N trials' by $\mathbf{X}_N$ then
>
> $$P(\mathbf{X}_N = j) = \frac{N!}{j!(N-j)!} p^j q^{N-j} \tag{2.4.5}$$
>
> This is the binomial distribution. It is a simple matter to show that $\mu = Np$ and $\sigma^2 = Npq$.
>
> Example 2. *The Gaussian Distribution.* The Gaussian, or normal, distribution is a purely continuous distribution for a random variable whose values may range from $-\infty$ to ∞. The distribution function is given by
>
> $$p(x) = (2\pi\sigma^2)^{-1/2} \exp[-(x - \mu)^2/2\sigma^2] \tag{2.4.6}$$
>
> The distribution function depends on two parameters, μ and σ, the symbols for which were chosen because they have the significance of the mean and standard deviation of the distribution respectively.

The Gaussian distribution is ubiquitous in probability and statistics. It occurs in the most varied places.[11] While it would take us too far afield to explore this in any detail, we shall show a relation between the binomial and the Gaussian distributions known as the de Moivre– Laplace limit theorem. In the limit as $N \to \infty$, $j \to \infty$, $N-j \to \infty$ (large numbers of trials, large numbers of successes, large numbers of failures)

$$\begin{aligned} b(j;N) &= \frac{N!}{j!(N-j)!} p^j q^{N-j} \\ &\approx (2\pi Npq)^{-1/2} \exp[-(x-Np)^2/2Npq] \end{aligned} \tag{2.4.7}$$

Since the binomial distribution is discrete and the Gaussian is continuous, eqn (2.4.7) cannot be literally true. What eqn (2.4.7) means is that the sum of the left hand side over some range of j, and the integral of the right hand side over the corresponding range of x are approximately equal, the approximation improving as N and j increase.

To see this, we use Stirling's approximation

$$\ln M! \sim M \ln M - M + \frac{1}{2}\ln(2\pi M) \tag{2.4.8}$$

on the right-hand side of (2.4.7), where the symbol $\sim$ means 'asymptotically equals.' This yields

$$\begin{aligned} &\ln b(j;N) \approx N\ln N - j\ln j - (N-j)\ln(N-j) + j\ln p \\ &+(N-j)\ln q + \frac{1}{2}\ln N + \frac{1}{2}\ln j - \frac{1}{2}\ln(N-j) - \frac{1}{2}\ln 2\pi \end{aligned} \tag{2.4.9}$$

Now introduce the deviation of the variable j from its mean value Np, $y_j = j - Np$. This yields

$$\begin{aligned} \ln b(j;N) &= -N[(p+\frac{y}{N})\ln(1+\frac{y}{Np}) + (q-\frac{y}{N})\ln(1-\frac{y}{Nq})] \\ &-\frac{1}{2}\ln(p+\frac{y}{N}) - \frac{1}{2}\ln 2\pi - \frac{1}{2}\ln(q-\frac{y}{N}) - \frac{1}{2}\ln N \end{aligned} \tag{2.4.10}$$

Let N and $y_j \to \infty$ in such a way that y_j^2/N stays fixed but $y_j^3/N^2 \to 0$ (which implies that $y_j/N \to 0$ also). The final result is

$$\ln b(j;N) \sim -y_j^2/2Npq - \frac{1}{2}\ln(2\pi Npq) \tag{2.4.11}$$

or

[11]'Everyone believes in it because the experimentalists imagime that it is a theorem of mathematics and the mathematicians that it is an experimental fact.' (Poincaré 1908)

$$b(j;N) \sim (2\pi Npq)^{-\frac{1}{2}} e^{-y_j^2/2Npq} \tag{2.4.12}$$

This limit theorem is a special case of a more general theorem, the *central limit theorem*, which we shall state in detail in Section 2.5.

Almost everything that has been put forward in this paragraph can easily be generalized to vector valued (also called *multivariate*) random variables. For an n-tuple random variable $(\mathbf{X}_1 \ldots \mathbf{X}_n)$, there is a multivariate distribution function, $p(x_1, \ldots x_n)$, and multivariate moments

$$\mu_{j_1 \ldots j_n} =< (\mathbf{X}_1 - < \mathbf{X}_1 >)^{j_1} \ldots (\mathbf{X}_n - < \mathbf{X}_n >)^{j_n} > \tag{2.4.13}$$

As a rule, the more variables there are to consider, the more complicated is the computation.

It is worthwhile recording the explicit form of a multivariate Gaussian distribution function, which occurs often in applications. We shall use a convenient matrix notation; $\mathbf{V}$ is a positive definite $n \times n$ matrix and $\mathbf{x}$ and μ are n dimensional column vectors. $\mid \mathbf{V} \mid$ is the determinant of $\mathbf{V}$. The superscript ^{T} denotes transpose. If we have n random variables, to say that they have a joint Gaussian distribution is to say that their joint distribution function is

$$p(x_1, x_2, \ldots, x_n) = (2\pi)^{-n/2} \mid \mathbf{V} \mid^{\frac{1}{2}} e^{-\frac{1}{2}(\mathbf{x}-\mu)^T \mathbf{V}(\mathbf{x}-\mu)} \tag{2.4.14}$$

The probabilistic significance of the matrix $\mathbf{V}$ is that $V_{ij}^{-1} = \mu_{ij}$ where μ_{ij} is defined by eqn (2.4.13). To see this, note that μ_{ij} can be computed as

$$\mu_{ij} = \int (x_i - \mu_i)(x_j - \mu_j) p(x_1, \ldots, x_n) d^n x = \frac{\partial \ln \mid \mathbf{V} \mid}{\partial V_{ij}} \tag{2.4.15}$$

Then use the rule for the differentiation of a determinant. Clearly the vector μ denotes the expected value of $\mathbf{X}$.

Example 3. *The Poisson Distribution.* The Poisson random variable is a random variable capable of taking on nonnegative integral values, k, from 0 to infinity with probability distribution function given by

$$P(\mathbf{N} = k) = e^{-\lambda} \frac{\lambda^k}{k!} \tag{2.4.16}$$

This distribution has been found empirically to describe many different phenomena, the number of disintegrations of an unstable nuclear species in a given time, the number of incoming calls to a telephone exchange in a given time, even the number of Prussian cavalrymen kicked by their horses in a given year.

It is easy to compute the mean of the Poisson random variable $\mathbf{N}$. It is just λ. The variance is also easy to calculate, and is also λ.

There is an interesting connection between the Poisson distribution and the binomial distribution. We have seen that the binomial distribution approaches the normal distribution in the limit that the number of trials, n becomes infinite, the probability of success, p, in any given trial remaining fixed. A fixed fraction of the trials lead to a success, on the average. A different question is, what happens when the number of trials, n, increases and the probability of success on a given trial decreases in such a way that the average number (not the average fraction) of successes is fixed? In symbols, what is the limit of the binomial distribution as $n \to \infty$, $p \to 0$, with $np = \lambda$ fixed? In this limit, the limiting distribution is the Poisson distribution. This may explain why the Poisson distribution is found so often in practice; it describes the istribution of a large number of rare events.

To see this we expand $\ln b(n,k)$ as in eqn (2.4.9) and write $\ln p = \ln \lambda - \ln n$, $\ln q = \ln(1 - \lambda n) \approx -\lambda n$. In contrast to eqn (2.4.10), however, one keeps k and λ fixed and lets n get large. This yields

$$\ln b(k;n) = k \ln \lambda - \ln k! - \lambda - k^2/n + (\lambda + \frac{1}{2})k/n + o(1/n) \tag{2.4.17}$$

Exponentiating yields the Poisson distribution.

2.5 Characteristic Function; Sums of Random Variables

A useful auxiliary function for studying a distribution function is its Fourier transform

$$\phi(t) =< e^{it\mathbf{X}} >= \int_{-\infty}^{\infty} e^{itx} p(x) dx \tag{2.5.1}$$

In probability theory, this function is given a special name, and called the *characteristic function.* Because p is a probability distribution function, and hence non negative and integrable from $-\infty$ to ∞, the characteristic function always exists. Furthermore, a distribution function is determined uniquely by its characteristic function.

The inversion formula for going from the characteristic function back to the distribution function involves a non-trivial limiting process. This is because it must take into account the atomic and singular parts of the cumulative distribution (cf. Section 2.3), as well as its non singular continuous part. Since we will ignore the possible existence of a singular continuous part, and treat the atomic part as though it were a sum of delta functions, we may invert (2.5.1) by the conventional Fourier inversion formula

$$p(x) = \frac{1}{2\pi} \int_{-\infty}^{\infty} e^{-itx} \phi(t) dt \tag{2.5.2}$$

For the atomic parts of p, we use the formulae

$$\int_{-\infty}^{\infty} e^{itx} \delta(x - x_0) dx = e^{itx_0}; \int_{-\infty}^{\infty} e^{itx_0} e^{-itx} dt = 2\pi \delta(x - x_0) \tag{2.5.3}$$

For a multivariate distribution function, the characteristic function is also multivariate. It is defined by

$$\phi(t_1, \ldots, t_n) = \int dx_1 \ldots dx_n \exp\left[i(t_1 x_1 + \ldots + t_n x_n)\right] p(x_1, \ldots, x_n) \qquad (2.5.4)$$

A very important property of characteristic functions is that the characteristic function of the sum of independent random variables (the sum is a new random variable in its own right) is the product of the characteristic functions of the individual random variable summands. In formulas

$$< \exp(it[\mathbf{X}_1 + \mathbf{X}_2 + \cdots + \mathbf{X}_n]) >= \prod_{j=1}^{n} \phi_j(t) \qquad (2.5.5)$$

where ϕ_i is the characteristic function of $\mathbf{X}_i$. To see this, write

$$\int_{-\infty}^{\infty} e^{it\Sigma x_i} p(x_1, \ldots, x_n) d^n x = \int e^{itx_1} p(x_1) e^{itx_2} p(x_2) \ldots d^n x = \Pi_{j=1}^{n} \phi_j(t) \qquad (2.5.6)$$

It is clear that the assumption of independence has played an essential role in this result, for only then does the joint distribution function of the random variables $\{\mathbf{X}_\mathrm{i}\}$ factor into a product of the distributions for the individual $\mathbf{X}$'s.

As a result of eqn (2.5.5), the distribution function of the sum of independent random variables can be determined in principle by inverse Fourier transformation of eqn (2.5.6). This inverse Fourier transform may not be easy to carry out explicitly, so that it is fortunate that there is a simple and useful formula which gives an asymptotic approximation to the result when the number of terms in the sum is large. This formula is known as the *central limit theorem.* Necessary and sufficient conditions for the validity of the central limit theorem are known (Feller 1966), but they are sometimes difficult to verify in practice. We give here a sufficient condition for the validity of the theorem which is easy to verify and sufficiently general for our purposes in this book.

Let $\{\mathbf{X}_i\}$ *be a sequence of independent random variables, all of which have the same distribution, with second moments* $< \infty$. *Let*

$$S_N = \sum \mathbf{X}_j - N\mu \qquad (2.5.7)$$

Then $P(s_N) \sim (2N\sigma^2)^{-\frac{1}{2}} \exp(-s_N^2/2N\sigma^2)$ *as* $N \to \infty$. We shall not give a proof of this theorem here. The characteristic function plays an important role in the proof.

Notice in particular the requirement that the random variables have finite second moments. If this is violated, the central limit theorem need not be valid. For example, consider the Cauchy distribution

$$p(x) = \frac{a}{\pi} \frac{1}{a^2 + x^2} \tag{2.5.8}$$

This is sometimes also called the Lorentz distribution because it arises in the Lorentz theory of the shape of spectral lines. It is easy to verify that the sum of two Cauchy random variables is also a Cauchy random variable (with a different parameter). Hence it is manifestly impossible for the sum of any finite number of Cauchy variables to have a distribution function approximating a Gaussian. But this does not contradict the central limit theorem because the second moment of a Cauchy variable diverges.

We close this section by exhibiting the characteristic function of a multivariate Gaussian distribution function. This can be obtained by taking the Fourier transform of eqn (2.4.14), and is

$$\phi(t_1, \ldots, t_n) = e^{i\mathbf{t}\cdot\mu - \frac{1}{2}\mathbf{t}^T\mathbf{V}^{-1}\mathbf{t}} \tag{2.5.9}$$

where $\mathbf{V}$ is the same matrix that appears in eqn (2.4.14).

2.6 Conclusion

The subject of probability theory fills many textbooks and monographs. We have just skimmed the surface, giving a few of the basic concepts and definitions, and even fewer of the principal results. For example, of the fundamental limit theorems of probability, we have given only one, the central limit theorem (the Laplace–de Moivre limit theorem can be considered a special case even though it came first, historically). I have tried to emphasize conceptual matters, and so may have slighted many technical points. Nevertheless, this is not a book on probability theory *per se* so that we shall eschew further explorations in this realm until and unless they are needed in later chapters.

In a sense, our view of probability has been essentially static. We have been describing situations containing randomness, but this randomness does not change. The distribution functions which we have introduced do not contain any information on how the randomness may or may not evolve. In the next chapter we pass on to a subarea of probability theory which tries to study exactly this possibility of evolution.

3

STOCHASTIC PROCESSES

3.1 Stochastic Processes

In the previous chapter we studied random variables as functions on a sample space without regard to how these might depend on parameters. We now want to enlarge the scope of our discussion by including parametric dependence. The notion which has been found useful to capture this idea is that of *stochastic process* or *random process*; the two names are used interchangeably. The term *random function* is sometimes used also.

Recall that a random variable is a function defined on the sample space. It assigns a number to an event. A *stochastic process* is a family of random variables depending on a real parameter. In other words, a stochastic process is a function of two variables, one of which is a point in the sample space, the other of which is a real variable.

There are three equivalent ways to look on a stochastic process. The first is that just given: as a function of two variables. If the stochastic process is called $\mathbf{X}$, we might denote it as $\mathbf{X}(\omega, t)$ to emphasize that it is being thought of as a function of two variables. For a fixed value of t, $\mathbf{X}$ is a random variable of the type we have discussed in Chapter 2. For each t, it is a different random variable; we can regard $\mathbf{X}$ as a family of random variables indexed by the variable t. If we want to emphasize this viewpoint, we might write the stochastic process as $\mathbf{X}_t$, with the sample space argument suppressed for simplicity. A third viewpoint is to consider $\mathbf{X}$ as a family of functions of t, one for each fixed ω. To emphasize this viewpoint, we write $\mathbf{X}$ as $\mathbf{X}_\omega(t)$ or just $\mathbf{X}(t)$, suppressing the sample space argument for compactness of notation. $\mathbf{X}(t)$ for a fixed sample space point is called a *realization*, or a *sample function*, of the process. Again, the sample space argument, ω, is usually omitted for notational simplicity.

These three ways of viewing a stochastic process are completely equivalent. In any specific application, the viewpoint and notation which best clarifies the case at hand should be adopted.

The parameter, t, can belong to an interval of the real line (i.e. be a continuous parameter) or belong to some countable subset such as the integers (i.e. be a discrete parameter). In most physical applications, t denotes the time. This is so common that t is often generically referred to as time, even when no specific timelike application is in mind. When t is restricted to a discrete set, it is conventional to speak of 'discrete time.'

Some examples will help to make these abstract considerations more concrete.

Example 1. Let $\mathbf{X}$ be a random variable and $f(t)$ a given function of time (with no randomness involved, sometimes called a *sure* function). Then $f(t)\mathbf{X}$ is a stochastic process. This is a rather simple case, but an important one.

Example 2. Let $\mathbf{X}$, ω, and δ be random variables. Then $\mathbf{X}\sin(\omega t + \delta)$ is a stochastic process. It corresponds to an oscillation with random amplitude, frequency, and phase. Of course, it will still be a stochastic process if only one of these three quantities is random.

Example 3. A coin is tossed n times. The accumulated number of heads is a random variable which depends on the real parameter n. It is thus a stochastic process in discrete time.

3.2 Distribution Functions

As we saw in Chapter 2, we usually cannot write explicit formulas for random variables. Since stochastic processes are sets of random variables, it follows that we shall usually not be able to write explicit expressions for them either. We shall characterize them in terms of their probability distribution functions.

Choose some finite set of ts, $t_1, t_2, \ldots, t_n$. Then $\mathbf{X}_{t_1}, \ldots, \mathbf{X}_{t_n}$ are a set of random variables (using the second notation introduced in Section 3.1) and we can specify their joint probability distribution function. So we have the distribution functions

$$\begin{gathered} p_1(x_1, t_1) \\ p_2(x_1, t_1; x_2, t_2) \\ p_3(x_1, t_1; x_2, t_2; x_3, t_3) \\ \cdots \end{gathered} \tag{3.2.1}$$

Here $p_1(x_1, t_1)dx_1$ is the probability that the random variable $\mathbf{X}_{t_1}$ has the value x_1 in the range dx_1. $p_2(x_1, t_1; x_2, t_2)dx_1dx_2$ is the joint probability that the two random variables $\mathbf{X}_{\mathbf{t}_1}$ and $\mathbf{X}_{t_2}$ have the values x_1 and x_2 in the ranges dx_1 and dx_2 respectively. The p_n with $n > 2$ have similar meanings. These are probability distribution function of exactly the type discussed in Section 2.3. The notation looks a bit more complicated only because we have put the t variables, specifying which random variables of the family are being considered, inside the parentheses along with the x variables, as is customary.

Naturally, the functions, p_j are not all independent. First of all, each function, p_j, must be symmetrical to any permutation of the j pairs of arguments (x_i, t_i)

$$p_j(x_1, t_1; x_2, t_2; \ldots; x_j, t_j) = p_j(x_{i_1}, t_{i_1}; \ldots; x_{i_j}, t_{i_j}) \tag{3.2.2}$$

Furthermore, the ps with lower j must be compatible with those of higher j, that is

$$p_j(x_1, t_1; \ldots; x_j, t_j) = \int p_{j+1}(x_1, t_1; \ldots; x_j, t_j; x_{j+1}, t_{j+1})dx_{j+1} \tag{3.2.3}$$

This just says that obliterating all information about time t_{j+1} (by integrating over x_{j+1}) yields the probability distribution for j times: an obvious requirement. Equation (3.2.3) will be called the compatibility requirement.

Since p_j is a probability distribution, there must also be the condition $p_j \geq 0$ for all j. By the symmetry of p_j with respect to its arguments, it does not matter in which order the arguments of a p_j are written. A common convention is to write them in the order of time increasing from right to left. For example, writing $p(x_1, t_1; x_2, t_2; x_3, t_3)$ implies that $t_1 \geq t_2 \geq t_3$, unless stated to the contrary.

Note that there are an infinite number of distribution functions for a single random process. This contrasts with the case of a random variable where there is only one distribution function for a given random variable. The reason for this difference is that a random or stochastic process is a family of infinitely many random variables. It is desirable to avoid, as far as possible, formulating problems in terms of p_js for large j because of the difficulty, both experimental and theoretical, of determining these functions.

Since the p_js are probability distribution functions for a finite number of random variables, the entire machinery discussed in Chapter 2 for treating multivariate probability densities is available for use in studying them. In particular, the concept of *conditional probability* plays an important role. Conditional probability densities are defined as in Section 2.2:

$$p_n(x_1, t_1; \ldots; x_n, t_n) = p_{k,n-k}(x_1, t_1; \ldots; x_k, t_k \mid x_{k+1}, t_{k+1}; \ldots; x_n, t_n) \times p_{n-k}(x_{k+1}, t_{k+1}; \ldots; x_n, t_n) \quad (3.2.4)$$

Note the order of the times in this equation.

Mean values and moments are defined just as in Section 2.4. The only new feature is that these mean values are no longer just numbers, but are now functions of the t values specifying which random variables (in the family comprising the random process) are being averaged. That is,

$$\begin{aligned} < x(t_1)x(t_2) > &= R(t_1, t_2) \\ < x(t_1)x(t_2)x(t_3) > &= S(t_1, t_2, t_3) \\ &\text{etc.} \end{aligned} \quad (3.2.5)$$

These functions are called *autocorrelation functions.* The prefix *auto* is used because $\mathbf{X}(t_1)$ and $\mathbf{X}(t_2)$ belong to the same random process. If two random processes are involved in some problem, say $\mathbf{X}$ and $\mathbf{Y}$, expectations such as $< \mathbf{X}(t_1)\mathbf{Y}(t_2) >$ may arise. The resulting function of t_1 and t_2 is called a *correlation function.*

If the distribution functions are independent of the origin of time, the stochastic process is called *stationary.* More formally, a stationary process is one for which

$$p_j(x_1, t_1 + s; x_2, t_2 + s; \ldots; x_j, t_j + s) = p_j(x_1, t_1; x_2, t_2; \ldots; x_j, t_j) \qquad (3.2.6)$$

for all j. This implies that all autocorrelation functions depend only on the time differences, t_k - t_{k-1}, for all the ts which occur in them. In particular, the mean value of a stationary stochastic process is independent of t: $< \mathbf{X}(t) >$ is constant.

If a process has a constant first moment (expectation), and an autocorrelation function $R(t_1, t_2)$ that depends only on $t_1 - t_2$, it is called *wide sense stationary.* This is a weaker condition than stationarity, since it imposes no restrictions on the distribution functions of order greater than two. It suffices for many applications.

In most applications, a two time correlation function approaches zero as the difference between its arguments approaches infinity. The time beyond which the correlation function becomes negligible is called the *correlation time* of the random process. Clearly this is not a very precisely defined concept; it depends on the definition of negligible. Nevertheless, it is often a useful qualitative characterization of the time interval over which a random process loses its coherence. For example, if the autocorrelation function is an exponential, $R(t_1, t_2) =< \mathbf{X}^2 > exp[-|t_1 - t_2|/\tau]$, then it would be reasonable to call τ the correlation time.

3.3 Classification of Stochastic Processes

There are many ways in which stochastic processes can be classified. For example, they could be classified in terms of the continuity, boundedness, etc., properties of their sample functions. They could be classified with respect to the properties of their distribution functions. However, the classification which has been found to be the most useful for our purposes is a classification according to their memory; we shall explain what this means.

The simplest kind of random process, according to this classification, is one which has no memory of the past. More precisely

$$p_n(x_1, t_1; \ldots; x_n, t_n) = p_1(x_1, t_1), p_1(x_2, t_2), \ldots, p_1(x_n, t_n) \qquad (3.3.1)$$

Such a process is called a *completely random* process. In terms of conditional probabilities, the definition of a completely random process can be written

$$p_{1,n-1}(x_1, t_1 \mid x_2, t_2; \ldots; x_n, t_n) = p_1(x_1, t_1) \qquad (3.3.2)$$

The random variables $\mathbf{X}(t_1)$ and $\mathbf{X}(t_2)$ are independent random variables when $t_1 \neq t_2$.

When t is a discrete variable, i.e. $\mathbf{X}(t)$ is a random sequence, there are numerous examples of such processes. A simple one is a sequence of throws of a fair coin. The ordinal number of the throw is the 'time' variable, t. Since the throws are, by hypothesis, independent, we have an instance of a completely independent process, in this case a sequence. When $\mathbf{X}$ is a continuous variable, no physical process will be completely random. Since it takes some time for the physical state of a system to change, there is always some kind of inertia, albeit small. However, there are physical processes that can be very closely approximated by completely random processes. One such is white noise in physical systems, which we shall meet again later in this chapter.

More complicated than completely random processes with no memory are processes with short memory. These processes are called *Markov processes.* The case of sequences was first introduced by A. A. Markov (1907). Their precise definition is that a process is a Markov process if its conditional probabilities for (x_1, t_1) given $(x_2, t_2; x_3, t_3; \ldots; x_n, t_n)$ only depends on $(x_1, t_1; x_2, t_2)$, where t_2 is the closest previous time to t_1. In formulas

$$p_{1,n-1}(x_1, t_1 \mid x_2, t_2; \ldots; x_n, t_n) = p_{1,1}(x_1, t_1 \mid x_2, t_2), t_1 > t_2 > \ldots > t_n \quad (3.3.3)$$

This means that the general n-time distribution function can be written as

$$p_n(x_1, t_1; x_2, t_2; \ldots; x_n, t_n) = \\ p_{1,1}(x_1, t_1 \mid x_2, t_2) p_{1,1}(x_2, t_2 \mid x_3, t_3) \ldots p_{1,1}(x_{n-1}, t_{n-1} \mid x_n, t_n) p_1(x_n, t_n) \quad (3.3.4)$$

The functions $p_{1,1}(x_1, t_1 \mid x_2, t_2)$ are called *transition probabilities.* The elementary events of the sample space of the process are often called *states.* A process whose transition probabilities depend on time only through $t_1 - t_2$ is called *homogeneous.*

The theory of Markov processes is very highly developed. We shall not attempt to give a description of the entire field here. Rather, we shall restrict our discussion to those aspects used later in this book. The reader desiring a more complete treatment of the subject may consult Van Kampen (1981a); Gardiner (1983); and Gillespie (1992). The sourcebook by Wax (Wax 1954) contains reprints of important articles, both original articles and reviews.

Example: One of the simplest examples of a Markov process in continuous time is the dichotomic Markov process. Not only does this process illustrate some of the concepts just introduced, but it also has some physical interest as an elementary model for two state systems, for instance relaxation of a spin-1/2 particle due to environmental perturbations, or the behavior of a molecule undergoing transitions between two conformational states. For clarity, we shall use a simpler notation than the general notation introduced above, and then indicate how the results would look in the more complicated general notation.

A system has two states a and b. Let $w_a(t)$ be the probability that the system is in the state a at time t, and $w_b = 1 - w_a$. If the system is in state a at time t, it has probability $\lambda dt + o(dt)$ of changing its state to state b in time interval dt. Similarly, if it is in state b, it has probability $\mu dt + o(dt)$ of changing its state to a. Thus it has probability $1 - \lambda dt$

while in state a and $1 - \mu dt$ while in state b, of not changing its state in time interval dt. (We omit writing the $o(dt)$ to simplify the notation.) The probability of making more than one transition in time interval dt is $o(dt)$. It is assumed from the start that the time development of the system is a Markov process; that is, once the system is in a given state its future development is independent of how it got into that state. We calculate the transition probabilities of the system between its two states. Thus

$$\begin{aligned} w_a(t + dt) &= w_a(t)(1 - \lambda dt) + w_b(t)\mu dt \\ w_b(t + dt) &= w_a(t)\lambda dt + w_b(t)(1 - \mu dt) \end{aligned} \tag{3.3.5}$$

Now divide both sides of eqn (3.3.5) by dt, which is a small but finite increment, and let $dt \to 0$. The result is the pair of ordinary differential equations

$$\begin{aligned} \frac{dw_a}{dt} &= -\lambda w_a + \mu w_b \\ \frac{dw_b}{dt} &= \lambda w_a - \mu w_b \end{aligned} \tag{3.3.6}$$

These equations are linear and easy to solve in terms of initial conditions. The solution is

$$w_a(t) = \frac{\mu}{\lambda + \mu} + (w_a(0) - \frac{\mu}{\lambda + \mu})e^{-(\lambda+\mu)t} \tag{3.3.7}$$

$$w_b(t) = \frac{\lambda}{\lambda + \mu} + (w_b(0) - \frac{\lambda}{\lambda + \mu})e^{-(\lambda+\mu)t} \tag{3.3.8}$$

w_a and w_b change monotonically from their initial values to final, time independent, values as $t \to \infty$. This does not mean that the system stops fluctuating between the two states at long times; there are always fluctuations between the states. It is just that the mean values around which the fluctuations occur settle down to a time independent value.

What we have here called w is, in the more explicit notation of Section 3.2, actually a p_1 function, the probability of finding the system in a certain state at a certain time. λdt is $p_{1,1}(b, t + dt \mid a,\, t)$, the transition probability, and the other $p_{1,1}$ functions can be identified easily from eqn (3.3.5). Since the transition probabilities are independent of t, this process is homogeneous. It is not necessarily stationary, since stationarity depends on p_1 as well as $p_{1,1}$

A reasonable question is, how do we know that this process is a Markov process? We have only determined p_1 and have no obvious assurance that the higher p_js have the property specified by eqn (3.3.4). We cannot derive conditions on the higher p_js from knowledge only of the lower ones. As stated at the beginning of the example, the assertion that the process is Markov is a hypothesis, not a conclusion. Of course, since we have assumed that the process is Markov, then all of the ps can be constructed from p_1 and $p_{1,1}$.

In this example, it was not necessary to know $p_{1,1}$ for all values of the time difference $t_1 - t_2$, but only when this difference is small, in fact in the limit $dt \to 0$. This is general for stochastic processes in continuous time. We have, from eqn (3.2.4)

$$p_2(x_1, t + dt \mid x_2, t) = p_{1,1}(x_1, t + dt \mid x_2, t) p_1(x_2, t) \tag{3.3.9}$$

Integrate (or sum in the case of a process with discrete states) over all values of x_2. Integration of the left-hand side gives $p_1(x_1, t_1)$, by the compatibility condition (3.2.3), so that

$$p_1(x_1, t + dt) = \int p_{1,1}(x_1, t + dt \mid x_2, t) p_1(x_2, t), dx_2 \tag{3.3.10}$$

Equation (3.3.5) is just a special case of this. In the example, the transition probability did not depend on the time, only on the time difference; in the general case it may, of course, depend on both the time and the time difference. The Markov property did not enter into this analysis at all.

There is a useful identity satisfied by Markov processes, called the *Chapman–Kolmogorov equation.* Here is the identity and its derivation. For the three time distribution function, we may write, using eqn (3.2.4)

$$\begin{aligned} p_3(x_1, t_1; , x_2, t_2; x_3, t_3) &= p_{1,2}(x_1, t_1 \mid, x_2, t_2; x_3, t_3) p_2(x_2, t_2; x_3, t_3) \\ &= p_{1,1}(x_1, t_1 \mid x_2, t_2) p_{1,1}(x_2, t_2 \mid x_3, t_3) p_1(x_3, t_3) \end{aligned} \tag{3.3.11}$$

The second step follows from the hypothesis that $\mathbf{X}$ is Markov. Now integrate (or sum, for discrete state processes) both sides of eqn (3.3.11) over all x_2. By the compatibility condition eqn (3.2.3), the left hand side is $p_2(x_1, t_1; x_3, t_3)$, which can be written $p_{1,1}(x_1, t_1 \mid x_3, t_3) p_1(x_3, t_3)$. Finally, on canceling $p_1(x_3, t_3)$, we obtain

$$p_{1,1}(x_1, t_1 \mid, x_3, t_3) = \int p_{1,1}(x_1, t_1 \mid, x_2, t_2) p_{1,1}(x_2, t_2 \mid, x_3, t_3) dx_2 \tag{3.3.12}$$

This is the Chapman–Kolmogorov equation. It is valid for discrete or continuous time, and for discrete or continuous state space. We shall see an application of this equation in Section 3.4.

Processes with longer memory, i.e. those for which $p_{1,j} \neq p_{1,1}$ for $j > 1$ are generically called non-Markovian. There is no general theory of non-Markovian processes, although a number of special cases have been successfully treated.

We close this section with an example of a Markov process which we shall use later, the *Poisson process.* A Poisson process, $\mathbf{N}(t)$, is a homogeneous stochastic process with independent increments[12] (and hence Markov). It takes on only positive integers as values, and the transition probability is

$$p_{1,1}(n_1, t_1 \mid n_2, t_2) = e^{-\nu(t_1-t_2)} \frac{(\nu(t_1 - t_2))^{(n_1-n_2)}}{(n_1 - n_2)!} \tag{3.3.13}$$

That is, the change of $\mathbf{N}(t)$ in any interval, Δt, of t is a Poisson random variable with mean $\nu\Delta t$ and variance $\nu\Delta t$. ν is a parameter.

This process could describe, for example, the number of nuclear disintegrations counted up to time t, or the number of random collisions occurring up to time t. The Poisson process describes the statistics of the situation when the time intervals between the events counted by $\mathbf{N}$ are exponentially distributed. This means that if s denotes the interval between the jth and $j+1$st event, then the probability distribution of s is

$$p(s) = \nu e^{-\nu s} \quad s > 0 \tag{3.3.14}$$

If the interevent times have a different distribution, then resulting process is no longer Poisson.

The Poisson process $\mathbf{N}(t)$ can be looked on as a sum of independent random variables as follows

$$\mathbf{N}(t) = [\mathbf{N}(t) - \mathbf{N}(t-\tau)] + [\mathbf{N}(t-\tau) - \mathbf{N}(t-2\tau)] + \ldots + \mathbf{N}(0)] \tag{3.3.15}$$

By virtue of the assumed independence of the increments, each term in square brackets is an independent random variable. τ is a fixed time chosen so that t is an integral multiple of τ.

Now let t increase by multiples of τ, kept fixed. The central limit theorem then asserts that the scaled Poisson process, $(N(t) - \nu t)/(\nu t)^{\frac{1}{2}}$, approaches a Gaussian, or normal, distribution as $t \to \infty$. i.e. the centered process, $\mathbf{N}(t) - \nu t$, approaches a Gaussian process with mean zero and variance proportional to t. This fact will be used in Chapter 5.

3.4 The Fokker–Planck Equation

The Chapman–Kolmogorov equation (3.3.12) is quite general. It holds for all Markov processes, and consequently does not yield much information about any particular process. However, for a restricted class of Markov processes there is a more useful equation derivable from the Chapman–Kolmogorov equation. Let us write eqn (3.3.12) for a short time interval, $t+s$ for a homogeneous process.

[12] The definition of independent increments is given in Section 3.5.

$$p_{1,1}(x,t+s \mid y,0) = \int p_{1,1}(x,t+s \mid z,t)p_{1,1}(z,t \mid y,0)dz \tag{3.4.1}$$

Furthermore, we suppose that $p_{1,1}(x,t+s \mid y,t)$ is a very sharply peaked function of $x-y$ when s is small. That is, the system cannot change its state very much in a short time; there are no 'jumps'. Then only values of z near x will contribute to the integral. Now take some arbitrary smooth function, $\phi(x)$ which vanishes as $x \to \pm\infty$, and multiply eqn (3.4.1) by this function; then integrate eqn (3.4.1) over x. Since ϕ is smooth, and $p_{1,1}(x,s \mid y,0)$ is short ranged, ϕ may be expanded in a Taylor series around z

$$\phi(x) = \phi(z) + (x-z)\phi'(z) + \frac{1}{2}(x-z)^2\phi''(z) + \ldots \tag{3.4.2}$$

Then

$$\begin{aligned} &\textstyle\int \phi(x)p_{1,1}(x,t+s \mid y,0)dx = \\ &\textstyle\int p_{1,1}(z,t \mid y,0)p_{1,1}(x,t+s \mid z,t)[\phi(z) + (x-z)\phi'(z) + \\ &\tfrac{1}{2}(x-z)^2\phi''(z) + \ldots]dzdx \end{aligned} \tag{3.4.3}$$

Now let us assume that the order of the x and z integrals in eqn (3.4.3) can be interchanged, and concentrate on the z integral. The legality of the interchange should be proved, of course; this it is a delicate technical point which we omit here (Rozanov 1987). Then

$$\begin{aligned} &\textstyle\int p_{1,1}(x,s \mid z,0)[\phi(z) + (x-z)\phi'(z) + \frac{1}{2}(x-z)^2\phi''(z) + \cdots]dx = \\ &\textstyle\phi(z) + \phi'(z)\int(x-z)p_{1,1}(x,s \mid z,0)dx + \frac{1}{2}\phi''(z)\int(x-z)^2 p_{1,1}(x,s \mid z,0)dx \end{aligned} \tag{3.4.4}$$

Now let us suppose that the following limits exist:

$$\begin{aligned} A(x) &= \lim_{\Delta t \to 0} \frac{1}{\Delta t} \int (x-z)p_{1,1}(x,\Delta t \mid z,0),dz \\ B(x) &= \lim_{\Delta t \to 0} \frac{1}{\Delta t} \int (x-z)^2 p_{1,1}(x,\Delta t \mid, z,0)dz \\ 0 &= \lim_{\Delta t \to 0} \frac{1}{\Delta t} \int (x-z)^n p_{1,1}(x,\Delta t \mid, z,0),dz; \quad n > 2 \end{aligned} \tag{3.4.5}$$

In fact, the third equation of eqn (3.4.5) follows from the first two (Pawula 1967) and is not an independent assumption. In this case, eqn (3.4.4) becomes, for short times

$$\begin{aligned} &\textstyle\int p_{1,1}(x,s \mid z,0)[\phi(z) + (x-z)\phi'(z) + \frac{1}{2}(x-z)^2\phi''(z) + \ldots]dx = \\ &\phi(z) + sA(z)\phi'(z) + \tfrac{1}{2}sB(z)\phi''(z) + o(s) \end{aligned} \tag{3.4.6}$$

A rearrangement and integration by parts yield

$$\int \frac{[p_{1,1}(x,t+s|y,0)-p_{1,1}(x,t|y,0)]}{s}\phi(x)dx =$$
$$\int \phi(x)[-\frac{\partial}{\partial x}(A(x)p_{1,1}(x,t \mid y,0)) + \frac{1}{2}\frac{\partial^2}{\partial x^2}(B(x)p_{1,1}(x,t \mid y,0) + o(1)]dx \tag{3.4.7}$$

Finally let $s \to 0$. The difference quotient on the left-hand side of eqn (3.4.7) becomes $\partial p_{1,1}/\partial t$ and the $o(1)$ terms on the right vanish. Since ϕ is an arbitrary smooth function, we conclude that

$$\frac{\partial p_{1,1}(x,t \mid y,0)}{\partial t} = -\frac{\partial}{\partial x}(A(x)p_{1,1}) + \frac{1}{2}\frac{\partial^2}{\partial x^2}(B(x)p_{1,1}) \tag{3.4.8}$$

This equation is called the *Fokker–Planck* equation (Fokker 1914), (Planck 1917); it is also known as Kolmogorov's first equation, or the *forward* equation.

Starting from the Chapman–Kolmogorov equation for $p_{1,1}(x,t \mid y,t-s)$ and carrying through the same procedure on it yields another equation

$$\frac{\partial p_{1,1}(x,t \mid y,s)}{\partial s} = -A(y)\frac{\partial p_{1,1}}{\partial y} - \frac{1}{2}B(y)\frac{\partial^2 p_{1,1}}{\partial y^2} \tag{3.4.9}$$

Since this is derived by differentiation with respect to the earlier time, s, it is called the *backwards* equation. It is the adjoint of the forward equation.

We can summarize the assumptions made in deriving the Fokker–Planck equation, eqn (3.4.8). We assumed that the process is Markov, homogeneous, that $p_{1,1}$ is twice continuously differentiable, and that its transition moments, A and B, are both proportional to the time difference. Clearly, not all Markov processes have these properties; those that do are called *diffusion* processes.

At first glance, it is not clear how to use the Fokker–Planck equation. It contains two functions A and B, whose definition depends on $p_{1,1}$. Yet it is an equation for $p_{1,1}$, so it looks like an identity for the class of diffusion processes. Indeed, it was derived from an identity, the Chapman–Kolmogoroff equation. Yet there is one consideration that makes it more than an identity. A and B depend on $p_{1,1}$ only for very small time differences, in particular as $\Delta t \to 0$. Once A and B are calculated using this short time information, eqn (3.4.8) determines $p_{1,1}$ for all times. The necessary short time information to determine A can often be obtained from a short time perturbation theory. For thermodynamic systems, B is most easily obtained by requiring that the long time asymptotic solution for p_1 corresponds to the known p_1 for thermal equilibrium. We shall see how this works in Chapter 7.

3.5 Some Special Processes

A stochastic process, $\mathbf{X}(t)$, is said to be a process with *independent increments* if, given $n+1$ times, $0 \le t_1 \le t_2 \le \ldots \le t_n$, the random variables $\mathbf{X}(0), \mathbf{X}(t_1) - \mathbf{X}(0), \ldots, \mathbf{X}(t_n) - \mathbf{X}(t_{n-1})$ are pairwise independent. A physical example could be the number of photons arriving at a detector in the time interval $[0, t)$. If the photons are emitted independently from their source, the number of photons arriving in non-overlapping time intervals are independent.

A stochastic process is said to be a *Gaussian process* if all of its distribution functions are multivariate Gaussian distributions(cf. eqn (2.4.14)).

$$p_n(x_1, t_1; , x_2, t_2; \ldots ; x_n, t_n) = (2\pi)^{-n/2} \mid \mathbf{V} \mid^{1/2} \exp\left(-\frac{1}{2}\mathbf{x}^T\mathbf{V}\mathbf{x}\right) \quad (3.5.1)$$

The matrix $\mathbf{V}$ is the inverse of the matrix $\mathbf{B}$, where

$$B_{i,j} = < \mathbf{X}(t_i)\mathbf{X}(t_j) > \quad (3.5.2)$$

The set of Gaussian distributions for all n with the matrix $\mathbf{V}$ determined by eqn (3.5.2) clearly obey the positivity and symmetry requirements (3.2.2), since the integral of a multivariable Gaussian function over one of its variables is also Gaussian. They also obey the compatibility condition, eqn (3.2.3). This can be shown by detailed analysis of the matrix remaining when one x is integrated out, but an easier way is to consider the characteristic function.

Let us compute the characteristic function $\phi_n(y_1, t_1; \ldots ; y_n, t_n)$ of p_n. We use the symbol y for the characteristic function independent variable, instead of t as in Section 2.5, since here t is already used for the time variable. Since the Fourier transform of a Gaussian function is also a Gaussian, the result is

$$\phi_n = \exp\left(-\frac{1}{2}\mathbf{y}^T\mathbf{B}\mathbf{y}\right) \quad (3.5.3)$$

But integrating over x_n and then calculating the characteristic function of the resulting p_{n-1} is the same as just taking $y_n = 0$ in eqn (3.5.3). Consequently, the characteristic function of p_{n-1} is

$$\phi_{n-1} = \exp\left(-\frac{1}{2}\mathbf{y}'^T\mathbf{B}'\mathbf{y}'\right) \quad (3.5.4)$$

where $\mathbf{y}'$ is the column vector $\mathbf{y}$ truncated by removing y_n, and $\mathbf{B}'$ is the matrix $\mathbf{B}$ truncated to an $n-1 \times n-1$ matrix by removing the nth row and column. Consequently the matrix elements of $\mathbf{B}'$ are the same correlation functions of $\mathbf{X}$ as those occurring in $\mathbf{B}$ (except for those discarded in the truncation).

The next step is to take the inverse Fourier transform of (3.5.4), which yields

$$p_{n-1}(x_1, t_1; \ldots ; x_n, t_n) = (2\pi)^{(n-1)/2} \mid \mathbf{V}' \mid^{1/2} \exp\left(-\frac{1}{2}\mathbf{x}'^T\mathbf{V}'\mathbf{x}'\right) \quad (3.5.5)$$

where $\mathbf{V}'$ is the inverse of $\mathbf{B}'$. Note that merely truncating $\mathbf{V}$ does not yield $\mathbf{V}'$; it is necessary to truncate $\mathbf{B}$ before inverting it, not to invert and then truncate.

Gaussian processes, and especially Gaussian Markov processes, will occur often in our future considerations. A particularly important Gaussian Markov process is the *Wiener process*. The Wiener process is a homogeneous Gaussian Markov process with independent increments whose transition probability is given by

$$p_{1,1}(x,t \mid y,0) = (4\pi Dt)^{-1/2} \exp[-(x-y)^2/2Dt] \tag{3.5.6}$$

In Chapter 4 we shall study in detail how this process arises in a physical context. For the time being we merely note that it is the solution of the Fokker–Planck equation (with zero drift coefficient, A, and constant diffusion coefficient, D) which approaches a delta function as $t \to 0$.

3.6 Calculus of Stochastic Processes

A stochastic, or random, process is not a single function, but a family of functions, its sample functions. In order to describe limiting operations on members of this family, we need to introduce the concept of *convergence* of sequences of random variables. This concept is relevant to sequences of arbitrary random variables, not only random processes; it could, in fact, have been introduced in the last chapter.

There are several types of convergence that arise for random variables. Let $\{\mathbf{X}_n\}$ be a sequence of random variables, and $\mathbf{X}$ some fixed random variable.

1) The sequence $\{\mathbf{X}_n\}$ is said to converge to $\mathbf{X}$ *almost certainly* if $\mid \mathbf{X}_n - \mathbf{X} \mid\to 0$ for all sufficiently large n, except on some set of events of probability zero. This mode of convergence is sometimes called *strong convergence*.

2) The sequence $\{\mathbf{X}\}$ converges to $\mathbf{X}$ *in probability* if $P(\mid \mathbf{X}_n - \mathbf{X} \mid> 0) \to 0$ as $n \to \infty$. This mode of convergence is sometimes called *weak convergence*.

The difference between almost certain convergence and convergence in probability is that in almost certain convergence, the set on which the sequence does not approach $\mathbf{X}$ settles down to some fixed set of zero probability. For convergence in probability, the set where $\mid \mathbf{X}_n - \mathbf{X} \mid$ is not zero also becomes one of zero probability, but it need not be fixed. It may move about the event space, Ω, as n increases, never settling down.

3) The sequence $\{\mathbf{X}_n\} \to \mathbf{X}$ in *mean square* if $<\mid \mathbf{X}_n - \mathbf{X} \mid^2>\to 0$ as $n \to \infty$. This mode of convergence is analogous to mean square convergence in vector spaces. It is the kind of convergence for which most of the results of stochastic calculus have been derived.

4) $\{\mathbf{X}_n\}$ is said to converge to $\mathbf{X}$ *in distribution* if the cumulative distribution functions of the $\mathbf{X}_n$ approach the cumulative distribution function of $\mathbf{X}$ at all continuity points of this latter. Since the cumulative distribution functions are ordinary sure functions, their convergence needs no special discussion here. The convergence in the de Moivre–Laplace limit theorem (Section 2.4) and the central limit theorem (Section 2.5) are examples of convergence in distribution.

Having the idea of convergence, we may now ask, what does it mean to say that a stochastic process is continuous? We shall say that a stochastic process is continuous if, with probability one, all of its sample functions are continuous functions of t. Recall that each sample function corresponds to a point in the space of events, or states, and so the statement that sample functions have a property with probability one makes sense.

The example given in Section 3.3 was not continuous, since the value of the random functions could only be one of two possibilities. A process can take values in a continuous range, and still not be continuous. For example, a process could have discontinuous jumps of arbitrary size occurring with a continuous probability distribution of the jump position and the jump size. Such a process would not be continuous even though the values could occur in a continuous range.

Sufficient conditions to determine whether a random process is or is not continuous are known, but are not easy to apply; it is often quite difficult to verify the hypotheses of the continuity criteria. However, for Markov processes there is a relatively simple criterion. It is, if

$$\lim_{\Delta t \to 0} \frac{1}{\Delta t} \int_{|x-y|>\epsilon} p_{1,1}(x, t + \Delta t \mid y, t) dx = 0 \tag{3.6.1}$$

for all $\epsilon > 0$, then the process $\mathbf{X}(t)$ is continuous (Ghikman and Skorohod 1974; 178). This condition means that finite jumps of arbitrarily small size become very improbable for sufficiently short time intervals. Gardiner (1983) illustrates this very nicely with two simulations: the transition probability of the Wiener process, eqn (3.5.6), satisfies eqn (3.6.1), and yields a simulated sample function that is very jagged, but continuous. (It is exactly this transition probability that arises in Einstein's theory of Brownian motion. We shall discuss in detail in the Chapter 4.) On the other hand,

$$p_{1,1}(x, t + \Delta t \mid y, t) = \frac{\Delta t}{\pi} \frac{1}{(x-y)^2 + (\Delta t)^2} \tag{3.6.2}$$

the Cauchy distribution, does not satisfy eqn (3.6.1) and yields a simulated sample function that has many discontinuous jumps of varying amplitudes.

There is another notion of continuity, called *stochastic continuity*, which we shall not discuss because we shall not need it in this book. We only mention it here to avoid possible confusion on the part of readers who encounter it in a further study of stochastic processes. Roughly speaking, stochastic continuity means that almost all sample functions are continuous at each t, but the exceptional sample functions that are discontinuous at t may be different for different t. Continuity, in the sense defined in the previous paragraphs is a stronger statement; the same set of sample functions are continuous over the same range of t. Still another notion of continuity, *mean-square continuity*, is based on the notion of mean square convergence. We shall not use this concept, and mention it only for those who wish to study the subject further.

Once there is a concept of continuity, the next step is to ask about differentiability. Although differentiability of stochastic processes can be defined in a similar fashion to continuity it is rarely done. Smoothness (differentiability) is a more delicate property than lack of jumps (continuity), and it is rarely necessary to enquire about the differentiability of a random process. Differentiability of sample functions can, of course, be studied by the ordinary methods of the calculus since sample functions are ordinary functions.

Integration is another matter. Integrals of random processes are very important, especially in the theory of Brownian motion, so we must devote some time to this topic. Indeed, it will be the main subject of Chapter 5. One kind of integral of random processes is easy to define. Let $\mathbf{X}(t)$ be a random process, $x(t)$, one of its sample functions, and $f(t)$ some fixed function. Then we define

$$\begin{aligned} &\textstyle\int_a^b f(t)\mathbf{X}(t)dt = \text{The family of sample functions} \\ &\{\lim \textstyle\sum f(t'_k)x(t'_k)(t_k - t_{k-1})\}; \quad t_{k-1} \leq t'_k \leq t_k \end{aligned} \tag{3.6.3}$$

The quantity in braces is the familiar Riemann sum defining the integral of the sample function, x. If the sample functions are integrable, for example if $\mathbf{X}$ is continuous, then the integral of $\mathbf{X}$ defined by (3.6.3) will exist. Similarly, if the sample functions of $\mathbf{X}$ are of bounded variation, one can define the Stieltjes integral $\int f(t)d\mathbf{X}(t)$ as the family of Stieltjes integrals of the sample functions.

$$\lim \sum_k f(t_k)(x(t_k) - x(t_{k-1}))$$

It is a common occurrence, however, that the processes we may want to integrate are not of bounded variation. Especially in Brownian motion theory, we need to integrate very wildly varying functions representing the random impacts of the surroundings on the Brownian particle. We would like to define integrals of the structure $\int \phi(\mathbf{X}(t))d\mathbf{X}(t)$ when $\mathbf{X}(t)$ is not of bounded variation. We might try to define them as limits of Riemann–Stieltjes sums

$$\lim \sum \phi(x(t'_k))(x(t_k) - x(t_{k-1}))$$

where the limit must be taken, as usual, as the number of subdivisions of the interval (a, b) becomes infinite, the size of the largest of them approaches zero, and $t_{k-1} \leq t'_k \leq t_k$. If the limit exists, is independent of the mode of subdivision, and of the choice of the t', then we say that the integral exists and is equal to the limit. Unfortunately, for the wildly varying functions arising in Brownian motion theory, while the limit often exists, it is *not* independent of the choice of the intermediate points, t'_k. If $x(t)$ varies very rapidly in the interval $t_k - t_{k-1}$, no matter how small that interval, then the limit of the Riemann–Stieltjes sum will clearly depend on how those intermediate points are chosen. Notice that this problem does not arise if the function ϕ does not depend on $\mathbf{X}$, but is merely a function of t.

This difficulty can be circumvented by specifying how the t'_k points are to be chosen in constructing the sums. There are many ways of doing this, and each of them will give rise to a different definition of the integral. There is no reasonable unique prescription.

There are two definitions that have been used (though many more are possible). The first in point of time, and the one almost exclusively used by the mathematical community is due to K. Ito (1944); see also Gardiner (1983). In Ito's definition, t'_k is always chosen to be t_k, the value at the end of the k'th interval. The second definition is due to R. Stratonovich (1966). According to this definition, one takes $t'_k = (t_k + t_{k-1})/2$, the midpoint of the kth interval.

Ito's integral has the advantage that the resulting integral belongs to a class of stochastic processes about which much is known. Rigorous proofs have been given for the properties of the integral. It has the disadvantage that it does not obey all of the rules of elementary calculus. As an example (Gardiner 1986), if $\mathbf{W}(t)$ is the Wiener process, then

$$\int_a^b \mathbf{W}(t)d\mathbf{W}(t) = \frac{1}{2}[\mathbf{W}^2(b) - \mathbf{W}^2(a) - (b-a)] \tag{3.6.4}$$

If this were an ordinary integral, the $b - a$ term would, of course, not be present. Stratonovich's integral, on the other hand, does obey the rules of elementary calculus. That is its advantage. Its disadvantage is that it is very difficult to prove theorems about its properties: existence, interchange of limits, convergence of integrals of sequences, etc.

Mathematicians find the advantages of the Ito definition overwhelming, and so use it almost exclusively. Physicists find the advantage of the Stratonovich definition compelling. Furthermore, they argue that physical processes are actually smooth on extremely short time scales, say the scales of molecular collision times. The wildly varying functions used to describe the observed motions are actually approximations valid on time scales which, while short in comparison to observation times, are still long compared to these microscopic times. As the divisions of the t axis used in defining the integral get finer and finer, the time scale is eventually reached where the sample functions should be looked on as ordinary functions to which the rules of calculus should apply. Then, when the actual system paths are replaced by their approximations by wild functions, the replacement should be done in such a way as to preserve these rules. This means: use the Stratonovich definition.

In a sense, however, it does not matter which definition is used, as long as it is used consistently! This means, not only sticking to one definition, but also using the proper associated formulas for change of variables, and the proper coefficients in the associated Fokker–Planck equation. We shall discuss this in detail in Chapter 5.

3.7 Fourier Analysis of Random Processes

In the analysis of ordinary functions, it is often useful to decompose the function into frequency components, that is, to construct either a Fourier series or a Fourier integral representation of the function. The same thing can be done with random processes. Since random processes are families of functions, not single functions, one will only be able to determine statistical properties of the

Fourier coefficients or transforms. Nevertheless, much useful information about the process can be obtained. This is especially true in applications such as communications engineering, where the problems of extracting signals (deterministic functions) from noise (random processes) are very important. We shall not treat communications problems in this book, but Fourier techniques are applicable to Brownian motion theory also.

The sample functions of a random process are generally not periodic functions. Therefore they cannot be developed in Fourier series in the conventional way. Nor do these functions vanish as $t \to \pm\infty$. Consequently they do not have Fourier integrals in the conventional sense. So how are we to apply Fourier - like ideas, that is, the decomposition of a function into harmonic components, to such functions? We treat the integral case first, and then return to sketch the series approach.

Let $x(t)$ be a real sample function of the stochastic process $\mathbf{X}(t)$. For the sake of simplicity, suppose $< x(t) >= 0$; if this is not the case, consider instead the function $x(t)- < x(t) >$, whose mean does vanish identically. Let T be some time (eventually to approach infinity, but for the moment a finite time), and define a new function

$$x_T(t) = \begin{cases} x(t) & -T/2 \leq t \leq T/2 \\ 0 & \text{otherwise} \end{cases} \tag{3.7.1}$$

In other words, x_T is a 'clipped' version of x. In the paragraphs to follow, we shall drop the subscript T for the sake of economy of notation, but with the understanding that, as long as T is finite, we really mean the clipped function. Now x, being zero outside of a finite interval, can be expressed as a Fourier integral in the ordinary way

$$\begin{aligned} A_T(\omega) &= \int_{-T/2}^{T/2} x(t)e^{-i\omega t}dt \\ x(t) &= \int_{-\infty}^{\infty} A_T(\omega)e^{i\omega t}\frac{d\omega}{2\pi} \end{aligned} \tag{3.7.2}$$

Since $x(t)$ is real, we must have $A(\omega) = A^*(-\omega)$, where * means complex conjugate. Form

$$x(t+s)x(t) = \int\int_{-\infty}^{\infty} A_T(\omega)A_T^*(\omega')e^{i[\omega(t+s)-\omega' t]}\frac{d\omega}{2\pi}\frac{d\omega'}{2\pi} \tag{3.7.3}$$

Now suppose that $\mathbf{X}$ is a homogeneous process. First take the expectation of eqn (3.7.3), and then take the time average of both sides

$$\tfrac{1}{T}\int_{-T/2}^{T/2} < x(t+s)x(t) > dt =< x(s)x(0) >=$$

$$\tfrac{1}{T}\int < A_T(\omega)A_T^*(\omega') > e^{i\omega s}\int_{-T/2}^{T/2} e^{i(\omega-\omega')t}\tfrac{d\omega}{2\pi}\tfrac{d\omega'}{2\pi}dt \tag{3.7.4}$$

As T approaches infinity,

$$\int_{-T/2}^{T/2} e^{i(\omega-\omega')t}dt \to 2\pi\delta(\omega-\omega')$$

where δ denotes the Dirac delta function. We also suppose that

$$\lim_{T\to\infty}\frac{|A_T(\omega)|^2}{T} \equiv G(\omega) \tag{3.7.5}$$

exists. Then eqn (3.7.4) becomes

$$< x(s)x(0) >= \int_{-\infty}^{\infty} G(\omega)e^{i\omega s}\frac{d\omega}{2\pi} = \frac{1}{\pi}\int_0^{\infty} G(\omega)\cos\omega s\, d\omega \tag{3.7.6}$$

The second equality in eqn (3.7.6) arises from the fact that $G(\omega)$ is an even function because $A(\omega) = A^*(-\omega)$.

The quantity $G(\omega)$ is called the *power spectrum* or just the *spectrum* of the process **X**. The name comes from communications engineering, where x is often a noise voltage or a noise current; then $< x^2 >$ is proportional to the power of the noise and this is just $\int G(\omega)d\omega$.

The normalized correlation function

$$\rho(t) = < x(s+t)x(s) > / < x^2 >$$
$$S(\omega) = G(\omega)/\int_0^{\infty} G(\omega)d\omega \tag{3.7.7}$$

is sometimes convenient. With these functions, eqn (3.7.6) takes the simple form

$$\rho(t) = \int_0^{\infty} S(\omega)\cos\omega t\, d\omega \tag{3.7.8}$$

Equation (3.7.6) or (3.7.8) is called the *Wiener–Khinchin* theorem. The demonstration of the theorem presented here is suggestive rather than compelling.[13] The order of sending T to infinity and integration has been interchanged. Limits have been assumed to exist without investigation. Furthermore, **X** is a stationary process by hypothesis, but the clipped **X** is not; a privileged time has been introduced by the clipping. Nevertheless, unless t is very near $\pm T/2$ and s is less than several characteristic times, **X** will behave as though it were stationary. In the limit as $T\to\infty$ the result will still be valid. Wiener and Khinchin proved the theorem with complete rigor.

[13] 'A demonstration will convince a reasonable person. A proof will convince a skeptic.' Mark Kac, remark during a lecture.

As mentioned above, the problem of a frequency decomposition of a random process can also be approached from the theory of Fourier series. Instead of clipping the sample functions so that they vanish outside of $(-T/2, T/2)$, we take the sample functions in this interval and continue them periodically for $|t| \geq T/2$. Then each sample function can be represented by a Fourier series

$$x(t) = \sum_{k=-\infty}^{\infty} a_k e^{i\omega_k t}; \quad \omega_k = \frac{2\pi k}{T} \tag{3.7.9}$$

As before, because x is real, $a_k = a^*_{-k}$.

Using the orthogonality relation

$$\int_{-T/2}^{T/2} e^{i(\omega_k - \omega_l)t} dt = T\delta_{k,l} \tag{3.7.10}$$

and the stationarity of x,

$$< x(t)x(0) >= \sum_{k=-\infty}^{\infty} | a_k |^2 e^{i\omega_k t} \tag{3.7.11}$$

Now assume that $\lim T \mid a_k^2 \mid = G(\omega_k)$ exists. Letting T increase without bound yields just eqn (3.7.6).

The frequency decomposition of random noise is used a great deal in communications engineering, where detection apparatus is often frequency sensitive. There are many sources of noise in communications devices. Among them is the Brownian motion of electrons or other charge carriers in circuits. That is one reason why the Fourier analysis of random processes is appropriate in this book.

3.8 White Noise

A random process whose spectrum, $G(\omega)$, is independent of frequency is called *white noise*. The name comes from an analogy with white light, whose spectrum is independent of frequency.

If $\mathbf{X}(t)$ is a white noise, then its correlation function $< x(t)x(0) >$ is a delta function, $\delta(t)$. That is, $\mathbf{X}(t+s)$ and $\mathbf{X}(t)$ are uncorrelated, no matter how small s is. There are no functions in the normal mathematical sense of the word that behave this way. In the physical sense also, this would be extraordinary behavior. All physical systems have some inertia, however small. Hence they cannot change infinitely rapidly, as would be required for a truly white spectrum.

Nevertheless, the concept of white noise is neither nonsense nor useless. If all times of physical interest are much longer than the correlation time of the random process, then the correlation function is essentially zero over all times of interest, and the process appears to be white (until measurements are made at much shorter times or much higher frequencies). Thus white noise can be considered to be a convenient model for processes with short correlation times. Similarly, from the mathematical point of view, white noise can be considered to be a limit of well defined stochastic processes. This outlook is similar to one way to look at the Dirac δ function. The δ function is not a function; it is a linear functional. But it can be considered as a limit of a sequence of sharply peaked proper functions.

White noise is often introduced as a limiting case of shot noise, a stochastic process introduced to model the result of a stream of particles arriving at random times at a target. It would take us too far afield to develop the theory of shot noise here, but we can illustrate how white noise arises from a limiting process by using the dichotomous model already introduced in Section 3.3.

We want to consider a stationary process; this means that p_1 should be independent of time. From eqn (3.3.7), this requirement holds only if $p_a = \mu/(\mu+\lambda)$ and $p_b = \lambda/(\mu+\lambda)$. Recall that the dichotomic random variable is defined only for the values a or b. Define a function $\phi(t)$ by

$$\begin{aligned} \phi(t) &= \begin{cases} 1 & \text{if the system is in state a} \\ -1 & \text{if the system is in state b} \end{cases} \\ \delta\phi(t) &= \begin{cases} \frac{2\lambda}{\lambda+\mu} & \text{if the system is in state a} \\ \frac{-2\mu}{\lambda+\mu} & \text{if the system is in state b} \end{cases} \end{aligned} \tag{3.8.1}$$

where $\delta\phi_a(t) = \phi_a(t) - <\phi>$. Clearly $\delta\phi$ has mean zero and variance $4\lambda\mu/(\lambda+\mu)^2$. It is now easy to compute $<\delta\phi(t)\delta\phi(s)>$, with the result

$$<\delta\phi(t)\delta\phi(s)> = \frac{4\lambda\mu}{(\lambda+\mu)^2} e^{-(\lambda+\mu)(t-s)} \quad t > s \tag{3.8.2}$$

The spectrum corresponding to this correlation function is

$$S(\omega) = \frac{2}{\pi} \frac{\lambda+\mu}{\omega^2 + (\lambda+\mu)^2} \tag{3.8.3}$$

It is clear from eqn (3.8.3) that $S(\omega)$ is very close to $S(0)$ when ω is much less than $\lambda+\mu$. When λ and μ become very large, S will be approximately constant over a very wide frequency interval. This corresponds to a very rapid rate of switching between state a and state b. In the limit as μ and λ become infinite, the spectrum becomes white and thus the correlation function becomes a delta function, $\delta(t-s)$.

Also, eqn (3.8.2) shows directly that the normalized correlation function $\rho(t)$, approaches a delta function as the rates approach infinity. The normalized correlation function is $\rho(t) = \frac{1}{2}(\mu + \lambda \exp[-(\mu+\lambda)t]$. If $f(t)$ be an arbitrary integrable function, then

$$\begin{aligned} \int_{-\infty}^{\infty} \frac{2}{\tau} e^{-t/\tau} f(t)dt &= 2\int_{-\infty}^{\infty} e^{-s} f(s\tau)ds \\ &\to 2f(0) \text{ as } t \to 0 \end{aligned} \tag{3.8.4}$$

Thus the stochastic process defined by (3.8.1) approaches white noise as the transition rates become unbounded.

Of course, in this simple example the stochastic process ϕ can take on only the two values ± 1. The process is certainly not a Gaussian process. In more realistic models of natural processes, the randomness is due to the cumulative effect of many small causes. Because of the central limit theorem, (Section 2.5), if each of these small causes is a dichotomic process we would expect the resulting process to be Gaussian. Indeed, Gaussian white noise is the common model for process with correlation times that are extremely short on the time scales relevant to the process being studied. Yet the simple two state model does illustrate the salient feature of white noise: many rapid transitions in short time intervals.

3.9 Conclusion

In this chapter, we have sketched some of the main features of the theory of stochastic processes. There is much more that could be said. The theory is quite rich, and we have consciously omitted many facets. In particular, we have left out all topics which did not bear directly on the subject of this book, Brownian motion. Furthermore, although we have tried to be precise in statements of mathematical fact, we have not attempted rigorous proofs here. The reader interested in rigor is referred to the following references: Iranpour and Chacon (1988), Ghikman and Skorokhod (1975), Doob (1953). Treatments more rigorous than given here, but less so than the preceding references can be found in Gardiner (1983) and Gillespie (1992).

There is also much more that can be said about Brownian motion from the mathematical point of view, and we shall return to mathematical topics in Chapters 5 and 6. For now, however, we have tarried long enough on the mathematical background. It is time to begin looking at the physical situation. This is the subject of the next chapter.

4

EINSTEIN–SMOLUCHOWSKI THEORY

4.1 What is Brownian Motion?

When mobile particles are immersed in an ambient medium, the particles undergo an incessant and irregular motion. This is the Brownian motion, whose history was outlined in Chapter 1. The most common case of Brownian motion is the one originally observed by Brown, particles suspended in a fluid. However, less commonly more exotic situations are encountered, for example, an electron immersed in a black body radiation field (a gas of photons at equilibrium). Also, the motion of the particles need not be translational; rotational motion can also partake of the incessant and irregular character which we call Brownian. Brownian motion is ubiquitous and universal. Thousands of cases had been examined, even by the beginning of the twentieth century. In each case the motion was found, and had the same general character. Some of the general properties are as follows: The rapidity of the motions are greater, the smaller the size of the suspended particles. Smoluchowski (1906a) uses the term motional velocity (*Bewegungsgeschwindigkeit*) for what we call here rapidity. As we shall see, the velocity of a Brownian particle is not a measurable quantity so that we have used a less definite term.

For a particle of radius 1 mm in water the motion can hardly be noticed, while for particles on the edge of microscopic visibility the motion is marked. There is a definite dependence on the nature of the fluid medium, and especially on its viscosity. Another characteristic of Brownian motion is its stability in time. The motion persists as long as the particles remain suspended in the fluid. This has been observed in preparations allowed to stand for over a year. A final very characteristic property is independence of most external influences. Electric fields, light (as long as it is not absorbed and does not heat the system), gravity (as long as the particles do not settle out) and similar disturbances from the outside seem to have no effect.

Temperature has a marked effect, however. This could be expected from the observation that the motion depends on the viscosity of the medium; as is well known, the viscosity of fluids is appreciably temperature dependent. Whether there is a residual temperature effect, above that due to the temperature dependence of viscosity, cannot be ascertained on experimental grounds alone. Without a theory to tell how to determine the effect due to viscosity it is not possible to see if there is additional temperature dependence.

What mechanism is universal, stable, and independent of most external influences? The first answer that comes to mind is molecular collisions. The kinetic theory of matter asserts that the molecules of a fluid are constantly in motion with a mean kinetic energy proportional to the temperature. For systems in equilibrium, this kinetic motion is stable in time, and is independent of external influences. Thus we conjecture that the observed irregular motions of a suspended particle are due to irregular transfers of energy and momentum from the fluid molecules to the particle due to irregularly occurring molecular collisions between the suspended particle and the medium particles.

The suspended particle is, as a rule, much heavier than the medium particles. Thus its velocity change upon collision with a medium particle will be very small. If V denotes the velocity of the heavy particle, and M its mass, while v and m refer to the same quantities for the light particle, then the change of V on collision with a light particle is of the order of magnitude $\Delta V \approx (m/M)v$. This follows from the laws of elastic collisions. Now v is, on average, of the order of several hundred meters per second for water at room temperature. m/M is of the order of 10^{-13}. To make this estimate I have assumed a particle with a radius of 10^{-3} cm and density 1 g cm^{-3}, roughly comparable to the particles of Robert Brown's original observations. Consequently,[14] ΔV is of the order of 10^{-8} cm s^{-1}. Of course, this is only a rough estimate, but even if it is off by a factor of 1000, we can still conclude that the velocity change of the heavy particle due to a single collision will be very small.

Furthermore, collisions increasing the velocity of the heavy particle will, on average, be balanced by collisions decreasing that velocity. Thus, the net mean change in velocity due to many collisions will be much smaller than that due to an individual collision. How then can the observed motion be due to collisions? Although the average effect of the collisions will indeed be small, there will be fluctuations about that average. Fluctuations large enough to lead to observable effects, while relatively rare, are still common enough to explain the phenomenon. A one-dimensional toy model, due to Smoluchowski (1906a), illustrates this nicely.

> A test particle moving on a line is bombarded by other particles from the right and the left. A particle hitting from the left gives the test particle a fixed velocity increment Δv towards the right; one hitting from the right gives an increment $-\Delta v$ (to the left). After a certain time, n collisions have occurred. The collisions are independent. What is the mean magnitude of the total velocity increment? One could equally well ask for the root-mean-square velocity increment; for qualitative purposes,the two are equivalent.
>
> Let us suppose that, of the n collisions, m are from the right, and $n-m$ from the left. Then the total velocity increment, in units of Δv, is $2m-n$. By the hypothesis that the collisions are independent, the distribution of the random variable $\mathbf{m}$ is given by the binomial distribution (Section 2.4)

[14] Smoluchowski estimates this number to be about 10^{-6}cm s^{-1}.The difference arises because he considers a slightly smaller particle, hence one of lower mass.

$$P_n(m) = \frac{1}{2^n}\frac{n!}{m!(n-m)!} \tag{4.1.1}$$

Consequently

$$<\mid 2n - m \mid> \sim \sqrt{\frac{2n}{\pi}} \tag{4.1.2}$$

$$<\mid 2n - m \mid> = \frac{2}{2^n}\sum_{m=n/2}^{n}\frac{(2m-n)n!}{m!(n-m)!} = \frac{n}{2^n}\frac{n!}{[(n/2)!]^2} \tag{4.1.3}$$

For large values of n, this approaches

$$<\mid 2n - m \mid> \approx \sqrt{\frac{2n}{\pi}} \tag{4.1.4}$$

Now, in air, a particle undergoes of the order of 10^{16} collisions per second; in water, this may be 10^{20} s^{-1}. Therefore, in a time of the order of one second, it is rather likely that a particle can attain a velocity of the order of 10^8–10^{10} of the change on one collision. The fluctuations in velocity due to fluctuation in collision numbers can therefore explain the observed Brownian motion qualitatively. The rest of this chapter will show how the explanation is not only qualitative, but can also be made quantitative.

A word of warning: do not take the detailed predictions of this toy model too seriously. It assumes that there is a unique velocity increment imparted by a collision. In actuality there is a distribution of such increments. Furthermore, it assumes that left and right collisions are equally likely, independent of the magnitude and direction of the velocity of the test particle. Clearly collisions giving a Δv in the same direction as v itself are less likely than those in the opposite direction, especially for large v. It is this latter assumption which prevents the mean v from settling down to an equilibrium value. Nevertheless, it is an instructive model which illustrates in a simple manner the importance of fluctuations.

4.2 Smoluchowski's Theory

We now begin to treat in some detail the quantitative theories of Brownian motion of Einstein and Smoluchowski, which were described in outline in Chapter 1. Although Einstein's work was first chronologically by some months, we shall reverse the order of appearance and discuss Smoluchowski's work first, for Smoluchowski's work is based on a specific detailed kinetic model, collisions of hard spheres. Thus it gives considerable physical insight into the mechanism behind the fluctuations. On the other hand, many approximations are needed to deduce the consequences of the model. Furthermore, the theory is not generalizable, at least not easily, to other situations.

Einstein's considerations, in contrast, are based on statistical assumptions of a general nature, not tied to a specific model, which is both its strength and its weakness. The strength is that it is applicable to a wide range of circumstances, and is generalizable. The weakness is that it yields little insight on what is happening on the microscopic dynamical time scale.

Time scales in physical systems can usually be divided into three ranges: the *microscopic*, *macroscopic*, and *mesoscopic*. The microscopic time scale is the scale appropriate for the description of elementary dynamical events, for example the duration of a collision, or the mean time between collisions. The macroscopic time scale is the scale appropriate for the description of laboratory scale phenomena, for example the relaxation time of a disturbance of measurable size. The mesoscopic time scale lies between these two (from the Greek *mesos*– middle), and refers to time intervals long enough to contain many elementary events, yet short enough to be effectively infinitesimal on an observational time scale. Using this terminology, Smoluchowski's theory is microscopic; Einstein's is mesoscopic.

So we consider a system made up of hard spheres—light ones with mass m and heavy ones with mass M. The heavy particles are so dilute that we can completely neglect any collision of heavy particles with other heavy particles. If the mean (actually root-mean-square) velocity of the light particles is c and that of the heavy ones is C, then the equipartition theorem states that

$$\frac{c}{C} = \left(\frac{m}{M}\right)^{\frac{1}{2}} \tag{4.2.1}$$

Let the vector velocities of the light and heavy particles be denoted by $\mathbf{v}$ and V respectively, so that $c^2 = < \mathbf{v} \cdot \mathbf{v} >$ and $C^2 = < \mathbf{V} \cdot \mathbf{V} >$. The velocities of two particles entering into a collision are indicated by unprimed letters, while those just after a collision are denoted by primed letters, i.e. $\mathbf{v}'$ and $\mathbf{V}'$. Let $\mathbf{g}$ denote the relative velocity $\mathbf{v} - \mathbf{V}$. The kinematics of collision then tell us that $g' = g$ and that

$$\mathbf{V}' = \mathbf{V} - 2\frac{m}{M_{12}}(\mathbf{g} \cdot \mathbf{k})\mathbf{k} \tag{4.2.2}$$

where $M_{12} = m + M$, and $\mathbf{k}$ is a unit vector normal to the common tangent plane of the spheres at their point of contact: compare Fig. 4.1.

Equation (4.2.2) shows that $C = C' + O((m/M)^2)$, on average. We shall follow Smoluchowski, and neglect effects of order $(m/M)^2$. Consequently, we may consider that the magnitude of the velocity of the heavy particle does not change on collision; only its direction of motion changes. Smoluchowski states that the angle, ϵ, between $\mathbf{V}'$ and $\mathbf{V}$ is given by $\sin \epsilon = (3/4)(m/M)(c/C)$ 'from the laws of collisions of elastic spheres.'[15] There are two limiting cases to be

[15] He does not explain this calculation. Fürth reports obtaining the value 0.806 instead of 3/4 in attempting to reproduce the calculation; see Brush (1968, footnote 83). My attempt to verify the formula yielded $\pi/4$, intermediate between the values of Smoluchowski and Fürth. The differences are small relative to the other approximations of the theory and it does not seem worthwhile to try to sort them out further.

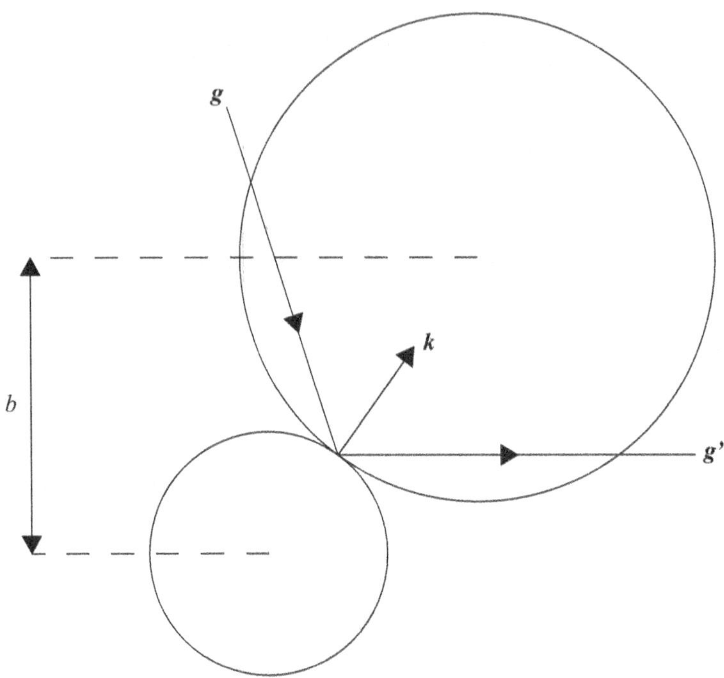

FIG. 4.1. The geometry of a collision of two hard spheres of differing radii.

considered. In the first, the radius of the heavy particle is much less than its mean free path, $R/l \ll 1$; in the second, the opposite holds, $R/l \gg 1$.

The first corresponds to the heavy particle being immersed in a dilute gas, the second to its being immersed in a dense gas or liquid. We consider the dilute gas case first. Because of the low density of the medium gas, a collision of a given light particle with the heavy one will be followed by many collisions of the heavy particle with other light molecules before it collides with the original light particle again. By that time the original light molecule will have undergone many collisions with other light molecules; its dynamical state will scarcely depend on its state before the previous collision with the heavy one. Thus we may consider successive collisions of the heavy particle with its surroundings to be independent events. Let us assume that the length of all trajectories of the heavy particle between collisions is the same; this will be l, the mean free path of the heavy particle. Now we come to the central question: the particle takes N steps of length l, each making an angle , ϵ, with the previous step. What is its root-mean-square distance from its starting point (which we can take to be the origin)? It is supposed that the end of the jth step can lie with equal probability anywhere on the base of the cone of angle ϵ whose axis is in the direction of the $j-1$th step.

Smoluchowski solved this problem by a rather complicated method. Nowadays, it is simpler to recognize that it is the same problem as that of finding the mean end-to-end distance of a polymer chain without excluded volume, but with fixed valence angle, ϵ. In the polymer case, angles of interest are of the order of 109°, while in the case of interest here the angle is very small. This does not affect the solution.

Let $\mathbf{l}_i$ be a vector of length l in the direction of the ith step. Then what we want is

$$< R_N^2 >=< \sum_{i,j=1}^{N} \mathbf{l_i} \cdot \mathbf{l_j} > \tag{4.2.3}$$

Consider a particular term $< \mathbf{l}_i \cdot \mathbf{l}_{i+p} >$. We compute this average by first averaging over the direction of $\mathbf{l}_{i+p}$ for fixed values of $\mathbf{l}_{i+p-1}$, $\mathbf{l}_{i+p-2}$, etc. Then we average over the result for fixed $\mathbf{l}_{i+p-2}$, etc., until we finally average over $\mathbf{l}_i$. The average of $\mathbf{l}_{i+p}$ given $\mathbf{l}_{i+p-1}$ is just $l_{i+p-1} \cos\epsilon$. The average of this gives $\mathbf{l}_{i+p-2} \cos^2\epsilon$. This process continues until the final $\mathbf{l}_i$ is averaged; this last average is $< \mathbf{l}_i \cdot \mathbf{l}_i >= l^2$. Consequently $< \mathbf{l}_i \cdot \mathbf{l}_{i+p} >= l^2 \cos^p \epsilon$.

There are $2(N-p)$ terms of this structure in the sum of eqn (4.2.3) for any given i. Hence eqn (4.2.3) reduces to

$$< R_N^2 >= l^2[N + 2\{(N-1)\cos\epsilon + (N-2)\cos^2\epsilon + \ldots + \cos^{N-1}\epsilon\}] \tag{4.2.4}$$

This sum can be simplified further, using the properties of geometric progressions, to yield, finally

$$< R_N^2 > /l^2 = N(\frac{1+\cos\epsilon}{1-\cos\epsilon}) - 2(\frac{\cos^2\epsilon - \cos^{N+1}\epsilon}{(1-\cos\epsilon)^2}) \tag{4.2.5}$$

The most interesting limiting case of this formula for the collision problem under discussion is: N very large and $N(1-\cos\epsilon)$ also large. In this limit

$$< R_N^2 >^{\frac{1}{2}}= l\sqrt{\frac{2N}{1-\cos\epsilon}} \tag{4.2.6}$$

Finally, if the heavy particle makes ν collisions per unit time, so that $N = \nu t$, we can write

$$< R_N^2 >^{\frac{1}{2}}= l\sqrt{\frac{2}{1-\cos\epsilon}}\, t^{\frac{1}{2}} \tag{4.2.7}$$

Since ϵ is very small, we may write $\epsilon = 3/4(mc/MC) = 3/4(C/c)$, and $l = C/\nu$. Thus, finally

$$< R_N^2 >^{\frac{1}{2}}= \frac{8}{3}\frac{c}{\sqrt{\nu}} t^{\frac{1}{2}} \quad \text{(t large)} \tag{4.2.8}$$

There are two important features of this result. First, the root-mean-square displacement of the heavy particle is independent of the heavy particle mass.

Second, it is proportional to the square root of the elapsed time. This means that if we attempt to compute the particle velocity by taking the limit of $< \Delta R^2 >^{\frac{1}{2}} / \Delta t$ (this is, after all, the definition), one would find that the limit does not exist. The trajectory of the heavy particle has no derivative; the particle does not have a well defined velocity.

How can this be? After all, we have assumed that the particle does have a velocity in carrying out the derivation. Recall that eqn (4.2.8) is a long time result; in deriving it, we assumed that $N\epsilon^2 \gg 1$. For $N\epsilon^2 \ll 1$ a quite different result follows from eqn (4.2.5)

$$< R_N^2 >^{\frac{1}{2}} = \nu l t \quad (t \text{ small}) \tag{4.2.9}$$

There is a crossover between the square root behavior and the linear behavior at some intermediate time very short compared to times of observation. The paradox of non differentiability only arises because we extrapolate the behavior of the function from macroscopic and mesoscopic times down to microscopic times. In the kinetic model we are considering, eqn (4.2.5) shows directly how the crossover occurs. It is important to keep this in mind because in the Einstein theory, which we shall treat a little later, the underlying microscopic dynamics are not explicitly used. The precise way that a crossover occurs is not so clear.

4.3 Smoluchowski Theory Continued

So far we have treated the case of a heavy particle in a dilute gas of light particle; the collisions of the light particles with the heavy ones may be considered to be independent events. In liquids by contrast, the heavy particle in interaction with many light particles simultaneously, and the collisions can be highly correlated. In this section we generalize the work of the previous section to cover this case, the one of primary interest. Again we follow Smoluchowski's arguments.

We cannot make detailed dynamical calculations in dense systems; it is too difficult. Consequently we look for a less exact but hopefully simpler method of analysis. If one places a particle with velocity V_0 in a viscous medium, the velocity in the direction of $\mathbf{V}_0$ decays according to the law

$$V_{\|} = V_0 e^{-t/\tau} \tag{4.3.1}$$

where τ is a relaxation time given by M/ζ. Here ζ is the friction coefficient. For spherical macroscopic particles, for example, when the velocity is not too high, ζ is given by Stokes' law, $\zeta = 6\pi\eta a$, η being the viscosity of the medium and a the radius of the particle. However, this cannot be the case for all time, for when $V_{\|}$ has decreased to the order of its equipartition value, it must remain there. When the particle is heavy, this equipartition value is small. The quantity τ may be taken as an estimate of the length of time before this decay occurs. That is, τ is a measure of the time during which the motion of the heavy particle can be considered to be rectilinear. Thus τV_0 is a measure of the length of the rectilinear section of the particle's path. Smoluchowski then makes the fundamental assumption that the motion of the heavy particle can be considered to be the motion of a gas particle with an apparent mean free path $\lambda = C\tau$ which makes irregular excursions at the beginning of each new free path segment. This assumption is substituted into the formula[16] $< R^2 >^{\frac{1}{2}} = \lambda\sqrt{(2\nu)}$, where ν is the number of free paths traveled; $\nu = 1/\tau$. This yields

$$< R_N^2 >^{\frac{1}{2}} = C(2\tau)^{\frac{1}{2}} = C(\frac{2M}{\zeta})^{\frac{1}{2}} = c(\frac{2m}{\zeta})^{\frac{1}{2}} \tag{4.3.2}$$

The friction constant, ζ, for a heavy hard sphere in a gas of light ones has the form

$$\zeta = \frac{4\pi}{3}a^2 mnc = \frac{4}{3}m\nu \tag{4.3.3}$$

where a is the collision diameter of the heavy particle, and ν is given by $\nu = \pi a^2 nc$. Here, n is the number density of the light gas.[17] The friction constant given by eqn (4.3.3) is twice as large as that given by Smoluchowski. The value quoted here is currently accepted as correct; we shall derive it later (Chapter 12). Consequently the numerical coefficients that will be derived here will differ from those in Smoluchowski's paper.

Putting together eqns (4.3.2) and (4.3.3) results in

$$< R_N^2 > = c\sqrt{\frac{3}{2\nu}} \tag{4.3.4}$$

Let us compare this result with eqn (4.2.8) for one unit of time (for which eqn (4.3.4) was derived). One sees immediately that eqn (4.2.8) is larger than eqn (4.3.4) by a factor of $(8/3)\sqrt{(2/3)}$. This is not too surprising, considering all of the rough estimates that entered into the derivation of eqn (4.3.4). Consequently, says Smoluchowski, let us correct eqn (4.3.2) for the dense fluid case by multiplying it by this same factor, $(8/3)\sqrt{(2/3)}$.

Of course, eqn (4.3.3) cannot be used for the dense fluid case; it is valid only for a very dilute ambient gas. However, for a dense fluid, we have the result from macroscopic hydrodynamics, $\zeta = 6\pi\eta a$, where η is the viscosity of the ambient fluid. This formula is called *Stokes' law.* Substitution of Stokes' law into eqn (4.3.2), corrected by the multiplicative factor given just above, yields

$$< R_N^2 > = \frac{8}{9}\sqrt{\frac{m}{2\pi\eta a}}c \tag{4.3.5}$$

This is for one unit of time; for an arbitrary time, t, it must be multiplied by t. We shall compare this result to that of Einstein and the experimental work of

[16] Smoluchowski refers to a previous paper (1906b) for the derivation. Although in French, it does not appear in Sommerfeld's list of Smoluchowski's papers in English, French, and German.

[17] Anyone who checks the original publication should note that Smoluchowski uses the symbol n for what we have denoted by ν

Perrin later in this chapter. For the present, we content ourselves with discussing the significance of Smoluchowski's work.

As we said earlier, the positive aspects of this work are related to the fact that a definite kinetic model is adopted. The fluctuations of the velocity of the heavy particle which engender its observed irregular path are laid out before the eyes of the reader. The analysis of a specific model enables one to get considerable physical insight into the mechanism of Brownian motion. Furthermore, Smoluchowski introduced, for the first time, the idea of *random walk* as a model for Brownian motion. The name *random walk* was first introduced by K. Pearson in 1905, one year before Smoluchowski's paper, and naturally, Smoluchowski did not use that terminology. Nevertheless, the idea was clearly there, and the solution was correct.

Random walk is the study of the statistical properties of a sum of random vectors. The derivation of eqn (4.2.5) is the solution of a particular problem of this genre. Other workers, particularly Rayleigh, had studied particular cases even earlier. We shall return to the study of random walk as a model for Brownian motion in Chapter 9.

One problem with Smoluchowski's derivation is that it is rather crude with respect to the numerical coefficients. He had to make many uncontrolled approximations, and there is, of course, the missing factor of two in his formula for the friction constant. Also, he uses the mean velocity and root-mean-square velocity interchangeably. As Smoluchowski remarks in a footnote, this makes little numerical difference. However, it does make comparing formulas with those of other workers difficult.

In my opinion, the greatest problem with Smoluchowski's work is that it is not easily generalizable. It was born of a specific kinetic model and the results are not immediately transferable to situations not encompassed by that model. Einstein's theory, published the year prior to Smoluchowski's, does not suffer from this difficulty. It is to Einstein's theory that we now turn.

4.4 Einstein's Theory

Particles suspended in a fluid undergo the same kind of motion that dissolved particles undergo. This is a fundamental tenet of the kinetic-molecular theory of matter. A suspended particle will usually have a much greater mass than a solvent molecule, while a dissolved molecule will usually have a mass of the same order of magnitude as that of the solvent. Thus the motions might appear quantitatively different, though qualitatively they are of the same general type. This is the situation discussed at the beginning of Section 4.2. We want to see what conclusions we can come to about the features of this motion from the principles of statistical mechanics, and some reasonable hypotheses, without adopting a detailed molecular kinetic model.

In observing a particle suspended in a fluid, one follows the motion of the particle as best one can. However, one does not follow the motions of the molecules of the ambient fluid; there are too many and they are too small. The number of collisions per second experienced by a particle in a dense fluid is enormous. Hence the suspended particle acts as though it were moving randomly through the medium since the individual collisions are not tracked. Consequently it is reasonable to assume that there is a time τ_r, short compared to macroscopically observable times yet long compares to the inverse collision frequency, such that the particle's motions in two consecutive time intervals of length τ_r are independent. In the language introduced in Chapter 3, the displacement is a process with independent increments. (Of course, Einstein did not use this terminology; it was not introduced until much later. But the idea is his.) In this time interval, τ_r, there will still be an enormous number of collisions, which will destroy the suspended particle's effective dependence on its initial position.

Let $\mathbf{r}$ denote the position of the suspended particle, let the position at time 0 be chosen as the origin of the coordinate system, and let $p(\mathbf{r}, t)$ be the probability density that the particle be at position $\mathbf{r}$ at time t. There is no external force, so our system is homogeneous. This means that $p(\mathbf{r}, t) = p(-\mathbf{r}, t)$. Let $\phi(\boldsymbol{\Delta}, \delta t)$ be the probability of the particle moving a distance $\boldsymbol{\Delta}$ in time δt. We assume that the process is a Markov process, so that p and ϕ are the only two quantities needed to specify the probability distribution. This was again the idea of Einstein, although he did not call it a Markov process. Indeed, Markov had not yet published his great work on Markov chains (Markov 1907); general Markov processes came even later.

Since the motion of the heavy particle is being treated as a Markov process, we may write down the Chapman–Kolmogorov equation for the process (eqn (3.3.12)). In the present notation, this is

$$p(\mathbf{r}, t + \delta t) = \int p(\mathbf{r} - \boldsymbol{\Delta}, t)\phi(\boldsymbol{\Delta}, \delta t) d\boldsymbol{\Delta} \tag{4.4.1}$$

From the Chapman - Kolmogorov equation, we may immediately go to the Fokker–Planck equation (3.4.8), obtaining

$$\frac{\partial p(\mathbf{r}, t)}{\partial t} = D\nabla^2 p \tag{4.4.2}$$

where

$$D = \lim_{t \to 0} \frac{1}{2\delta t} \int \Delta^2 \phi(\boldsymbol{\Delta}, \delta t) d\boldsymbol{\Delta} \tag{4.4.3}$$

The term in the first spatial derivative of p vanishes because of the assumed even parity of ϕ.

Equation (4.4.2) is the well known diffusion equation, and D has the physical significance of the self-diffusion coefficient of the heavy particle.

The derivation of eqn (4.4.2) given here follows Einstein's very closely. The only difference is that we have written down the equations for the distribution of the position vector, $\mathbf{r}$, while Einstein considered only the distribution of one Cartesian component, x. Of course, Einstein could not just call upon the

Chapman–Kolmogorov equation and the Fokker–Planck equation in his derivation. These equations had not yet been derived by the authors for whom they are named. What Einstein did was to derive eqn (4.4.2) directly for the special case at hand, using a simpler form of the same technique that we used in Chapter 3 to derive the general case.

Multiply both sides of eqn (4.4.2) by r^2 and integrate over all space. The left-hand side yields, by definition

$$\int r^2 \frac{\partial}{\partial t} p(\mathbf{r}, t) d\mathbf{r} = \frac{d < r^2 >}{dt} \tag{4.4.4}$$

The integral of the right-hand side yields $6D$, by Green's theorem. Consequently

$$\frac{d < r^2 >}{dt} = 6D$$

or

$$< r^2 >= 6Dt \tag{4.4.5}$$

The constant of integration must vanish, because $< r^2 >$ must vanish at $t = 0$. Since the starting time for observation, the time at which we took the particle position to be the origin, was arbitrary, this result shows that the sample paths of the random motion are differentiable nowhere because $\Delta r \sim t^{1/2}$. This conclusion is applies for all times, not just $t = 0$, because the process has independent increments. Of course, the conclusion is absurd when considered as a microscopic description of the path. It is a result of extrapolating a mesoscopic description down to the microscopic level.

The same reasoning that we have just used for translational motion can also be used to sudy the rotational Brownian motion of a spherical particle. Following Einstein (1906a), let us consider a spherical particle constrained to rotate about an axis fixed in space. The only variable describing the orientation of the particle is the angle, θ, of this rotation. The distribution function of the orientation of the particle will obey a Fokker–Planck equation of the form

$$\frac{\partial}{\partial t} p(\theta, t) = D_R \frac{\partial^2 p}{\partial \theta^2} \tag{4.4.6}$$

where D_R is the rotational diffusion coefficient. This equation is the same as eqn (4.4.2), specialized to one dimension, but the boundary conditions are very different. The angular variable, θ, is only defined up to an additive integral multiple of 2π. Furthermore, the distribution function, p, must be a periodic function of θ, $p(0, t) = p(2\pi, t)$.

Consequently, the root-mean-square angle $< \Delta\theta^2 >$ cannot increase like t indefinitely, since θ is bounded. It is more useful to investigate the behaviour of the average of some periodic function that behaves like θ^2 for small θ, yet remains bounded for all time. $\sin^2 \theta$ is such a function. Multiply eqn (4.4.6) by $\sin^2 \theta$ and integrate over θ. Integration by parts yields the differential equation

$$\frac{d}{dt} < \sin^2\theta >= D_R(1 - 2 < \sin^2\theta >) \tag{4.4.7}$$

The solution that vanishes at $t = 0$ is

$$< \sin^2\theta >= \frac{1}{2}(1 - e^{2D_R t}) \tag{4.4.8}$$

Clearly, for short times, when only small θ can contribute to the average, one obtains $< \Delta\theta^2 >= D_R t$. At times comparable to $1/D_R$, this linear dependence on time crosses over into an asymptotically constant behavior.

Clearly this description of rotational Brownian motion is incomplete. The model of a spherical particle constrained to rotate about a single fixed axis does not correspond to a realizable experimental situation. However, consider the case of a small mirror attached to a torsion fiber. Small rotations of the mirror can be amplified by reflecting light from it. Such an arrangement is used in sensitive galvanometers, strain gauges, and other instruments. Any 'jiggling' of the mirror due to Brownian motion will serve to reduce the precision of the measurement being attempted. Thus the study of the effect of Brownian motion on uniaxial rotation has considerable physical interest; we shall return to it when we discuss applications in Chapter 7 and in Chapter 13.

4.5 Diffusion Coefficient and Friction Constant

At first, the result eqn (4.3.5) obtained by Smoluchowski, and the result $< r^2 >= 6Dt$ obtained by Einstein look very different. They are even expressed in terms of different variables, the diffusion coefficient in Einstein's case, and the viscosity in Smoluchowski's. Nevertheless, the dependence on physical parameters is exactly the same for both theories because there is a close relation between the diffusion coefficient and the viscosity. This relation was found by Einstein in his 1905 paper on Brownian motion.

Consider a suspension of particles in a fluid with a spatially constant external field, e.g. gravity, imposed upon it. We denote the external force by F, and choose the z axis of the coordinate system in the direction of $\mathbf{F}$. The potential of the external field is $-Fz$. The suspended particle will move in the z direction and attain a terminal velocity, v, given by

$$v = \frac{F}{\zeta} \tag{4.5.1}$$

where ζ is the friction constant. The suspension is supposed to be sufficiently dilute that the individual particles in it do not interact with each other, but only with the constituents of the surrounding medium. The system is bounded in the z direction; the container has a bottom, for example. Consequently the motion will build up a concentration gradient in the z direction, and the concentration gradient will induce a diffusion current in the opposite direction to the current induced by the external force.

Eventually, the concentration gradient will become large enough that the two currents will cancel each other, and the system will reach equilibrium. If we denote the local concentration of suspended particles by $n(z)$, then the particle current induced by the external force is nv; that induced by the concentration gradient is $-Ddn/dz$. We have here used Fick's law for the diffusion current. At equilibrium

$$\frac{nF}{\zeta} = -D\frac{dn}{dz} \tag{4.5.2}$$

But at equilibrium, $n(z)$ is given by the Boltzmann law

$$n(z) \sim e^{-Fz/kT} \tag{4.5.3}$$

Putting this in eqn (4.5.2) yields

$$D = \frac{kT}{\zeta} \tag{4.5.4}$$

This is the Einstein relation between the friction coefficient and the diffusion coefficient. Einstein actually did not write it down in this form, but immediately assumed that the friction constant was given by Stokes' law, $\zeta = 6\pi\eta a$, where η is the viscosity of the medium, and a the radius of the suspended particles. This is the same assumption that was made by Smoluchowski. If we use Stokes' law for ζ, then (4.5.4) becomes

$$D = \frac{kT}{6\pi\eta a} \tag{4.5.5}$$

Equation (4.5.5) is called the *Stokes–Einstein* relationship, and is widely used in the interpretation of data since the viscosity is easily measurable.

A similar result holds for the rotational diffusion of a sphere. A torque, Q, applied to a sphere induces a steady state angular velocity, ω, given by

$$\omega = \frac{Q}{\zeta_R} \tag{4.5.6}$$

where ζ_R is the rotational friction constant (Landau and Lifshitz 1959). A similar argument to that just given for translation then yields

$$D_R = \frac{kT}{\zeta_R} \tag{4.5.7}$$

The equivalent of Stokes' law for the rotation of a sphere in a viscous liquid is $\zeta_R = 8\pi\eta a^3$, so that the Stokes–Einstein law for rotation is

$$D_R = \frac{kT}{8\pi\eta a^3} \tag{4.5.8}$$

In spite of the appealing nature of the use of Stokes' law, H. A. Lorentz (1921) has argued that Stokes' law is an approximation to the motion of spheres through a viscous fluid that is valid for slow steady motions. He investigated the first correction to Stokes' law when the motion is not steady, but still slow, and found that it was not negligible. He concludes that 'the details of the Brownian motion cannot be calculated with Stokes' law.' In spite of this, almost all workers use Stokes' law in discussing Brownian motion. Lorentz' considerations do not seem to be well known; they appear in a textbook based on his lectures on kinetic problems in 1911–1912. I have therefore translated the relevant portion from the original Dutch. The translation appears in Appendix A.

Let us now compare the results of Einstein and Smoluchowski. Recall that Smoluchowski's result is (4.3.5)

$$< R_N^2 >= \frac{8}{9}\sqrt{\frac{m}{2\pi\eta a}}c \tag{4.5.9}$$

Using eqn (4.5.5), Einstein's result, $< r^2 >= 6Dt$ can be written, for unit time

$$< r^2 >^{\frac{1}{2}}= \sqrt{\frac{kT}{\pi\eta a}} \tag{4.5.10}$$

Here we have used the equipartition theorem to write $c^2 = 3kT/m$. Therefore, the Smoluchowski result is larger than the Einstein result by a factor of $\sqrt{(32/27)}$; this is remarkably close considering all of the approximations that went into the theory. Smoluchowski and many subsequent commentators claim that his result was, in fact, $\sqrt{(64/32)}$ larger, but this was because of the error of a factor of two in his formula for the friction constant of a heavy particle in a dilute gas. His result was actually closer than he thought. It is now universally agreed, and was agreed by Smoluchowski, that Einstein's result is the correct one.

4.6 The Langevin Theory

The Einstein and Smoluchowski theories look very different on the surface. One utilizes the dynamics of the particle motion in an essential way, while the other is a purely statistical theory. A link between the two conceptions was provided by P. Langevin (1908).[18] A suspended particle in a fluid is acted upon by forces due to the molecules of the solvent. This force can be expressed as a sum of its average value and a fluctuation about this averagevalue. Langevin's idea was to treat the mean force dynamically and the residual fluctuating part of the force probabilistically.

We assume that the mean force on a particle moving slowly in a viscous medium is given by

[18] In this article Langevin also claims to have redone Smoluchowski's calculation more exactly and to find precise agreement with Einstein's result. He gives no indication, however, of the differences between his calculation and that of Smoluchowski.

$$\mathbf{F}_{avg} = -\zeta \mathbf{v} \tag{4.6.1}$$

where $\mathbf{v}$ is the velocity of the particle relative to the resting fluid. The difference in sign between eqn (4.6.1) and eqn (4.5.1) is that the former refers to the velocity induced by an external force, while the latter refers to the force caused by a given velocity. The two are equal in magnitude and opposite in sign.

We shall denote the residual fluctuating force by $\mathbf{X}$. We know little about $\mathbf{X}$ in detail, so we shall make only a few statistical hypotheses about its properties. First of all, $\mathbf{X}$ is a fluctuation about a mean, so it must, itself, have zero mean.

$$< \mathbf{X} >= 0 \tag{4.6.2}$$

Second, we assume that $\mathbf{X}$ is a stationary process with a very short correlation time

$$< \mathbf{X}(t) \cdot \mathbf{X}(t+s) >=< \mathbf{X}^2 > \phi(s) \tag{4.6.3}$$

where $\phi(s)$ is a function that is very sharply peaked about $s = 0$. The correlation time is so short compared to M/ζ (the only characteristic time of the system) that ϕ may be taken to be a Dirac delta function, $\phi(s) \propto \delta(s)$. Moreover, $\mathbf{X}(t)$ is not correlated with the position of the particle at time t, nor with the velocity at any previous time.

$$\begin{aligned} &< \mathbf{X}(t) \cdot \mathbf{r}(s) > = 0 \\ &< \mathbf{X}(t) \cdot \mathbf{v}(s) > = 0, \quad t > s \end{aligned} \tag{4.6.4}$$

Newton's second law of motion for this system them reads

$$M \frac{d\mathbf{v}}{dt} = -\zeta \mathbf{v} + \mathbf{X}(t) \tag{4.6.5}$$

This differential equation cannot be solved in the usual sense because we do not know enough about $\mathbf{X}(t)$. Furthermore, in order that its correlation time be as short as assumed $\mathbf{X}$ must be a wildly fluctuating function . It is not clear that solutions of (4.6.5) exist, even in principle, for such functions. We shall come back to this point in more detail in Chapter 5. For the present, we shall proceed entirely formally, assuming that the differential equation eqn (4.6.5) can be treated as a normal differential equation.

Take the scalar product of eqn (4.6.5) with $\mathbf{r}(t)$, and take the mean value of the result. Since $< \mathbf{X} \cdot \mathbf{r} >= 0$ by hypothesis, we get

$$M < \mathbf{r} \cdot \frac{d\mathbf{v}}{dt} >= -\zeta < \mathbf{r} \cdot \mathbf{v} > \tag{4.6.6}$$

Since $\mathbf{v} = d\mathbf{r}/dt$, this equation can be put into the form

$$\frac{M}{2}\frac{d^2 < r^2 >}{dt^2} + \frac{\zeta}{2}\frac{d < r^2 >}{dt} = 3kT \qquad (4.6.7)$$

where we have used the equipartition theorem $< v^2 >= 3kT/M$. The solution to this equation that vanishes at $t = 0$ is

$$< r^2(t) >= \frac{6kT}{\zeta}t + B(e^{-\zeta t/M} - 1) \qquad (4.6.8)$$

where B is an integration constant specified by $< \mathbf{r} \cdot \mathbf{v} >= 0$ at time zero. Although we do not need to know it, the value of B can easily be computed to be $6MkT/\zeta^2$. Thus, after a time of the order of ζ/M, Einstein's result is valid. This time is indeed quite short, of the order of 10^{-7} s for typical situations. Consequently, Einstein's result may be considered to be valid for practical purposes for all time.

Why is this result equal to Einstein's for practical purposes only, and not for all time? Einstein worked completely in the configuration space of the Brownian particle; he never introduced the velocity of the particle. Thus he completely neglected the inertia of the particle and the possibility of persistence of velocity. In other words, the short time, τ, after which the displacements of the particle should be independent, should be longer than M/ζ. Langevin, on the other hand, worked in the particle's phase space and was able to treat the velocity relaxation. Langevin's description is on a finer scale than that of Einstein.

If we are only interested in the configuration space properties of the Brownian particle, for example the mean-square displacement, it makes no practical difference which description we adopt. However, for the study of the Brownian particle's velocity space properties, for example the velocity correlation function, it is important to have the phase space description available. We shall treat this question in detail in Section 8.1.

5

STOCHASTIC DIFFERENTIAL EQUATIONS AND INTEGRALS

5.1 The Langevin Equation Revisited

At the end of the last chapter, we used the differential equation proposed by Langevin to study the mean square displacement of a Brownian particle.

$$M\frac{dv}{dt} = -\zeta v + F(t) \tag{5.1.1}$$

This equation contains one function, the fluctuating force F, whose functional form is unknown; all we postulated were some statistical properties of this function. Whatever the form of this term, it must be very wild in order that the function have the postulated statistical property of being delta function correlated (eqn (4.6.3) *et seq.*). It is not at all obvious that the differential equation has a unique solution for a given initial condition.

There is, in fact, a standard existence theorem for differential equations which guarantees the existence of a *local* solution if the function F is continuous. A local solution is one which exists in some neighborhood of the point at which the initial value is given. But even if a solution exists it may be only local, or it may not be unique, unless some stronger conditions are imposed on F.

Perhaps an existence theorem is irrelevant since an explicit solution of eqn (5.1.1) can be exhibited, showing directly the existence of a solution for the case at hand. This explicit solution is

$$v(t) = v(0)e^{-\zeta t/M} + \frac{1}{M}\int_0^t F(s)e^{-\zeta(t-s)/M}ds \tag{5.1.2}$$

But this only transfers the problem elsewhere. How do we know that the integral in eqn (5.1.2) exists, that it is more than just a formal symbol? We have already discussed this problem in Section 3.6. To go any further in this direction we must specify the stochastic properties of the function F in more detail than we have yet done.

Since $F(t)$ is delta correlated, it is a white noise. We assume that it is due to the random collisions of the ambient fluid particles with the Brownian particle. Each collision lasts a very short time on the mesoscopic time scale. We may therefore assume that a collision results in an exchange of momentum described by a delta function in time, $\delta(t - t_k)$, where t_k is the time of occurrence of the kth collision. There are ν collisions per unit time. The time between successive collisions, the interarrival time, is exponentially distributed.

Since the delta function is the derivative of the unit step function, the $F(t)$ process is the derivative of a series of unit steps occurring at exponentially distributed times. In other words, $F(t)$ is the derivative of a Poisson process. This is not a white noise process, but becomes one in the limit that $\nu \to \infty$. Also, in this same limit the Poisson process (scaled to zero mean) approaches a Gaussian process with variance proportional to t. But a Gaussian process with variance linear in t is a Wiener process. Thus we have made it plausible that our white noise process should be looked upon as the derivative of a Wiener process.

This is an important conclusion, so let us look at it from another point of view to get additional insight. Let us define $\mathbf{B}(t)$ by

$$\mathbf{B}(t) = \int_a^t F(s)ds \tag{5.1.3}$$

However the integral is to be defined, we should insist that it have the following two properties, at least, in order that it be worthy of the name of integral.

a) The integral should be a continuous function of its upper limit.
b) The integral should be additive, in the sense that

$$\int_a^t F(s)ds = \int_a^b F(s)ds + \int_b^t F(s)ds; \quad a \le b \le t \tag{5.1.4}$$

Condition a) implies that B is a continuous stochastic process, and condition b) implies that it is a Markov process.

We can compute the mean and variance of the $\mathbf{B}$ process from our hypotheses about the F process.

$$\begin{gathered} < \mathbf{B}(t) >= 0 \\ < \mathbf{B}^2(t) >= \int_0^t \int_0^t < F(s)F(s') > dsds' =< F^2 > t \end{gathered} \tag{5.1.5}$$

Since $\mathbf{B}(t)$ is a continuous Markov process, its transition probability will obey a Fokker– Planck equation. Let us compute the A and B coefficients of this equation. We have

$$\begin{gathered} < \mathbf{B}(t+\Delta t) - \mathbf{B}(t) \mid \mathbf{B}(t) >= \int_t^{t+\Delta t} < F(s) > ds = 0 \\ < [\mathbf{B}(t+\Delta t) - \mathbf{B}(t)]^2 \mid \mathbf{B}(t) >= \int_t^{t+\Delta t} ds \int_t^{t+\Delta t} ds' < F(s)F(s') >= \Delta t \end{gathered} \tag{5.1.6}$$

Consequently, returning to the definitions of A and B, eqn (3.4.5), we find $A = 0$, $B = 1$. Hence the Fokker–Planck equation describing $\mathbf{B}(t)$ is precisely that for the Wiener process, and we shall henceforth denote $\mathbf{B}(t)$ by $\mathbf{W(t)}$.

The conclusion that $\mathbf{W}(t)$ defined by eqn (5.1.3) (with the just mentioned change in notation) is a Wiener process seems a bit paradoxical. $\mathbf{W}$ is the integral of F, so the temptation would be to write $F = d\mathbf{W}/dt$. But, with probability one, the sample functions of a Wiener process are not differentiable. This is because the white noise process, F, is not a stochastic process in the ordinary sense that we have discussed in Chapter 3. Rather, it is the limit of such a process as certain parameters, in this case the frequency of collisions, approach infinity. But let us ignore this problem for a while longer, while we use heuristic means to reformulate the concept of a stochastic differential equation in such a way that it makes mathematical sense.

5.2 Stochastic Differential Equations

Instead of confining our attention to the simple Langevin equation (5.1.1), let us consider the more general stochastic differential equation

$$\frac{d\mathbf{X}}{dt} = f(\mathbf{X}, t) + g(\mathbf{X}, t)\frac{d\mathbf{W}}{dt} \tag{5.2.1}$$

Of course, this is not the most general type of first order equation, but it is sufficiently general for our considerations in this book. It may be a nonlinear equation, and is said to have a *multiplicative noise* because the fluctuating force, $d\mathbf{W}/dt$, is multiplied by a function of the unknown process, $\mathbf{X}$. In contrast, eqn (5.1.1) is said to have *additive noise* because the coefficient of the fluctuating term does not depend on the unknown. f and g are given functions of their arguments.

For the reason discussed in Section 5.1, eqn (5.2.1) has no a priori meaning. Let us nevertheless formally integrate both sides from zero to t.

$$\mathbf{X}(t) - \mathbf{X}_0 = \int_0^t f(\mathbf{X}(s), s)ds + \int_0^t g(\mathbf{X}(s), s)d\mathbf{W}(s) \tag{5.2.2}$$

In this form, the equation does make sense, provided that the integrals appearing on the right hand side are unambiguously defined. We know how to give a meaning to such integrals; this was discussed in Section 3.6. In fact, we outlined there two definitions, those of Ito and of Stratonovich. Each of these definitions, or rules, will give rise to a different stochastic solution process, $\mathbf{X}$.

In the additive noise case, for example eqn (5.1.1), both rules lead to the same result, and it is a matter of indifference which is picked. In the case of multiplicative noise, this is not the case. The question of which rule to use is of the utmost importance for applications. In discussing this question we shall follow the excellent treatment of Van Kampen (Van Kampen 1981b). We first determine the Fokker–Planck equation for the two cases of the Ito and Stratonovich rules.

Again, recall the definitions of the drift and diffusion coefficients from eqn (3.4.5). We treat the Ito case first. We shall write eqn (5.2.2) to first order in Δt, omitting all higher order terms for the sake of brevity.

$$< \mathbf{X}(t + \Delta t) - x_0 >_{Ito} \approx f(x_0, t)\Delta t + g(x_0, t) < \Delta\mathbf{W} > \tag{5.2.3}$$

Since $< \Delta\mathbf{W} >= 0$, we get $A = f(x, t)$. Similarly

$$< [\mathbf{X}(t + \Delta t) - x_0]^2 > \approx g^2(x_0, t) < [\Delta \mathbf{W}]^2 > \tag{5.2.4}$$

Since $< [\Delta \mathbf{W}]^2 >= t$, we immediately conclude that $B = g^2$. We have here assumed that the variance of the Wiener process, $\mathbf{W}$, is t; all the constant coefficients have been lumped into the function g. Thus the Fokker–Planck equation for the Ito interpretation of eqn (5.2.2) is

$$\frac{\partial p}{\partial t} = \frac{\partial}{\partial x}[-f(x,t)p + \frac{g^2}{2}\frac{\partial p}{\partial x}] \text{ (Ito)} \tag{5.2.5}$$

The analogous result for the Stratonovich equation is more complicated. In place of eqn (5.2.3),the change in $\mathbf{X}$ is

$$< \mathbf{X}(t+\Delta t) - x_0 > \approx < f((\mathbf{X}+x_0)/2, t) > \Delta t + < g((\mathbf{X}+x_0)/2, t)\Delta \mathbf{W} > \tag{5.2.6}$$

Since $f(x + x_0)/2, t) = f(x_0, t) + o(1)$, we may write the first term of eqn (5.2.6) as merely $f(x_0, t)\Delta t$. The second term is

$$< [g(x_0, t) + (\frac{\partial g}{\partial x})_{x_0} \Delta \mathbf{X}]\Delta \mathbf{W} > \tag{5.2.7}$$

The term $< g(x_0, t)\Delta \mathbf{W} >$ vanishes, and we have to substitute eqn (5.2.3) for $\Delta \mathbf{X}$ in the second term. This yields $\frac{1}{2}gg' < (\Delta \mathbf{W})^2 >$, and so we find that

$$A_s = f(x,t) - \frac{1}{2}gg' \tag{5.2.8}$$

The B coefficient is the same as that already found for the Ito rule. Hence the Fokker– Planck equation for the Stratonovich rule is

$$\begin{aligned} \frac{\partial p}{\partial t} &= \frac{\partial}{\partial x}(-(f - \frac{gg'}{2})p + \frac{g^2}{2}\frac{\partial p}{\partial x}) \\ &= \frac{\partial}{\partial x}(-fp + \frac{g}{2}\frac{\partial gp}{\partial x}) \text{ (Stratonovich)} \end{aligned} \tag{5.2.9}$$

The extra drift term in the Fokker–Planck equation is sometimes called *anomalous drift.*

Clearly, the Ito and Stratonovich rules applied to the same formal stochastic differential equation give rise to different stochastic processes. Notice, however, that the stochastic differential equation with drift coefficient f and diffusion coefficient g, together with the Stratonovich definition for the integral, yields the same Fokker–Planck equation as the Ito definition for the integral with the different drift coefficient $f - \frac{1}{2}gg'$.

Another difference between the stochastic differential equations we have been discussing and ordinary differential equations lies in the rules for changing variables. Since the stochastic differential equations are really integral equations with special rules for interpreting the integral, this should not come as a surprise. Were we dealing with normal differential equations, given eqn (5.2.1) we would feel free to introduce a new independent variable $y = h(x)$ and obtain

$$\frac{dy}{dt} = F(y,t) + G(y,t)\frac{d\mathbf{W}}{dt} \tag{5.2.10}$$

where

$$\begin{aligned} F(y,t) &= h'(x)f(x,t) \\ G(y,t) &= h'(x)g(x,t) \end{aligned} \tag{5.2.11}$$

The coefficients in the Fokker–Planck equation arising from eqn (5.2.10) can now be calculated. Taking into account that the distribution function for the y variable is $P(y,t) = p(x,t)/h'(x)$ the Stratonovich rule yields

$$\frac{\partial P}{\partial t} = \frac{\partial}{\partial y}(-FP + \frac{1}{2}G\frac{\partial}{\partial y}GP) \text{ (Stratonovich)} \tag{5.2.12}$$

which is also the result of changing variables in the Fokker–Planck equation (5.2.9) in the normal way, by the rules of ordinary calculus.

The Fokker–Planck equation corresponding to the Ito interpretation and the same change of variables might be expected to be

$$\frac{\partial P}{\partial t} = \frac{\partial}{\partial y}(-FP + \frac{\partial G^2 P}{\partial y}) \text{ (Ito ?)} \tag{5.2.13}$$

but this is not the case. When the variable in eqn (5.2.5) is transformed, the result is

$$\frac{\partial P}{\partial t} = \frac{\partial}{\partial y}([-F - \frac{1}{2}G^2\frac{h''}{h'^2}]P + \frac{1}{2}\frac{\partial G^2 P}{\partial y}) \tag{5.2.14}$$

Thus, in the case of a change of variable, the Ito equation seems to have picked up some anomalous drift. This is inherent in the Ito definition of the stochastic integral. The stochastic differential equation, eqn (5.2.11), cannot be treated as though it were a differential equation of the usual kind with respect to change of variable. The chain rule for differentiation, which must be used in transforming a differential equation by a change of independent variable, picks up an extra term (Schuss 1980).

$$d\phi(x,t) = \frac{\partial\phi}{\partial x}(fdt + gd\mathbf{W}) + (\frac{\partial\phi}{\partial t} + \frac{1}{2}g^2\frac{\partial^2\phi}{\partial x^2})dt \tag{5.2.15}$$

where $x(t)$ is supposed to satisfy eqn (5.2.1). Equation (5.2.15) is called *Ito's rule.* It is clear that if Ito's rule is used instead of the chain rule of ordinary calculus, we do not obtain eqn (5.2.10). Instead, we obtain a stochastic differential equation whose associated Fokker-Planck equation has the structure of eqn (5.2.14) and eqn (5.2.13) is incorrect in the Ito interpretation.

5.3 Which Rule Should Be Used?

Since the Ito and Stratonovich interpretations (or definitions) of the stochastic integral give rise to different stochastic processes as solutions of the stochastic differential equation, which one should be used to model a given physical situation? The answer that is often given is that one should use the Stratonovich interpretation.

The reason given for this choice is that the use of the Wiener process is in fact an idealized version of the fluctuations in real physical systems. In real systems, the fluctuations have a finite, albeit small, correlation time, not a zero correlation time as in the Wiener process. Consequently, the argument goes, any results that only appear when the correlation time is strictly zero are artifacts of the limit. This occurs in the Ito interpretation because the 'differential' $d\mathbf{W}$ is of the order of magnitude $(dt)^{\frac{1}{2}}$, not dt as for an ordinary differential. Of course, this is also true for the Stratonovich interpretation, but its effects are ameliorated by Stratonovich's rule for choosing the evaluation points in the stochastic integral; the ordinary rules of calculus hold. Indeed, this was Stratonovich's motivation for introducing his rule.

For stochastic processes involving noise with a finite correlation time, it does not matter which rule is chosen for picking the evaluation points in the finite sum approximation to the stochastic integral. Then, in the zero correlation time limit, the Stratonovich integral goes smoothly into its limiting form, while the Ito integral picks up the anomalous properties we have discussed in the previous section. Thus the Stratonovich interpretation should be used.

In one restricted sense, however, it does not matter which interpretation is used. A Stratonovich Langevin equation with drift term f yields a Fokker–Planck equation with drift term $f - \frac{1}{2}gg'$. An Ito Langevin equation with drift term $f - \frac{1}{2}gg'$ gives rise to the same Fokker–Planck equation. This assumes that both Langevin equations have the same noise term, g. Thus, if the drift term is modified appropriately the same result will arise from both the Ito and Stratonovich forms. On the other hand, the usual situation is that the drift term is thought to be known on physical grounds; a choice is not available. Then the question of whether the Ito or Stratonovich rules should be used must be answered before any solution is possible.

The arguments of Van Kampen (1981a, 1981b) on this subject are the clearest and most convincing. The problem arises because the form of the drift term is believed known a priori. It is usually assumed to be the macroscopic drift that would occur if there were no fluctuations at all. If a physical argument leading to a Langevin equation is sound, it ought not only lead to the coefficients in the equation, for example the drift term, but also to the interpretation rule. It is the interpretation rule which translates a string of symbols into a meaningful mathematical statement.

There are two cases to consider. The first is when the noise is due to external sources, unaffected by the system. An example of this is noise in radio propagation caused by electrical fluctuations in the atmosphere that are independent of the transmitter or receiver. Another example is fluctuations in crop production caused by the weather. In such cases the external influence can, in principle at least, be turned off (grow the plants in a greenhouse). The system can be studied in the absence of fluctuations and the proper description of the drift determined. To describe the system with fluctuations included, merely add in the fluctuation term. As discussed above, if the correlation time of a real system is short enough that it may be considered to be zero on the time scale being studies, the Stratonovich interpretation should be adopted for the stochastic differential equation.

Let us remark that it may be useful to subtract off the 'anomalous drift' in order to treat the equation as an Ito equation, just because so much more is known about the mathematical properties of Ito integrals than of Stratonovich integrals. But this is a matter of technique, not of fundamental principle.

The second case is that of systems whose fluctuations are internal, caused by properties of the system itself. Brownian motion is a case in point here. The fluctuations are due, as we have seen, to the thermal motions of the particles that constitute the system. These motions cannot be turned off, so that the system cannot be studied in the absence of fluctuations. We must therefore turn to a more fundamental description of the system. If the system is known to be Markovian, Van Kampen recommends starting from the Master Equation.[19] Whatever this more fundamental description may be, it should contain both the average behavior of the system under study and its fluctuations as well.

In many cases the more fundamental governing equation contains a small parameter; $1/\Omega$. For Brownian motion the small parameter will be the mass ratio m/M.[20] In the limit of large Ω, the fluctuations become negligible relative to the deterministic change; that is, the probability distribution has a delta function peak at the deterministic value. This suggests that we should attempt a systematic expansion of the equation in powers of $1/\Omega$; the zeroth order term will give the drift. The next term, of order $1/\Omega$, will give the diffusion term of a well defined Fokker–Planck equation.

The question of the interpretation of the Langevin equation can thus be bypassed. If we wish to work with a Langevin equation, we may chose either the Ito form or the Stratonovich form, according to convenience, as long as we pick the Langevin drift term corresponding to the interpretation we have adopted.

Carrying out the expansion in the small parameter to higher order terms in $1/\Omega$ will generate terms more complicated than those appearing in the Fokker–Planck equation, and there will be no equivalent Langevin equation of either Ito or Stratonovich form. The noise term need not be Gaussian white noise.

[19]We have not previously introduced this equation, and it would take us too far afield to do so now. Suffice to say that it is an equation for the transition probability of a Markov process; the Fokker–Planck equation is a special case (Van Kampen 1981a; Oppenheim *et al.* 1977).

[20]In other contexts Ω is often the volume of the system, and so the expansion is sometimes called the system-size expansion.

5.4 Some Examples

The foregoing paragraphs are irrelevant for the Langevin equation eqn (5.1.1), for here the Ito and Stratonovich rules are equivalent. In order to justify spending so much space on this point, we give three examples of situations giving rise to a Langevin equation with multiplicative noise.

Example 1. A molecule with dipole moment μ rotating in a fluid: The molecule has dipole moment μ and angular velocity ω. There is an electric field $\mathbf{E}$ present also. The equations of motion for this system are

$$\begin{aligned} \frac{d\mu}{dt} &= \omega \times \mu \\ I\frac{d\omega}{dt} &= -\zeta_R \omega + \mu \times \mathbf{E} + \tau(t) \end{aligned} \tag{5.4.1}$$

I is the moment of inertia of the molecule, ζ_R the rotational friction constant, and τ the fluctuating torque. The first of these equations is a general kinematic relation expressing the rate of change of any vector fixed in a rotating frame. The second says that the rate of change of the angular momentum of the molecule is given by the sum of the systematic frictional torque, the torque due to the external field, and the random fluctuating torque, τ. In the limit that the rotational inertia of the molecule is negligible, small moment of inertia or large friction, one can neglect the $d\omega/dt$ term. Then combining both eqns (5.4.1) yields

$$\begin{aligned} \frac{d\mu}{dt} &= \zeta_R^{-1}(\mu \times \mathbf{E}) \times \mu + \tau \times \mu \\ &= \zeta_R^{-1}\mu^2(\mathbf{I} - \frac{\mu\mu}{\mu^2}) \cdot \mathbf{E} + \tau \times \mu \end{aligned} \tag{5.4.2}$$

In this equation, $\mathbf{I}$ is the unit tensor, not the moment of inertia.

The fluctuating torque is supposed to have zero mean and to be a delta function correlated Gaussian white noise. This is just the kind of Langevin equation with multiplicative noise that we have been discussing.

Example 2. An oscillator with random frequency modulation: The equations of motion of an ordinary harmonic oscillator can be written

$$\dot{z} = i\omega z \tag{5.4.3}$$

where z is the complex variable $x + iv$ where x is the position of the oscillator and v its velocity. ω is the frequency of oscillation. We modify this equation to

$$\dot{z} = i(\omega + \varepsilon\xi(t))z \tag{5.4.4}$$

where ϵ is a strength parameter, and ξ is Gaussian white noise. This clearly corresponds to an oscillator whose frequency is constantly changing due to random perturbations from the environment.

Example 3. Stock market prices: We digress here to mention an example from the domain of finance, of more pedagogical than practical interest. An elementary model for fluctuations in equity prices is the stochastic process

$$\mathbf{X} = \mathbf{X}_0 e^{\mu t + \sigma \mathbf{W}(t)} \tag{5.4.5}$$

Here μ, σ, and $\mathbf{X}_0$ are constants and $\mathbf{W}$ is, as before, a Wiener process. This is sometimes called exponential Brownian motion. The random variable $\mathbf{X}$ undergoes exponential growth with Gaussian fluctuations of the growth rate. It is of interest as an elementary example of stock price changes. It has fluctuations added to a drift (the μt term), and the magnitude of the fluctuations in price are proportional to the price itself as we shall see below. The question at issue is, which Ito equation does this function solve?

Clearly $\mathbf{Y} = \ln \mathbf{X}$ can be written

$$\mathbf{Y} = \mathbf{Y}_0 + \mu t + \sigma \mathbf{W}(t) \tag{5.4.6}$$

Consequently

$$d\mathbf{Y} = \mu dt + \sigma d\mathbf{W} \tag{5.4.7}$$

Using Ito's rule to change variables back to X, we find

$$d\mathbf{X} = (\mu + \frac{1}{2}\sigma^2)\mathbf{X}dt + \sigma \mathbf{X} d\mathbf{W} \tag{5.4.8}$$

Note that we did not have to use Ito's rule to obtain eqn (5.4.6) since that was an ordinary equation relating sample functions, not an integral equation. Equations between stochastic differentials are really just shorthand notation for integral equations.

Thus we see that the stock prices purportedly described by this simple model have a noise induced drift and the fluctuation term is indeed proportional to the solution. We have thus another example of a multiplicative Langevin equation, albeit in a field far removed from the rest of this book.

While it is true that most of the stochastic differential equations that occur in physics and chemistry are equations with additive noise, the above examples illustrate that there are interesting problems which are best described by a multiplicative noise model. When such a situation arises, the complications outlined in this chapter cannot be avoided. For the reader interested in studying the subject further, Gardiner (1983) has a good chapter on the subject. The more mathematical books by Schuss (Schuss 1980) and Arnold (Arnold 1973) may also be consulted; there are even more technical treatments from the mathematical side, to which we need not make reference here.

6

FUNCTIONAL INTEGRALS

6.1 Functional Integrals

In Chapter 2 we discussed the mean value, or mathematical expectation, of a random variable. It is an integral of the random variable, weighted by the probability of occurrence of the various values of its argument. In fact, in probability theory, the phrases 'integral with respect to a probability measure' and 'mathematical expectation' are synonymous. Such integrals are ordinary integrals of the type familiar from ordinary calculus. They have nothing to do with the Ito and Stratonovich integrals discussed in Chapters 3 and 5. We now want to discuss a new type of integral that is closely related conceptually to the first, or 'expectation' kind of integral, and only very indirectly related to the stochastic integral.

Suppose now that we ask a question such as, 'A Brownian particle in one dimension starts at the origin at $t = 0$. What is the mean number of times it returns to the origin up to time T?' This is a more complicated question than those we have been considering up until now because the elementary events, the elements of the space on which the probability is defined, are all the possible paths of Brownian particles in the time interval $[0, T)$. Since a path is a function, $x(t)$, one should say that the probability space is the space of functions defined on $[0, T)$. So, answering questions having to do with sample function properties fundamentally reduces to being able to define a probability measure on a space of functions.

Mathematicians had considered the problem of defining a measure on function spaces for some time. Several solutions had been proposed prior to 1921, but none of them captured the concept in a completely satisfactory way. In 1921, N. Wiener showed how to do it (Paley and Wiener 1934). Of course, once there is a measure on function space, an integral over function space can be defined. Years later, R. P. Feynman showed, in his 1942 Ph.D. thesis, how nonrelativistic quantum mechanics could be formulated without introducing the Schrödinger equation, in terms of an integral over the classical paths of particles (Feynman 1950; Feynman and Hibbs 1965). Feynman did not know of Wiener's work at the time, and indeed, the problems were different enough technically that a knowledge of Wiener's work probably would have been of very little help to him. Nevertheless, there are strong conceptual analogies between the two independent developments. Integrals of this type are nowadays also called functional integrals or path integrals.

In this chapter, we shall give an intuitive constuction of the Wiener functional integral. Then we shall give a brief description of the rigorous mathematical development due to Wiener and refined later by others; this description can be skipped without detriment to reading the rest of the book. A description of the Feynman path integral will follow. This is an excursion from the theme of Brownian motion, but the formal analogies are so strong, and the path integral approach to quantum mechanics so common nowadays, that I cannot let pass the opportunity to present it.

Then we shall address the question of how to evaluate functional integrals in practice. In ordinary calculus, integrals are often evaluated by seeing what function can be differentiated to give the integrand. That procedure is not available for functional integrals, since there is no analogy in this domain to the fundamental theorem of the calculus.

The Wiener integral, the subject of this chapter, has not turned out to be very useful for the solution of problems of direct interest for applications in the physical sciences. Its use has been in the insight it has given into the mathematical nature of stochastic processes. Consequently this chapter may be skipped by any reader anxious to get more quickly to physically relevant results.

6.2 The Wiener Integral

In this section, we work in one space dimension and choose units in which the diffusion coefficient, D is unity; this is only for simplicity. Suppose $x(t)$ is the path of a Brownian particle that starts at the origin, 0, at $t = 0$. Now suppose we have a functional $F_n[x(t)]$ that depends on the value of x at only a finite number of points, $t_1, t_2, \ldots, t_n$. Of course, this is a very special type of functional; it is merely a function of n variables. According to the discussion in Chapters 2 and 3, the mean value of this type of functional is given by

$$
\begin{gathered}
< F_n >= \frac{1}{(2\pi)}^{n/2} \prod_{j=0}^{n} \frac{1}{(t_j - t_{j-1})}^{1/2} \times \\
\int_{-\infty}^{\infty} e^{-x_1^2/2t} e^{-(x_1-x_2)^2/2(t_2-t_1)} \ldots e^{-(x_n-x_{n-1})^2/2(t_n-t_{n-1})} \times \\
F(x_1, \ldots, x_n) dx_1 \ldots dx_n
\end{gathered} \tag{6.2.1}
$$

As stated above, we have taken $x_0 = 0$, $t_0 = 0$. The integral is an ordinary Riemann integral; it may easy to evaluate or it may be difficult, but there should be no problem in understanding its definition.

Such functionals, however, are the exception rather than the norm. We are usually interested in functionals that depend on the values of a function over its entire domain, not just at a finite number of points. Let us divide the interval $[0, T)$ into n subintervals of equal length. The division points are $t_j = j/n$, and we write $\epsilon = 1/n$ for short. Then we approximate $F[x]$ by an ordinary function, $F_n(x_1, \ldots, x_n)$ depending on the values of x at $\epsilon, 2\epsilon, \ldots, n\epsilon; \epsilon = T/n$. The mean of this function can be calculated according to eqn (6.2.1).

Now let n increase; F_n depends on more and more values of x. In the limit $n \to \infty$, F_n approaches the functional F, and our definition of the average of F becomes

$$< F[x(t)] >= \lim_{n\to\infty} \frac{1}{(2\pi\epsilon)^{n/2}} \int dx_1 \dots dx_n \times$$
$$F(x_1, \dots, x_n) e^{-\sum (x_j - x_{j-1})^2/2\epsilon} \qquad (6.2.2)$$

This is the Wiener integral, or integral over paths, or functional integral. The hard work consists of proving that the limit exists for some interesting class of functionals, and that it is independent of the method of subdivision of the interval. We shall proceed purely formally, and shall not attempt to study these questions.

Let us rewrite eqn (6.2.2) in a different way. Notice that the sum in the exponent can be written

$$\frac{1}{2\epsilon}\sum_{j=1}^{n}(x_j - x_{j-1})^2 = \frac{1}{2}\sum_{j=1}^{n}(\frac{\Delta x}{\Delta t})^2 \Delta t \to \frac{1}{2}\int_0^T (\frac{dx}{dt})^2 dt \qquad (6.2.3)$$

This suggests that $< F >$ be written

$$< F >= \int F[x(t)] e^{-\frac{1}{2}\int_0^T (dx/dt)^2 dt} \mathcal{D}x(t) \qquad (6.2.4)$$

where the symbol $\mathcal{D}x(t)$ stands for an integration 'volume element' for an integration over all paths. The normalization factor is not written explicitly, being included in the 'volume element' symbol. This equation is very suggestive, but it can in no way be taken literally. As we mentioned in Section 2.4, almost all of the sample paths of a Brownian particle are nowhere differentiable. Consequently the integral in eqn (6.2.3) only makes sense (i.e. has a well defined integrand) for a subset of sample paths whose contribution to $< F >$ have zero statistical weight. Thus eqn (6.2.4) must be regarded as a shorthand notation for eqn (6.2.2), nothing more and nothing less.

It is sometimes convenient to define *conditional Wiener integrals*. Here, all the sample paths begin at $x(0) = 0$ and end at $x(T) = x$. The definition is almost the same as that given by eqn (6.2.2) except that the last integral over x_n is not carried out. This has the consequence that the integral over all space (for the $n - 1$ variables, and for $F \equiv 1$) is not normalized to unity; however, if the final integral is carried out, the result will again be unity.

Thus we have seen how to average functionals that can be approximated by functions of a finite number of variables over the space of Brownian paths. But we may want to compute quantities which are not easily treated by this method. For example, we may want to ask for the probability that a Brownian path is differentiable almost everywhere. We have stated above that this is zero, but suppose we didn't know it. The question can be formulated as the computation of the average of a functional by the following artifice. Define the functional $G[x(t)]$ by

$$G[x(t)] = \begin{cases} 1 \text{ if x(t) is differentiable almost everywhere} \\ 0 \text{ otherwise} \end{cases} \tag{6.2.5}$$

Then $< G >$ is the required probability. On the other hand, it does not seem possible to express G as the limit of a sequence of functions of a finite number of variables, so that we cannot calculate $< G >$ using eqn (6.2.2). We have to introduce a probability measure directly in the space of Brownian paths.

6.3 Wiener Measure

In this section we shall attempt to explain in qualitative terms how a measure can be introduced on a space of functions. In order to follow this exposition, some previous knowledge of measure theory on the part of the reader will be necessary. It would be too much of a digression from the main theme of this book to develop measure theory, which will, in any case, not be needed after this chapter. Readers with no previous knowledge of measure theory are advised to skip this section. This presentation owes much to the exposition by Kac (1959, 1966).

Let us recall how Lebesgue measure on the real line is introduced. The starting point is to define the measure of an interval (a, b) as its length, $b - a$. Then a σ-field of sets is generated from the collection of intervals by the usual set theoretic union and set difference operations. The measure is extended from the intervals to the σ-field by requiring countable additivity. Unfortunately, the σ-field so generated does not contain all sets of interest but it is possible to extend the measure to a larger σ-field as follows. The measure just constructed can be used to define an *outer measure*, and concomitantly an *inner measure*, on the set of all subsets of the line. The sets for which the outer measure and inner measure are equal are the Lebesgue measurable sets of the line; the common value of the inner and outer measures is their measure. This thumbnail description encapsulates much work, and many theorems must be proved in order to show that all this can be done in a consistent way and to demonstrate all the properties of the measure. This development is beyond our scope.

Wiener actually gave several different, but equivalent constructions, of his measure on the space of paths. In one of these, he constructed an explicit mapping between functions on the line and the real numbers. The measure of a set of functions was then defined as the Lebesgue measure of the subset of $(0, 1)$ onto which these functions get mapped. Another approach involved the use of Fourier analysis. Here we shall sketch an approach which brings out the analogy with the construction of Lebesgue measure, just described. First consider the set of all real valued functions (continuous or not) on the interval $[0, T)$ with $x(t) = 0$. Select n points $t_1 < t_2 < \ldots < t_n$, and n pairs of numbers a_i and b_i, with $a_i < b_i$. That set of functions, $x(t)$, for which

$$a_i \leq x(t_i) \leq b_i \quad i = 1, \ldots, n \tag{6.3.1}$$

is called a *quasiinterval.* Quasiintervals play a role in Wiener measure similar to that which ordinary intervals play in Lebesgue measure. By definition, the measure of the quasiinterval eqn (6.3.1) is

$$\text{measure} = \prod_{i=1}^{n} \frac{1}{(2\pi(t_{i+1} - t_i))^{1/2}} \times$$
$$\int_{a_1}^{b_1} \cdots \int_{a_n}^{b_n} e^{-\frac{1}{2}\Sigma(x_i - x_{i-1})^2/(t_i - t_{i-1})} dx_1 \ldots dx_n \tag{6.3.2}$$

From the set of all quasiintervals, i.e. all n, all $\{t_i\}$, all $\{a_i,\ b_i\}$, construct a σ-field. There is a general theorem of Kolmogorov that states that a completely additive measure on the σ-field so generated can be constructed (Kolmogorov 1950).

Unfortunately things are not this simple. This is not Wiener measure. The σ-field constructed from the quasiintervals is too small; it does not contain most of the sets of interest (note the analogy to the corresponding phenomenon in the construction of Lebesgue measure). In particular, the set of continuous functions is not a measurable set with respect to this measure. Its outer measure is one and its inner measure is zero. Therefore, its measure is not defined. The question of what is the probability that as sample path is continuous, for example, cannot even be asked! This measure is therefore not very interesting.

To construct an useful measure, let us restrict ourselves for the time being to the class of *continuous* functions. Adopt the same quasiinterval restrictions as given in eqn (6.3.1) and the same measures for quasiintervals as in eqn (6.3.2), now only including continuous functions. Then the same argument as in two paragraphs above guarantees that one can construct a countably additive measure on the space of continuous functions on $[0, T)$. This is almost, but not quite, Wiener measure, for now sets of discontinuous functions are not measurable.

To finish the story, adjoin to the σ-field defined on the set of continuous functions the sets generated by the original set of quasiintervals, containing both continuous and discontinuous functions. Doob has shown that it is possible to extend the measure in a consistent way to this larger class of sets (Doob 1953). Thus one ends up with a measure on the space of functions on $[0, T)$ which is entirely concentrated on the continuous functions. The discontinuous (almost everywhere) functions have measure (probability) zero. This is very satisfying from the physical point of view for one would certainly have to reject as incompatible with experiment a model in which a Brownian particle could disappear and instantaneously reappear at a different position with non-zero probability.

Of course, once there is a measure, an integral can be defined and functionals can be integrated. For example, we could evaluate the integral of $G[x(t)]$ defined in eqn (6.2.5), should we desire.

It can be shown that the set of functions which are differentiable almost everywhere are a set of Wiener measure zero. We have alluded to this result several times already, and it is not at all surprising in light of the fact that $< \Delta x^2 >\sim t$, as we have already discussed in Chapter 4. Nevertheless, it is satisfying to know that there is a rigorous proof of the assertion On the other hand, the rigorous proof is very technical and difficult, and we shall not discuss it any further here.

There are many other results of the same type that have been proved. For example, for every $\epsilon > 0$, almost every continuous function satisfies a Lipshitz condition of order $\frac{1}{2} - \epsilon$, and almost none satisfy a Lipshitz condition of order $\frac{1}{2} + \epsilon$. Such theorems, interesting as they are, have not been of use in applications and are beyond the scope of this book.

This has all been in the nature of an existence discussion. We have not entered into the question of the actual evaluation of functional integrals. One technique of evaluation is discussed in the following section.

6.4 The Feynman–Kac Formula

The evaluation of some interesting functional integrals can be reduced to the solution of a differential or integral equation. Kac relates that he once attended a lecture by R. P. Feynman on his space-time approach to quantum mechanics. At that time, he was interested in the problem of the evaluation of Wiener integrals, and he got the idea to look for a relation between Wiener integrals and differential equations (Kac 1966). The result of his investigations are now known as the Feynman–Kac formula.

Let $V(x)$ be a continuous function of x, and suppose we want to evaluate the mean of the functional

$$F[x] = \exp(-\int_0^t V(x(s))ds) \tag{6.4.1}$$

conditional on $x(t) = x_{\text{final}}$. $V(x)$ might, for instance, represent the rate of annihilation (by absorbtion, say) of a Brownian particle at point x. Then eqn (6.4.1) would represent the probability that the particle has not been annihilated and is at x at time t if it has traversed the path $x(t)$. Using the notation of eqn (6.2.4), we may write

$$p(x,t) =< \exp(-\int_0^t V(x(s))ds) >_C= \int e^{-\int_0^t [\frac{1}{2}(dx/ds)^2 + V(x(s))]ds} \mathcal{D}_C x(t) \tag{6.4.2}$$

The formula of Feynman and Kac asserts that the functional integral eqn (6.4.2) is given by the solution to the partial differential equation

$$\frac{\partial p(x,t \mid x_0,t_0)}{\partial t} = (\frac{\partial^2}{\partial x^2} - V(x))p(x,t \mid x_0,t_0)$$

$$\lim_{t \to t_0} p(x, t \mid x_0, t_0) = \delta(x - x_0\ )\quad t > t_0 \tag{6.4.3}$$

We give a heuristic derivation of this following the path of Martin (1996). An alternate derivation, closer to the original one of Kac, can be found in Schulman (1981); see also Kac (1959).

We begin by introducing what seems to be a complication. The transition probability, $p(x, t \mid x_0, t_0)$, for a free Brownian particle can be looked on as the matrix element of an operator $\exp[-K(t - t_0)]$

$$\frac{1}{(2\pi\epsilon)^{\frac{1}{2}}} e^{-(x-x_0)^2/2(t-t_0)} =< x_0 \mid e^{-K(t-t_0)} \mid x > \tag{6.4.4}$$

We have introduced standard Dirac notation for the matrix element. In fact, the operator K is just $-\partial^2/\partial x^2$. To see this, note that eqn (6.4.4) is the operational solution of the Fokker–Planck equation eqn (4.4.2) (for the one dimensional case, of course). $V(x)$ is a multiplication operator, and hence diagonal in the $\mid x >$ basis.

Turning to the discrete approximation eqn (6.2.2), we may write

$$\begin{gathered}< \exp(-\textstyle\int_{t_0}^{t} V(x(s))ds) >\approx \\ \frac{1}{(2\pi\epsilon)}^{n/2} \int \cdots \int e^{-(x-x_n)^2/2\epsilon - V(x_n)\epsilon} e^{-(x_n - x_{n-1})^2/2\epsilon - V(x_{n-1})\epsilon} \cdots \\ e^{-(x_1-x_0)^2/2\epsilon - V(x_0)\epsilon} dx_1 \cdots dx_n \\ = \int dx_1 \cdots dx_n < x \mid e^{-K\epsilon} e^{-V\epsilon} \mid x_n >< x_n \mid e^{-K\epsilon} e^{-V\epsilon} \mid x_{n-1} > \cdots \\ < x_1 \mid e^{-K\epsilon} e^{-V\epsilon} \mid x_0 >=< x \mid (e^{-K\epsilon} e^{-V\epsilon})^{n+1} \mid x_0 >\end{gathered}$$

The exponentials of K and V cannot simply be combined into a single exponent because K and V do not commute. There is a way around this. Remember that $\epsilon = (t - t_0)/(N + 1)$. Consequently

$$< \exp(-\int_{t_0}^{t} V(x(s))ds >_C =< x \mid (e^{-K(t-t_0)/(n+1)} e^{-V(t-t_0)/(n+1)})^{n+1} \mid x_0 > \tag{6.4.5}$$

Now, of course, we must pass to the limit $n \to \infty$. To do this we use the Trotter product formula

$$\lim_{n \to \infty} (e^{A/n} e^{B/n})^n = e^{A+B} \tag{6.4.6}$$

(Nelson 1964) so that

$$< \exp(-\int_{t_0}^{t} V(s)ds) >_C =< x \mid e^{-(K+V)(t-t_0)} \mid x_0 > \tag{6.4.7}$$

This clearly is the solution of the equation

$$\frac{\partial p}{\partial t} = \frac{\partial^2 p}{\partial x^2} - V(x)p \tag{6.4.8}$$

Thus we have finished the task of showing that the evaluation of the functional integral eqn (6.4.2) can be accomplished by solving a partial differential equation.

However, the question is often reversed. We are presented with a partial differential equation that has come up in some application, and whose solution is difficult. Can we transform the equation into some equivalent problem whose solution is easier than the problem in its original form? In particular, we now know that if we have a differential equation of the form eqn (6.4.8), we can write the solution as a functional integral. Can we use the functional integral to help obtain the solution, or at least some of its properties? Perhaps it might suggest a worthwhile approximation method, for example. Consideration of the functional integral formulation has sometimes proved a worthwhile way to proceed; we shall see several examples later in this book.

6.5 Feynman Path Integrals

The aim of Feynman's space-time approach to quantum mechanics is to formulate an expression for the transition amplitude of a quantum mechanical system to go to state B from state A in terms of possible classical paths. As the reader will see, there are very strong analogies to the case of the Wiener integral, but these analogies cannot be pushed too far. In the Brownian motion case, we were interested in computing a probability measure in the space of paths, a non-negative set function. In the quantum case, the quantity of interest is a probability amplitude, which may be negative; in fact, it may even be complex! In the Brownian motion case, all paths give a positive contribution to the measure, and their contributions add. In the quantum mechanical case, the paths give contributions of both signs. Furthermore, the integrands are rapidly oscillating and there is a great deal of cancellation between the contributions of neighboring paths. Nevertheless, the two problems have many features in common.

Were this a treatise on quantum mechanics, the proper way to introduce the Feynman path integral would be to begin from basic physical notions, in particular the principle that probability amplitudes must be added before squaring to form a probability. It would then be shown how these concepts lead to the idea of a 'sum over histories', or an integral over possible classical paths. This is the trail followed by Feynman himself. But it would take us too far afield to present enough quantum mechanics here to make following such a trail possible. We shall take the more direct, albeit less physically appealing, route of starting from the Wiener functional integral and analytically continuing it into the complex plane.

In the previous section, for simplicity of notation we set $D = 1$, where D is the diffusion coefficient of the Brownian particle. Let us now change units so as to restore the D. Then eqn (6.2.4) can be written

$$< x \mid e^{-(K+V)(t-t_0)} \mid x_0 >= \int e^{-\int_{t_0}^{t}(\frac{D}{2}(dx/ds)^2+V(x(s)))ds}\mathcal{D}x(t) \qquad (6.5.1)$$

Now let V be a pure imaginary quantity $V(x) = iU(x)/\hbar$; U is then a quantity with the dimensions of energy. In addition, let D be the imaginary quantity $-im/\hbar$. Then the right-hand side of (6.5.1) can be written

$$\int e^{\frac{i}{\hbar}S(x,\dot{x})}\mathcal{D}x(t) \qquad (6.5.2)$$

Here $S(x, \dot{x})$ is the *action* along the path

$$S = \int_{t_0}^{t} (\frac{m}{2}(dx/ds)^2 - U)ds \qquad (6.5.3)$$

The action is the time integral of the Lagrangian, $T - U$, where T is the kinetic and U the potential energy; m is the particle mass, and $\hbar$ is Planck's constant.

The functional integral eqn (6.5.2) is then the solution of the partial differential equation

$$i\hbar\frac{\partial P}{\partial t} = -\frac{\hbar^2}{2m}\frac{\partial^2 P}{\partial x^2} + U(x)P \qquad (6.5.4)$$

which is immediately recognizable as the Schrödinger equation of a particle of mass m moving in a potential field $U(x)$.

This method, of course, shows nothing of the physical content of the path integral eqn (6.5.2). The reader interested in this physical content is referred to the many excellent books devoted to the Feynman integral. Perhaps the best of these for orientation purposes is that by Feynman himself (Feynman and Hibbs 1965). Neither the method of presentation used here nor any other simple method shows much of the mathematical content either, for it is not at all clear in what sense the path integral converges, if at all. One might attempt to make sense of it by giving $\hbar$ an imaginary part, thus giving a convergence factor to the integrals in the finite approximations to the path integral. After evaluating the integral by passing to the limit $n \to \infty$, one then should effect another passage to the limit $\operatorname{Im} \hbar \to 0$. If that limit exists, it is the value of the path integral. The quantity so constructed may indeed satisfy the Schrödinger equation, and may be of real physical interest because of that. Nevertheless, the resulting quantity is not an integral over a countably additive measure. This was conjectured by Gel'fand and Yaglom (1960); detailed technical discussion can be found in papers by Cameron and Nelson (Cameron 1960/61; Nelson 1964). A review of the mathematical status of Feynman integrals has been given by Cartier (1997). Montroll (1952) has also discussed the relation between Wiener integrals and quantum theory.

6.6 Evaluation of Wiener Integrals

So far, we have seen two methods of evaluating Wiener integrals. The first is to evaluate the finite dimensional integrals eqn (6.2.1), and then pass to the limit $n \to \infty$. When this can actually be carried out, it is a perfectly legitimate method. The second method is that of Section 6.4. The Wiener integral is equivalent to a solution of a certain differential equation, so solve the differential equation. This has the advantage of replacing unfamiliar mathematics (integration in spaces of infinite numbers of dimensions) with more familiar mathematics (differential equations), but very often the differential equation can't be solved explicitly either. Usually the functional integral has to be evaluated by finding a clever trick, by a numerical method, or by perturbation theory.

As an example of evaluation by passage to the limit, let us compute the mean value of

$$F[x] = \exp(-\int_0^t x^2 ds) \tag{6.6.1}$$

This presentation follows the method of Gel'fand and Yaglom (1960) In the discretized form of eqn (6.2.1) or eqn (6.4.5), the exponent of the approximating discretized integral is clearly a positive definite quadratic form. For integrals of this form, there is the well known formula

$$\int_{-\infty}^{\infty} e^{-\Sigma a_{ij} x_i x_j} dx_1 \ldots dx_n = \pi^{n/2} [\det A]^{-\frac{1}{2}} \tag{6.6.2}$$

Therefore we have that

$$< \exp(-\int_0^t x^2 ds) >= \lim_{n \to \infty} D^{(n)\,-\frac{1}{2}} \tag{6.6.3}$$

where $D^{(n)}$ is the determinant

$$D^{(n)} = \begin{vmatrix} 2+(t/n)^2 & -1 & 0 & \cdots & 0 & 0 \\ -1 & 2+(t/n)^2 & -1 & \cdots & 0 & 0 \\ & & & \cdots & & \\ 0 & 0 & 0 & \cdots & 2+(t/n)^2 & -1 \\ 0 & 0 & 0 & \cdots & -1 & 2+(t/n)^2 \end{vmatrix} \tag{6.6.4}$$

The notation $D_k^{(n)}$ will denote the principal minor of this determinant of order $n-k+1$ situated in the lower right-hand corner of $D^{(n)}$. If we expand the determinant $D_k^{(n)}$ by the elements of the first row, we obtain the recurrence relation

$$D_k^{(n)} = (2 + (\frac{t}{n})^2) D_{k+1}^{(n)} - D_{k+2}^{(n)} \quad 1 \le k \le n-2 \tag{6.6.5}$$

Writing $t/n = \Delta t$, this equation may be transcribed as

$$\frac{D_k^{(n)} - 2D_{k+1}^{(n)} + D_{k+2}^{(n)}}{(\Delta t)^2} = 2D_{k+1}^{(n)} \tag{6.6.6}$$

Clearly, the left-hand side begins to look like a second derivative. Now let $k \to \infty$, $\Delta t \to 0$ in such a way that $k\Delta t = s$ is fixed. Then eqn (6.6.6) goes over into

$$\frac{d^2 D}{ds^2} = 2D, \quad D(t) = 1; D'(t) = 0 \tag{6.6.7}$$

The conditions on $D(t)$ and $D'(t)$ come from examining the recursion relation for $k = n$ and $k = n - 1$. The quantity we want is

$$< \exp(-\int_0^t x^2 ds) >= [D(0)]^{-\frac{1}{2}} \tag{6.6.8}$$

From the solution to eqn (6.6.7), which is elementary, the final result is

$$< \exp(-\int_0^t x^2 ds) >= \text{sech}(\sqrt{2}t) \tag{6.6.9}$$

Another method for evaluation of functional integrals is to expand the functions, $x(t)$, representing the Brownian path, in some orthonormal set of functions between 0 and T:

$$x(t) = \Sigma a_n \omega_n(t) \tag{6.6.10}$$

The functional $F[x]$ can then be looked on as a function of a denumerable set of variables $\{a_k\}$, which can then be integrated over the a_ks. This technique has not often been applied to Wiener integrals; we shall merely refer to Brush (1961) who reviews these methods for Feynman integrals.

We also refer to Brush for a review of numerical methods vaguely analogous to the trapezoid rule or Simpson's rule for ordinary integrals. These methods have proved to be of limited accuracy. There is, however, a numerical method which is tailor made for the approximate evaluation of functionals of Brownian motion paths. This is the Monte Carlo method.

We shall see in Chapter 9 that Brownian paths can be modeled quite well by random walks. In fact this is basic in Smoluchowski's theory, (Chapter 4). Random walks are easy to generate on a computer. To evaluate the mean of a functional, generate a random walk, evaluate the functional for that random walk as its argument, and do this over and over again. When enough samples have been accumulated, the computer can take their average. Of course, there are often tricks that can be applied to lessen the labor or shorten the computer time, but this is the basis of the method. The Monte Carlo method is inherently a statistical method, and here it is being used to answer a statistical question; the method seems well adapted to the question.

6.7 Applications of Functional Integrals

The theory of the Wiener integral is a beautiful one, generalizing the theory of integration over regions of finite dimensional spaces to infinite dimensional regions of infinite dimensional spaces (function spaces). But is it useful as a computational tool, or as a tool in deriving new results for physical applications? Such uses have, to date, been very limited.

The Feynman integral has, indeed, proved very efficacious. Most modern texts on quantum field theory and quantum many body theory use the Feynman integral formulation, rather than the canonical Hamiltonian formalism, to present the theory and to derive the perturbation theory which has been essential for extracting experimentally verifiable results from the formalism. In quantum statistical mechanics too, the Feynman integral arises. If the time, t , is allowed to become complex, the Feynman integral is just the partition function of the system (Brush 1961; Gel'fand and Yaglom 1960). Here also, perturbation theory is of the utmost importance.

As far as I know, there have been no comparably important applications of the Wiener integral. The Wiener integral has remained a tool in the mathematical theory of stochastic processes, primarily useful in the formulation and proof of theorems. We shall therefore now leave this topic and go on in the next chapter to our first serious study of physical systems undergoing Brownian motion.

7

SOME IMPORTANT SPECIAL CASES

7.1 Several Cases of Interest

In this chapter we shall treat several special cases of particular interest. These cases are, the freely translating particle, the harmonically bound particle, the particle in a constant field of force, and the uniaxial rotor. These models can be attacked either from the Langevin equation point of view or from the Fokker–Planck equation point of view. We shall approach the freely translating particle via the Langevin route, and the other models via the Fokker–Planck route. The reason for choosing different methods is merely to illustrate the different techniques.

We have already discussed the free particle and a uniaxial rotor in Chapter 4. We did this using Einstein's methods, treating the description of the particles in position space, and completely neglecting their inertia This is, in fact, the origin of the peculiar non-differentiability of the sample paths. We now intend to treat the Brownian motion of these models in phase space, with the particle's velocity being taken into account. Of course, the collisions with the ambient medium are still considered as random events, so we still have a stochastic theory and not a purely dynamical one.

7.2 The Free Particle

We consider a particle of mass M immersed in an ambient medium. The interaction between the particle and the medium is characterized by a systematic linear frictional force, $-\zeta u$, and a random force, $F(t)$. u is the velocity of the particle. The random force is assumed to be a Gaussian stochastic process with

$$
\begin{aligned}
< F(t) > &= 0 \\
< F(t_1)F(t_2) > &= 2kT\zeta\delta(t_1 - t_2)
\end{aligned}
\tag{7.2.1}
$$

We shall treat here only the case of motion in one dimension. The extension to three dimensions is simple, and is left as an exercise for the reader.

The equation of motion is the Langevin equation, eqn (5.5.1), which we rewrite here for ease of reference.

$$M\frac{du}{dt} = -\zeta u + F(t) \tag{7.2.2}$$

We also rewrite the formal solution

$$u = u_0 e^{-\beta t} + \frac{1}{M}\int_0^t F(s) e^{-\beta(t-s)} ds \tag{7.2.3}$$

where we have introduced the conventional abbreviation $\beta = \zeta/M$.

The reader may wish to review the material of Section 5.2 to refresh his or her memory of the meaning of eqn (7.2.1). Because we are dealing with the case of additive noise, the distinction between the Ito and Stratonovich equations is immaterial in the present chapter.

Since $F(t)$ is a Gaussian random process, and since the sum of Gaussian random processes is Gaussian, so is the integral in eqn (7.2.2). We therefore conclude that

$$u - u_0 e^{-\beta t}$$

is a Gaussian random process. To characterize the process completely we need its distribution functions. Since the process is Gaussian, the first and second moments suffice to determine the distribution functions.

From $< F >= 0$, it follows from eqn (7.2.1) that

$$< u >= u_0 e^{-\beta t} \tag{7.2.4}$$

The variance of u can also be obtained from eqn (7.2.1). It is

$$< (u - u_0 e^{-\beta t})^2 >= \frac{2kT}{M}[1 - e^{-2\beta t}] \tag{7.2.5}$$

Thus, since we know that the distribution function of u is Gaussian, we can immediately write, for the probability of u conditional on $u(0) = u_0$

$$p(u, t \mid u_0, 0) = [\frac{M}{2\pi kT(1 - e^{-2\beta t})}]^{\frac{1}{2}} e^{-\frac{M}{2kT}\frac{(u-u_0 e^{-\beta t})^2}{(1-e^{-2\beta t})}} \tag{7.2.6}$$

Note that the argument starting from the Langevin equation gives the conditional probability because the Langevin equation is a first order differential equation. Its solution is determined by solving an initial value problem. Alternatively you may note that the integral equation (cf. eqn (5.2.2)) that expresses the differential equation in rigorous form, contains u_0 as a parameter. The reader can easily verify that, if the velocity is u_0 at time t_0 instead of time 0, the only changes are to replace the lower limit on the integral by t_0 and $u_0 \exp(-\beta t)$ by $u_0 \exp(-\beta(t-t_0))$ in eqns (7.2.3)–(7.2.6). That is, the process is stationary. The essential reason for this is that F has been assumed to be stationary.

So far our considerations tell us nothing about the higher order distribution functions. However, we can assert that the stochastic process describing the motion is a Markov process; hence the conditional probability eqn (7.2.6) uniquely determines all of the higher distribution functions. Why can we assert that the process is Markov? There is no mathematical fact that requires this conclusion, thought it is permissible mathematically. The reason is physical. The Langevin equation describes the velocity change of the Brownian particle as depending only on its current state, i.e. the current velocity and the current force. The previous history of the particle is assumed irrelevant. This is the essence of the Markov property, and is why we are justified in making the assumption that the process is Markov.

The stochastic process whose transition probability we have just derived in called the *Ornstein–Uhlenbeck process* or often just the O–U process (Uhlenbeck and Ornstein 1930). It is a Gaussian Markov process with a correlation function that decays exponentially in time; this follows directly from eqn (7.2.4). There is a theorem due to Doob (1942) which states that the Ornstein–Uhlenbeck process is, with one trivial exception, the only stochastic process with these three properties. The exception is the completely independent Gaussian random process (cf. eqn (3.3.1)).

We have carried out the calculations here only for the one-dimensional case. It is very easy to extend them to a particle moving in three dimensional space. Essentially all one has to do is replace the scalars u and F by the vectors $\mathbf{u}$ and $\mathbf{F}$. This is left as an exercise for the reader.

7.3 The Distribution of Displacements

Einstein's theory of Brownian motion, which gives rise to the Wiener process, neglects the inertia of the Brownian particle. The Ornstein–Uhlenbeck process takes the inertia into account. The Einstein development is formulated in the configuration space of the particle, while the O–U development is formulated in the particle's velocity space. In order to compare the results of the two developments we have to find the distribution of displacements of the O–U process. As in the previous section, we shall only consider the one-dimensional case explicitly.

To get an expression for the displacement of the Brownian particle, integrate eqn (7.2.3) with respect to time. The result is

$$x - x_0 = \frac{u_0}{\beta}(1 - e^{-\beta t}) + \frac{1}{M}\int_0^t dy \int_0^y dsF(s)e^{-\beta(y-s)} \tag{7.3.1}$$

The same argument as in the previous section shows that $\mathbf{x}$ is a Gaussian random process. Consequently all we have to do is find its mean and variance. From eqn (7.3.1), it follows immediately that

$$< x - x_0 >= \frac{u_0}{\beta}(1 - e^{-\beta t}) \tag{7.3.2}$$

and

$$< (x- < x >)^2 >= \frac{kT}{M\beta^2}(2\beta t - 3 + 4e^{-\beta t} - e^{-2\beta t}) \tag{7.3.3}$$

Therefore we can immediately write down the conditional distribution function

$$p(x,t \mid x_0,0) = \left(\frac{M\beta^2}{2\pi kT\sigma^2}\right)^{\frac{1}{2}} \exp\left[-\frac{\beta^2}{2mkT}\frac{(x - x_0 - \frac{u_0}{\beta}(1-e^{-\beta t}))^2}{\sigma^2}\right]$$
$$\sigma^2 = 2\beta t - 3 + 4e^{-\beta t} - e^{-2\beta t} \tag{7.3.4}$$

These are the formulae that should be compared with those of Einstein (cf. Section 4.4).

For large times, the O–U process yields

$$< x > \approx x_0 + \frac{u_0}{\beta}$$
$$< (x - x_0)^2 > \approx \frac{2kT}{\zeta}t \tag{7.3.5}$$

The first of these equations can be interpreted as saying that the particle remembers its initial velocity for a time of the order of $u_0 M/\zeta$. Thereafter it wanders with no propensity for any given direction. The second formula, for the mean square displacement, is exactly the same as that of Einstein. Equation (4.4.5) has a coefficient of 6, while we here obtain a coefficient of 2, but this is merely a manifestation of the difference between three and one dimensions. In any given number of dimensions the asymptotic O–U results and the Einstein results are the same.

What happens for short times? For times short compared to M/ζ, eqns (7.3.2) and (7.3.3) show that

$$x \approx x_0 + u_0 t \tag{7.3.6}$$

and

$$< (x - x_0)^2 > \approx u_0^2 t^2 \tag{7.3.7}$$

Consequently it seems intuitively clear that the paths of particles undergoing the Ornstein-Uhlenbeck process are differentiable. It is the velocity that is nondifferentiable, inheriting this property from the wild irregularities of F. This can be proved rigorously. There are thus two reasons that the O–U process is a better model for the physical phenomenon of Brownian motion than is the Wiener process. These are (a) The O–U gives a description of the process in phase space, while the Wiener process is restricted to coordinate space, and (b) the path is described more realistically.

So far, we have only determined the velocity space distribution and the coordinate space distribution, the marginal distributions. The joint distribution of u and x still has to be determined. We shall do this as a byproduct of our treatment of the harmonically bound particle in the next section. Although the joint distribution can be obtained by the methods we have just used, the method has already been amply illustrated, and it is worthwhile introducing a different technique.

7.4 The Harmonically Bound Particle

The free particle, which we have just treated, is a model for the Brownian motion of a particle in a medium with no external forces. We now want to study the case of a particle subject to a linear restoring force as well as the frictional force and the random force due to the medium. Although this could be studied using the Langevin methods of the last two paragraphs, we prefer to use the method of the Fokker–Planck equation.

The harmonically bound particle is interesting because it illustrates the effect that external forces can have on Brownian motion, it is simple enough to be solved exactly, and it approximates some actual physical situations: an atom in a crystal lattice, an atom adsorbed on a surface, or an atom trapped in a potential well.

We need to have the Langevin equation for this model at hand, however, in order to evaluate the coefficients in the Fokker–Planck equation. We take as origin the point to which the particle is bound; the restoring force is then

$$K = -\omega_0^2 x \tag{7.4.1}$$

The Langevin equation is

$$\begin{aligned}\frac{du}{dt} &= -\frac{\zeta}{M}u - \omega_0^2 x + \frac{1}{M}F(t)\\ \frac{dx}{dt} &= u\end{aligned} \tag{7.4.2}$$

and the Fokker–Planck equation is

$$\begin{aligned}\frac{\partial p}{\partial t} = -\frac{\partial}{\partial x}(A_1 p) - \frac{\partial}{\partial u}(A_2 p) + \frac{1}{2}(\frac{\partial^2}{\partial x^2}(B_{11}p) + \\ \frac{\partial^2}{\partial u^2}(B_{22}p) + 2\frac{\partial^2}{\partial x \partial y}(B_{12}p))\end{aligned} \tag{7.4.3}$$

We need to determine the A and B coefficients (cf. eqn (3.4.5)).

Using eqn (7.4.2), we see immediately that

$$\begin{aligned}A_1 &= u\\ A_2 &= -\beta u - \omega_0^2 x\end{aligned} \tag{7.4.4}$$

For the B coefficients, since Δu and Δx are both linear in Δt for small Δt, it follows that B_{11} and B_{12} both vanish. Although the same argument superficially seems to say that B_{22} vanishes also, this is not the case because of the delta correlation property of $F(t)$. In fact, from eqn (7.4.2)

$$< (\Delta u)^2 >= \frac{1}{M^2} \int_0^{\Delta t} ds \int_0^{\Delta t} dt < F(t)F(s) > +O((\Delta t)^2) \tag{7.4.5}$$

The integral is easily evaluated leading to

$$\begin{aligned} B_{11} &= 0 \\ B_{22} &= \frac{kT\beta}{M} \equiv D \\ B_{12} &= 0 \end{aligned} \tag{7.4.6}$$

Note that D is *not* the self-diffusion coefficient of the particle; it even has the wrong dimensions for that. However, it is the conventional symbol in this context.

The Fokker–Planck equation for the harmonically bound particle is thus

$$\frac{\partial p}{\partial t} = -\frac{\partial}{\partial x}(up) + \frac{\partial}{\partial u}([\beta u + \omega_0^2 x]p) + D\frac{\partial^2 p}{\partial u^2} \tag{7.4.7}$$

It now remains to solve this equation, subject to the initial condition

$$p(u, x, t \mid u_0, x_0, 0) \to \delta(x - x_0)\delta(u - u_0) \text{ as } t \to 0 \tag{7.4.8}$$

This is a necessary condition for the interpretation of p as the transition probability of a Markov process.

It is convenient to begin the solution by a change of variables. Experience with the harmonic oscillator as a mechanical system, one without the random force, suggests that useful variable for the integration of its equations of motion are

$$\begin{aligned} z_1 &= u + ax \\ z_2 &= u + bx \end{aligned} \tag{7.4.9}$$

where

$$\begin{aligned} a &= \frac{\beta}{2} + i\omega_1 \\ b &= \frac{\beta}{2} - i\omega_1 \\ \omega_1^2 &= \omega_0^2 - \frac{\beta^2}{4} \end{aligned} \tag{7.4.10}$$

Note that ω_1 need not be a real frequency; it is imaginary when the system is overdamped.

With this change of variables, the Fokker–Planck equation becomes

$$\frac{\partial p}{\partial t} = b\frac{\partial}{\partial z_1}(z_1 p) + a\frac{\partial}{\partial z_2}(z_2 p) + D(\frac{\partial}{\partial z_1} + \frac{\partial}{\partial z_2})^2 p \tag{7.4.11}$$

a more symmetrical form than eqn (7.4.7).

The next step is to Fourier transform eqn (7.4.11). Because the independent variables only enter into the coefficients of the equation to, at most, the first power, the Fourier transform yields a first order partial differential equation:

$$\frac{\partial \tilde{p}}{\partial t} = -bk_1\frac{\partial \tilde{p}}{\partial k_1} - ak_2\frac{\partial \tilde{p}}{\partial k_2} - D(k_1 + k_2)^2\tilde{p} \tag{7.4.12}$$

Here $\tilde{p}$ is the Fourier transform of p, and $k_{1,2}$ are the Fourier transform variables.

The standard way to attempt to solve such a first order equation is to consider the auxiliary set of ordinary differential equations

$$\frac{dt}{1} = \frac{dk_1}{bk_1} = \frac{dk_2}{ak_2} = -\frac{d\tilde{p}}{D(k_1 + k_2)^2\tilde{p}} \tag{7.4.13}$$

This means, introduce an auxiliary variable, s, and consider the equations $dt/ds = 1$, $dk_1/ds = ak_1$, etc. There will be three first integrals of this set of equations that are independent of r. For the present case, these first integrals can be taken to be

$$\begin{aligned} k_1 e^{-bt} &= c_1 \\ k_2 e^{-at} &= c_2 \\ \tilde{p}\exp\left[D(\frac{k_1^2}{2b} + \frac{k_2^2}{2a} + \frac{2k_1k_2}{a+b})\right] &= c_3 \end{aligned} \tag{7.4.14}$$

Now the general theory of first order linear partial differential equations asserts (Courant and Hilbert 1937; Sneddon 1957) that the general solution to eqn (7.4.12) is of the form

$$\Phi(c_1, c_2, c_3) = 0 \tag{7.4.15}$$

where Φ is an arbitrary function of its arguments. In principle, eqn (7.14.15) can be solved for $\tilde{p}$, yielding

$$\tilde{p} = \phi(k_1 e^{-bt}, k_2 e^{-at})e^{-D(\frac{k_1^2}{2b} + \frac{k_2^2}{2a} + \frac{2k_1k_2}{a+b})} \tag{7.4.16}$$

where ϕ is another arbitrary function. The remaining problem is to determine ϕ.

ϕ is determined using the initial condition eqn (7.4.8) which, when Fourier transformed, reads

$$\tilde{p} \to e^{i(k_1 x_0 + k_2 u_0)} \quad \text{as } t \to 0 \tag{7.4.17}$$

This tells us that

$$\phi(t=0) = e^{D(\frac{k_1^2}{2b}+\frac{k_2^2}{2a}+\frac{2k_1k_2}{a+b})+i(k_1x_0+k_2u_0)} \tag{7.4.18}$$

For $t > 0$, therefore, we need only substitute $k_1\exp(-bt)$ for k_1 and $k_2exp(-at)$ for k_2 in ϕ. The final result is

$$\tilde{p}(k_1, k_2, t) = \exp[i(k_1x_0e^{-bt} + k_2u_0e^{-at})] \times$$
$$\exp\left[-D(\frac{k_1^2}{2b}(1-e^{-2bt}) + \frac{k_2^2}{2a}(1-e^{-2at}) + \frac{2k_1k_2}{a+b}(1-e^{-(a+b)t}))\right] \tag{7.4.19}$$

Completing the square in the exponent of eqn (7.4.18) shows that $\tilde{p}$ is a Gaussian function of the ks. Since the Fourier transform of a Gaussian function is a Gaussian function, the final conclusion is that $p(x, u, t \mid x_0, u_0, 0)$ is Gaussian.

The only remaining problem is to find the means and correlation functions, the coefficients in the exponent of the Gaussian function. In principle, this can be done as follows

1. Read off the coefficients of the quadratic form in the exponent of the Gaussian in k space.

2. Form the matrix of these coefficients and calculate its inverse. This gives the coefficients of the quadratic form in the exponent of the Gaussian in z space.

3. Make the inverse linear transformation to eqn (7.4.9) to get the quadratic form in (x, u) space.

This is a feasible procedure, but involves much tedious algebra. An alternative procedure is to compute the means and correlation functions of x and u directly from the Langevin equation. Since the Langevin equation for the case at hand is a system of linear differential equations with constant coefficients, this is straightforward. The results are

$$
\begin{aligned}
< u > &= e^{-\beta t/2}[-x_0\frac{\omega_0^2}{\omega_1}\sin\omega_1 t + \\
&\quad u_0(\cos\omega_1 t - \frac{\beta}{2\omega_1}\sin\omega_1 t)] \\
< x > &= e^{-\beta t/2}[x_0(\cos\omega_1 t + \frac{\beta}{2\omega_1}\sin\omega_1 t) + \\
&\quad u_0\frac{\sin\omega_1 t}{\omega_1}] \\
< (u- < u >)^2 > &= \frac{D}{\beta}[1 - \frac{e^{-\beta t}}{\omega_1^2}(\omega_1^2 + \frac{1}{2}\beta^2\sin^2\omega_1 t - \\
&\beta\omega_1\sin\omega_1 t\cos\omega_1 t)] \\
< (x- < x >)^2 > &= \frac{D}{\beta\omega_0^2}[1 - \frac{e^{-\beta t}}{\omega_1^2}(\omega_1^2 + \frac{1}{2}\beta^2\sin^2\omega_1 t + \\
&\beta\omega_1\sin\omega_1 t\cos\omega_1 t)] \\
< (x- < x >)(u- < u >) > &= \frac{D}{\omega_1^2}e^{-\beta t}\sin^2\omega_1 t
\end{aligned}
\tag{7.4.20}
$$

As might be expected, the mean position and velocity decay to zero in a time of the order of $2M/\zeta$; because there is a restoring force, the approach to zero is oscillatory (except when the motion is overdamped). By the equipartition theorem, the long time asymptotic values of $< (u- < u >)^2 >$ and $< (x- < x >)^2 >$ are kT/M and kT/ω_0^2 respectively. Recalling the definition of D from eqn (7.4.6), one can see that these limits are also reached in a damped oscillatory manner.

We shall not write down the explicit formula for $p(x, u, t \mid x_0, u_0, 0)$; it is clearly a very long expression. The situation can be summarized by stating that the conditional distribution function is a bivariate Gaussian. The matrix $\mathbf{V}$ (cf. eqn (2.4.14)) is the inverse matrix to the matrix of correlation functions given in eqn (7.4.20).

We are now in a position to compute the bivariate conditional distribution function in position and velocity for the case of the free particle. The marginal distributions were already computed in Sections 7.2 and 7.3. We need only take the distribution for the harmonically bound particle and let $\omega_0 \to 0$; this entails $\omega_1 \to i\beta/2$. $< (u- < u >)^2 >$ and $< (x- < x >)^2 >$ are as given in eqns (7.2.5) and (7.3.3) respectively. The cross correlation is given by

$$
< (x- < x >)(u- < u >) >= \frac{D}{4\beta^2}(1 - e^{-\beta t})^2 \tag{7.4.21}
$$

As in the previous paragraph, the conditional distribution for a free particle is a bivariate Gaussian distribution. The matrix of the quadratic form in the numerator is the inverse of the matrix of the correlation functions of x and u.

7.5 A Particle in a Constant Force Field

Another case that is of interest and is simple enough to be solved exactly is that of a particle in a constant field of force. Examples of this are particles in the Earth's gravitational field, or charged particles in a constant electric field. As is well known, the Earth's field may be considered to be constant if the distances involved are small enough, say of laboratory size.

The method of treatment of this problem is so similar to that of the harmonically bound particle that we merely intend to outline the derivation, leaving the details as an exercise for the reader. This will therefore be a short section.

Again, we treat only the one-dimensional case. The extension to three dimensions is easy. The particle is moving in an external field of force, K, constant in space and time. There are also the frictional force and the random force, both due to the medium. These forces are assumed to have the same properties as in previous sections. The Langevin equation for the system is

$$\begin{aligned} \frac{du}{dt} &= -\zeta u + K + F(t) \\ \frac{dx}{dt} &= u \end{aligned} \tag{7.5.1}$$

From this, the methods used earlier show that

$$\begin{aligned} < u > &= u_0 e^{-\beta t} + \frac{K}{M\beta}(1 - e^{-\beta t}) \\ < x > &= x_0 + \frac{Kt}{M\beta} - \frac{1}{\beta}(u_0 - \frac{K}{M\beta})e^{-\beta t} \\ < (u- < u >)^2 > &= \frac{kT}{M}(1 - e^{-\beta t}) \\ < (x- < x >)^2 > &= \frac{kT}{M\beta^2}(2\beta t - 3 + 4e^{-\beta t} + e^{-2\beta t}) \end{aligned} \tag{7.5.2}$$

Note that the variances of x and u are the same as in the absence of the external field. The Fokker–Planck equation for this model is

$$\frac{\partial p}{\partial t} = -\frac{\partial}{\partial x}(up) + \frac{\partial}{\partial u}(\beta u - K)p + D\frac{\partial^2 p}{\partial u^2} \tag{7.5.3}$$

This equation can be solved by the method used in Section 7.4; the preliminary transformation of variable to z_1 and z_2 is not needed in the present case.

The final result for the bivariate distribution of x and u is a bivariate Gaussian whose parameters are determined by the correlation functions given by eqn (7.5.2). Recall that thequadratic form in the exponent of the Gaussian has as coefficients the matrix elements of the inverse matrix to the matrix of correlation functions.

7.6 The Uniaxial Rotor

The problem of the Brownian motion of a general rotating body is technically a complicated one. The reason for this is at least three-fold: (1) rotations about different axes do not commute, (2) the range of position variables, the angles specifying the body's orientation, is finite, (3) the relation between angular velocity and angular momentum is a tensorial one, not a vectorial one as in translation. Nevertheless, we shall have to study rotational Brownian motion because of its importance for molecular rotations in dense media. We shall do this in Chapter 15.

For the present chapter we confine ourselves to the case of a body restricted to rotate about a single axis. There may be a restoring torque hindering the rotation. We have in mind the example of a mirror mounted on a torsion fiber, such as is found in galvanometers. The mirror undergoes rotational Brownian motion due to collisions with molecules of the surrounding gas, while the fiber provides a restoring torque.

The orientation of the body is characterized by a single angle, θ. The body has a moment of inertia I about the unique axis. In analogy to the case of translation, the frictional torque due to the medium is proportional to the angular velocity, ω; the proportionality constant is ζ_R, the rotational friction constant. The restoring torque is taken to be proportional to the angle of rotational displacement, θ. Of course, for large displacements this will not be an accurate representation of the actual torque. On the other hand, for all cases of interest, the rotations are small, and the linear approximation to the resorting torque is adequate.

The Langevin equation for this system is

$$\begin{aligned} \frac{d\omega}{dt} &= -\frac{\zeta_R I}{\omega} - \omega_0^2\theta + \frac{1}{I}N(t) \\ \frac{d\theta}{dt} &= \omega \end{aligned} \tag{7.6.1}$$

$N(t)$ is the random torque due to the ambient medium, and has the same properties as the random force in the translational case

$$\begin{aligned} < N(t) > &= 0 \\ < N(s)N(t) > &= 2kT\zeta_R\delta(t-s) \end{aligned} \tag{7.6.2}$$

It is clear that eqn (7.6.2) has precisely the same structure as eqn (7.4.2) for the harmonically bound particle. Only the symbols are different.

Without doing any further calculation, we can carry over the results of Section 7.4, merely making the change of notation

$$\omega \leftrightarrow u \qquad\qquad M \leftrightarrow I$$

$$\theta \leftrightarrow x \qquad \zeta \leftrightarrow \zeta_R \tag{7.6.3}$$

The quantity of main interest here is the mean square amplitude of the 'jiggle', that is, of the angular displacement. This is, from eqn (7.4.20)

$$< (\theta - < \theta >)^2 > = \frac{kT}{I\omega^2}[1 - \frac{e^{-\beta_R t}}{\omega_1^2}(\omega_1^2 + \frac{1}{2}\beta_R^2 \sin^2 \omega_1 t + \beta_R \omega_1 \sin \omega_1 t \cos \omega_1 t)]$$

$$\beta_R = \frac{\zeta_R}{I}; \quad \omega_1^2 = \omega_0^2 - \frac{\beta_R^2}{4} \tag{7.6.4}$$

After long times, the mean square angular displacement is just $kT/I\omega_0^2$ as the equipartition theorem predicts.

7.7 An Equation for the Distribution of Displacements

We have derived the distribution of displacements of a free Brownian particle in Section 7.3. The derivation was based on noting that $x - < x >$ must have a Gaussian distribution, and calculating its moments from the Langevin equation. Wang and Uhlenbeck (1945) asked the question, whether it was possible to derive it differently: to find some differential equation whose solution is the distribution in question? They did not answer this question, but the answer is 'yes.'

Since we already have the solution, eqn (7.3.4), it might seem that having the equation in addition is redundant. But the solution which we have obtained is for infinite space. If we need to consider a bounded system for some application, it would be desirable to have the equation satisfied by the distribution function in question, which we would attempt to solve subject to appropriate boundary conditions. We shall not discuss what are appropriate boundary conditions in this section, confining our interest for the time being, to determining the evolution equation for the probability of the displacements.

Let us calculate the characteristic function of the Gaussian random variable $x - < x >$, whose distribution is given by eqn (7.3.4). By eqn (7.2.6), $\rho(q)$ is

$$\rho(q) = \exp[-\frac{kT}{2M\beta^2}q^2(2\beta t - 3 + 4e^{-\beta t} - e^{-2\beta t}) + iq\frac{u_0}{\beta}(1 - e^{-\beta t})] \tag{7.7.1}$$

We have used q as the Fourier transform variable since k is already used as the Boltzmann constant. Taking the time derivative of eqn (7.7.1) yields

$$\frac{\partial \rho}{\partial t} = -\frac{kT}{M\beta}q^2(1 - 2e^{-\beta t} + 2e^{-2\beta t})\rho + iqu_0 e^{-\beta t}\rho \tag{7.7.2}$$

Finally, the inverse Fourier transform of eqn (7.7.2) yields

$$\frac{\partial p}{\partial t} = \frac{kT}{M\beta}q^2(1 - 2e^{-\beta t} + 2e^{-2\beta t})\frac{\partial^2 p}{\partial x^2} - u_0 e^{-\beta t}\frac{\partial p}{\partial x} \tag{7.7.3}$$

Clearly, the evolution equation for p depends on u_0 as a parameter. A dependence on u_0 is to be expected, since it takes some time for the influence of the initial velocity to die out. Equation (7.7.3) shows how fast it dies out. As might be expected from all of our previous examples the falloff time for the influence of the initial velocity is of the order of M/ζ.

Often, however, we do not know the initial velocity of the Brownian particle. Hence we should like, somehow, to average eqn (7.7.3) over u_0. This cannot be done by a straightforward averaging of eqn (7.7.3) since the final term in that equation contains u_0 both in the coefficient of p and implicitly in p itself. We can, however, average eqn (7.7.1) over the Maxwell distribution, φ_M

$$\varphi_M(u_0) = (\frac{M}{2\pi kT})^{\frac{1}{2}} e^{-Mu_0^2/2kT} \tag{7.7.4}$$

The integration of eqn (7.7.1) involves a standard Gaussian integral; completing the square in the exponent yields

$$\int \rho(q)\varphi_M(u_0)du_0 = \exp[-\frac{kT}{2M\beta^2}q^2(2\beta t - 3 + 4e^{-\beta t} - e^{-2\beta t} + (1 - e^{-\beta t})^2)] \tag{7.7.5}$$

Consequently, using the notation $< ... >_0$ to denote averaging over u_0

$$\frac{\partial < \rho >_0}{\partial t} = -\frac{kT}{M\beta} q^2 (1 - e^{-\beta t}) < \rho >_0 \tag{7.7.6}$$

and so

$$\frac{\partial < p >_0}{\partial t} = \frac{kT}{M\beta}(1 - e^{-\beta t})\frac{\partial^2 < p >_0}{\partial x^2} \tag{7.7.7}$$

Thus the displacement evolution equation, averaged over initial velocities, approaches the conventional diffusion equation after a time of order M/ζ. Of course, this is just the same order of magnitude of time that it takes for the velocity distribution to relax to its equilibrium form. In dense systems diffusion is quite slow, while velocity relaxation is rapid. Consequently, the ordinary diffusion equation should be an adequate description except at extremely short times.

7.8 Discussion

All of the results of this chapter depend in a very direct way on the postulated form ofthe random force correlation functions, eqn (7.2.1) and its analog for rotation. Why have we postulated this particular form?

That the mean of the random force vanishes is trivial. The random force is supposed to represent the deviation of the actual force from its average; the mean of this difference must vanish by definition. For the two time correlation function, there are actually two questions. The first is why do we assume a delta function correlation in time? The physical assumption is that the random force has a very short correlation time, shorter than any other time of interest in the problem. Therefore we may as well assume that the correlation time is, in fact, zero; this is easier to treat, technically, and does not discard any important physics of the problem.

The second question is why we chose the specific numerical coefficient of the delta function in eqn (7.2.1), $2kT\zeta$? This is the only choice of constant that leads to the Maxwell distribution of velocities in the limit as $t \to \infty$. We can see this with a minimum of additional computation. First of all, a glance at eqn (7.2.6) shows that the Maxwell distribution is indeed reached in the limit $t \to \infty$. Thus this choice of the constant is sufficient.

It is also necessary since the entire calculation could have been carried out with an arbitrary coefficient, say A instead of $2kT\zeta$. Since the temperature, T, only enters the computation through the coefficient of the delta function, the result will be a formula that only differs from eqn (7.2.6) by having $2A/\zeta$ wherever kT stands in eqn (7.2.6). Therefore the Maxwell distribution will not be recovered in the long time (equilibrium) limit unless A is, in fact $2kT\zeta$.

The problems considered in this chapter were all exactly soluble. The feature that makes them so is linearity. The Langevin equation for all of the cases considered in this chapter are linear equations. The Fokker–Planck equation is always a linear partial differential equation, but in the present cases has an additional, different, kind of linearity. The A and B coefficients of the Fokker–Planck equation are linear functions of the independent variables. Of course, this is intimately tied up, through the definitions of the A and B coefficients, with the linearity of the Langevin equation.

Let us mention one more problem which is linear and exactly soluble. This is a charged Brownian particle in a uniform magnetic field. The reader is invited to work out the distribution function for him- or herself. We make only several introductory remarks. (1) Since a magnetic field makes charged particles move in circular orbits in a plane perpendicular to the field, the problem is a two-dimensional one; this is in contrast to the one-dimensional problems we have considered heretofore. The motion in the direction of the field is independent and may be treated separately; it is just that of a free particle, already discussed in Section 7.2. (2) The external force on the charged particle is the Lorentz force: $(q/c)\mathbf{u} \times \mathbf{B}$, where q is the charge of the particle and c is the velocity of light. (3) The problem is conveniently treated by the method used for the case of the harmonically bound particle.

As is the case in most branches of science, many problems of interest do not fall into the category of 'exactly soluble.' To make progress with them, we must resort to approximation methods. Much of the rest of this book will deal with examples of such methods. Nevertheless, a reservoir of exactly solved problems is useful for developing intuition about the nature of solutions and for the testing of approximation methods. Furthermore, many approximate methods require the solution of some known problem as their initial step. The results of this chapter will prove useful in our future work.

8

THE SMOLUCHOWSKI EQUATION

8.1 The Kramers–Klein Equation

In the previous chapter we studied some special cases of the Fokker–Planck equation for a Brownian particle in an external field of force: the harmonically bound particle, for which the force is linear in the displacement, and a constant force. While these are important examples, they are too special to serve in all cases of interest. In this chapter we shall extend our considerations to the case of a Brownian particle in a general field of force.

We begin with a remark about nomenclature, which is not standardized and can be confusing. Fokker and Planck independently studied the motion of a free Brownian particle, and, in doing so, derived the equation which bears their names. The name Fokker–Planck equation was then enlarged to mean any equation of the form eqn (3.4.8) and its extension to more than one independent stochastic variable. O. Klein (1922) proposed an extension of this work to the case when an external force was present, but this work was largely ignored. Perhaps this was because Klein's main suggested application was to the theory of solutions of electrolytes. The explosive development of electrolyte theory due to the work of Debye, Hückel, Onsager, and Falkenhagen starting in 1923 may have pushed Klein's work into obscurity.

Many years later, H. A. Kramers (1940) independently rederived this equation and applied it to the problem of the passage of molecules over potential barriers, an important question in chemical kinetics. Kramers' paper attracted much attention, and the Fokker–Planck equation in an external field is called the *Kramers equation* by many authors. In the heading to this section, we have called it the *Kramers–Klein equation* to give some recognition to the earlier derivation by Klein.

Because of the experience obtained in the previous chapter, we shall find it an easy task to derive the Kramers–Klein equation. As before, we shall confine our detailed discussion to a two-dimensional phase space, one dimension for the coordinate x and one for the momentum Mu. The generalization to higher dimensions is obvious once the two dimensional case is understood. The Fokker–Planck equation is given by eqn (3.4.8) and the coefficients A and B are defined by eqn (3.4.5). In the present case, of course, A is a two-dimensional vector and B is a two by two matrix.

To compute A and B, we have to solve the Langevin equation. The Langevin equation is

$$\begin{aligned} M\frac{du}{dt} &= -\beta u + K(x) + F(t) \\ \frac{dx}{dt} &= u \end{aligned} \tag{8.1.1}$$

Here, $K(x)$is the external force, assumed to be time independent; $F(t)$ is the random force with the statistical properties assumed in Chapter 7. eqn (8.1.1) is a nonlinear differential equation except when $K$is a linear function of x, and there is no hope of obtaining a formal explicit solution in the general case. However, in order to compute A and B, we only need the solution to first order in Δt, as indicated in eqn (3.4.5).

The solution to eqn (8.1.1) to first order in Δt is

$$\begin{aligned} x(t+\Delta t) &= x(t) + u(t)\Delta t \\ u(t+\Delta t) &= u(t) - \frac{\beta}{M}u(t)\Delta t + \frac{K(x(t))}{M}\Delta t + \frac{1}{M}\int_t^{t+\Delta t} F(s)ds \end{aligned} \tag{8.1.2}$$

From this, we can immediately read off the values of the A and B coefficients. They are:

$$\begin{cases} A_x = u \\ A_u = -\beta u/M + \frac{K(x)}{M} \\ B_{uu} = \frac{2kT\beta}{M} \\ \text{All other elements of } B \text{ are zero} \end{cases} \tag{8.1.3}$$

Consequently the distribution function $p(x, u, t)$satisfies

$$\frac{\partial p}{\partial t} + u\frac{\partial p}{\partial x} = \beta\frac{\partial}{\partial u}(u - \frac{K}{\zeta} + \frac{kT}{M}\frac{\partial}{\partial u})p \tag{8.1.4}$$

This is the Kramers–Klein equation. It is a linear partial differential equation. However, since $K(x)$ will not in general be a linear function of x, it is usually not possible to find a solution in closed form, as we did in Chapter 7. The generalization to six dimensions (three space, three velocity) just amounts to turning eqn (8.1.4) into a vector equation:

$$\frac{\partial p}{\partial t} + \mathbf{u}\cdot\nabla_x p = \beta\nabla_u\cdot(\mathbf{u} - \frac{\mathbf{K}}{\zeta} + \frac{kT}{M}\nabla_u)p \tag{8.1.5}$$

8.2 The Smoluchowski Equation

In a large number of examples that arise in practice, the velocity distribution of the Brownian particle rapidly thermalizes while the position distribution remains far from equilibrium for a much longer time.. This means that the velocity distribution is close to a Maxwell distribution while the position distribution still has not evolved too far from the initial distribution. Experiments on such a system probe aspects of the position distribution, and the velocity distribution is irrelevant. There are many interesting phenomena that take place on the time scale between that of thermalization and that of uniformization in space.

An example of such a phenomenon is diffusion in a fluid. If a concentration gradient of a solute is set up in a fluid and then allowed to approach a uniform distribution by free diffusion, the approach to uniformity in space will take much longer than the time necessary for the velocity distribution to thermalize. The relevant time scale for thermalization is β^{-1}, while that for uniformization in space is $l^2/2D$, where l is the macroscopic distance of uniformization, and D is the diffusion coefficient. The ratio of these two times is

$$\frac{t(\text{thermalization})}{t(\text{uniformization})} = \frac{2MD^2}{kTl^2} \tag{8.2.1}$$

If we use the same numbers for the mass and radius of the particle as were used in Section 4.1, take l as 1 mm, and use the Stokes–Einstein relation to estimate the diffusion coefficient, the ratio in eqn (8.2.1) is about 10^{-26}. Even for particles of molecular size and mass, the ratio is small, of order 10^{-19}.

If we would like to study the position distribution alone, it would be desirable to have an equation whose solution gives that distribution directly. At present, the only way we know to obtain the position space distribution is to first solve the Kramers–Klein equation for the phase space distribution function and then integrate over the velocity variables. This is a correct procedure but is a lot of work. Is all this work necessary? We saw in Chapter 7 that the position distribution function for a free particle satisfies a closed partial differential equation that approaches the diffusion equation for $t \gg \beta^{-1}$. Unfortunately we do not know how to generalize this exact result to the case when forces are present. The approximate result can be generalized, however. The resulting equation is known as the *Smoluchowski equation* (Smoluchowski 1915).

Let us use $n(x,t)$as the symbol for the position distribution function:

$$n(x,t) = \int p(x,u,t)du \tag{8.2.2}$$

We might try to obtain an equation for n by integrating eqn (8.1.4) over all u, assuming that the distribution function vanishes at infinity in velocity space. This procedure yields

$$\begin{aligned} \frac{\partial n}{\partial t} + \frac{\partial j}{\partial x} &= 0 \\ j(x,t) &= \int u\, p(x,u,t)du \end{aligned} \tag{8.2.3}$$

This equation is not very useful. In order to calculate j one must first know p. but this is the very quantity that we are trying to avoid computing. Because the velocity distribution is assumed to be close to Maxwellian, we might try to compute j by assuming that it is actually Maxwellian, $p \approx n(x,t)\varphi_M(u)$, where φ_M is defined by eqn (7.7.4). But this would yield $j = 0$. It is precisely the deviation, albeit small, of the velocity distribution from the Maxwellian distribution that is responsible for the uniformization in position space.

We shall first give a heuristic derivation of the Smoluchowski equation. This derivation has several unsatisfactory features, but has the advantage of coming to the correct answer rapidly and intuitively. We shall then go on to derive the equation on a more satisfactory basis. The basis assumption of both approaches is that thermalization occurs on a time scale short with respect to the time for appreciable changes in the positional distribution. This will occur when the friction constant is large, and is often called the *high friction limit.* We have just seen above that this limit is very common in practice.

The assumption of rapid thermalization means that after a short transient period, the inertial term in its equation of motion is very small, the frictional forces being balanced by the external (and random) forces. This means that we can write the Langevin equation, eqn (8.1.1), as

$$\begin{aligned} 0 &= -\beta u + K(x) + F(t) \\ \dot{x} &= u \end{aligned} \tag{8.2.4}$$

or

$$\dot{x} = \beta^{-1}(K + F) \tag{8.2.5}$$

Now there is a Fokker–Planck equation of the form eqn (3.4.8) for the x variable alone; we merely have to find the appropriate A and B coefficients. This is done, as in the previous paragraph and in Chapter 7, by solving eqn (8.2.5) to first order in Δt. The result is that

$$\begin{aligned} A &= K(x) \\ B &= \frac{kT}{\zeta} \equiv D \end{aligned} \tag{8.2.6}$$

Consequently

$$\frac{\partial n}{\partial t} + \frac{1}{\zeta}\frac{\partial K n}{\partial x} = D\frac{\partial^2 n}{\partial x^2} \tag{8.2.7}$$

This is similar to the diffusion equation except for the additional drift term involving the external force. Equation (8.2.7) is the Smoluchowski equation.

Note that in the case of the free particle, ($K = 0$), the Smoluchowski equation reduces to the equation derived by Einstein, eqn (4.4.2). For historical interest, we remark that Smoluchowski did not derive his equation by reduction from phase space to configuration space, the phase space version not being known at that time. Rather he followed a path similar in concept to Einstein's development for the field-free case (Smoluchowski 1915).

The derivation just given is unsatisfactory for several reasons. First, although it is true that the frictional forces will balance the external forces *on average* after thermalization, eqn (8.2.4) does not say that. It says that the frictional forces (unaveraged) balance the external force plus the random force. Although the inertial term may be negligibly small on the average, there is no compelling reason to assume that it is always negligible. In fact, one might expect it to be of the same order as F, the random force, a term which is kept in the analysis.

Secondly, throwing away the inertial term means throwing away the highest order derivative in the equation of motion, eqn (8.1.1). This is fraught with danger. A second order ordinary differential equation requires two initial conditions to specify a unique solution. a first order equation only requires one condition. Consequently the second initial condition that must be imposed on the original second order equation need not be satisfied by the solution of the truncated first order equation. This problem remains even when the neglected term is small. It is known as the problem of singular perturbations. We cannot go into singular perturbation theory in this volume; it would take us too far afield. Nevertheless, the existence of these difficulties means that we must search for a more satisfactory method of deriving eqn (8.2.7). This is the subject of the following section.

8.3 Elimination of Fast Variables

The situation where the variables describing a phenomenon can be divided into two sets, one evolving on a rapid time scale and one evolving on a slow time scale, is of frequent occurrence. It is often desirable to eliminate or average over the rapid variables in order to study the dynamics of the slow variables. The problem of eliminating the velocity variables from the Kramers–Klein equation is only one example of this genre of problem. Some general methods of attack are described in a review article by Van Kampen (1985). Because the problem is so common, we shall present here one method that we have found useful in several contexts. We shall then apply it to the Kramers–Klein equation.

We start by assuming that we have an evolution equation of the form

$$\frac{\partial \psi}{\partial t} = (L + \Lambda)\psi \tag{8.3.1}$$

where L and Λ are linear operators. The variables x are the slowly varying variables, and the variables ξ are the fast variables. Λ operates only on functions of the fast variables, ξ, while L may operate on functions of both sets of variables. We assume that Λ has the property that

$$\int \Lambda \varphi(x, \xi, t) d\xi = 0 \tag{8.3.2}$$

We call an operator with this property 'divergence - like' because the divergence operator of vector analysis (with suitable boundary conditions) has this property. We also assume that there is a small parameter in the problem, μ, and that

$$L = \mu L' + \mu^2 L'' + \cdots \tag{8.3.3}$$

while Λ is of order μ^0.

The above assumptions specify the problem mathematically; they are restrictions on the nature of the problem. The final assumption introduces an approximation. We assume that, after some short induction time $\psi(x, \xi, t)$ depends on t only as a functional[21] of $n(x,t)$, where

$$n(x,t) = \int \psi(x, \xi, t) d\xi \tag{8.3.4}$$

that is, $\psi(x, \xi, t) = \psi(x, \xi \mid n(\cdot, t))$. This means that the time dependence of the distribution in the combined (x, ξ) space is entirely due to the time dependence of the reduced distribution in the x space, the ξ variable distribution having relaxed to (almost) a value consistent with n. In this form, the hypothesis is usually attributed to Bogoliubov in the kinetic theory of gases (Bogoliubov 1946); the idea without the detailed elaboration of Bogoliubov can also be found in the treatise of Chapman and Cowling (1952).

Integrating eqn (8.3.1) over ξ yields

$$\frac{\partial n}{\partial t} = \int L\psi d\xi \tag{8.3.5}$$

But

$$\begin{aligned} \frac{\partial \psi}{\partial t} &= \int dy \frac{\delta \psi(x, \xi \mid n)}{\delta n(y,t)} \frac{\partial n(y,t)}{\partial t} \\ &= \int dy \frac{\delta \psi(x, \xi \mid n)}{\delta n(y,t)} \int d\eta L\psi(y, \eta, t) \end{aligned} \tag{8.3.6}$$

where the symbol $\delta/\delta n(y,t)$ signifies the functional derivative with respect to $n(y,t)$. Now let us write

$$\psi = \psi^{(0)} + \mu\psi^{(1)} + \mu^2\psi^{(2)} + \cdots \tag{8.3.7}$$

This is not necessarily a Taylor series expansion in powers of μ; it is an ordering of the terms according to their order of magnitude in μ. For example, a term $\mu \exp(-\mu t)$ is of first order in μ as far as its amplitude is concerned. One would not want to expand it further in powers of μ because, for $t \gg \mu^{-1}$, a very large number of terms would be necessary for acceptable accuracy. Long times are, of

[21] A brief description of functionals is given in Appendix B.

course, the time regime we are after, so we must avoid the temptation to expand everything we can in powers of the small parameter μ.

We are going to insert the representation given by eqn (8.3.7) into eqn (8.3.1) and equate terms of the same order in μ. This is a matter of our own choice in organization; it is not a mathematical theorem, since the μ expansion is not a Taylor expansion. Inserting eqn (8.3.7) into eqn (8.3.1), we see that $\partial\psi/\partial t$ is of order μ^0. Consequently

$$\Lambda\psi^{(0)} = 0 \tag{8.3.8}$$

Since Λ operates only on the ξ variables, the solution to (8.3.8) is of the form $f(\xi)g(x,t)$, where f satisfies $\Lambda f = 0$.

Now let us insist that

$$\int \psi^{(0)}(x,\xi,t)d\xi = n(x,t) \tag{8.3.9}$$

That is, $n(x,t)$ is of order μ^0; we have the freedom to require this as an extra condition on $\psi^{(0)}$. Thus

$$\int f(\xi)g(x,t)d\xi = n(x,t) \tag{8.3.10}$$

or

$$\begin{aligned}\psi^{(0)} &= \bar{f}(\xi)n(x,t)\\ \bar{f}(\xi) &= \frac{f(\xi)}{\int d\xi f(\xi)}\end{aligned} \tag{8.3.11}$$

This completes the determination of $\psi^{(0)}$.

We now have to determine $\psi^{(1)}$ which satisfies

$$\frac{\partial\psi^{(1)}}{\partial t} = \Lambda\psi^{(1)} + L'\psi^{(0)} \tag{8.3.12}$$

By eqn (8.3.6), this can be written as

$$\frac{\partial\psi^{(1)}}{\partial t} = \int dy \frac{\delta\psi^{(0)}(x,\xi \mid n)}{\delta n(y,t)} \int d\eta L'\psi^{(0)}(y,\eta,t) \tag{8.3.13}$$

Note that $\psi^{(0)}$ or n is of order $\mu^{(0)}$, but $\partial n/\partial t$ is of order μ. Equation (8.3.13) can be simplified by realizing that

$$\frac{\delta\psi^{(0)}(x,\xi \mid n)}{\delta n(y,t)} = \bar{f}(\xi)\delta(x-y)$$

Consequently

$$\frac{\partial \psi^{(1)}}{\partial t} = \int d\eta L' \psi^{(0)}(x, \eta \mid n) \bar{f}(\xi) \tag{8.3.14}$$

Thus, from eqn (8.3.12)

$$\Lambda \psi^{(1)} = \bar{f}(\xi) \int d\eta L' \bar{f}(\eta) n(x,t) - L' \bar{f}(\xi) n(x,t) \tag{8.3.15}$$

Equation (8.3.15) is an inhomogeneous linear equation for $\psi^{(1)}$; everything on the right- hand side is known. It can be solved by constructing the Green's function for the operator Λ. Note, however, that the equation $\Lambda f = 0$ has a non-trivial solution (cf. eqn (8.3.8)); that is, Λ has a zero eigenvalue. The Green's function in the usual sense, the inverse of Λ, does not exist. What is needed here is the Green's function in the extended sense (Courant and Hilbert 1937), the inverse of Λ defined on the space remaining after the subspace corresponding to the zero eigenvalue has been removed.

A solution of the homogeneous equation, a multiple of $\psi^{(0)}$, should also be added to give the general solution. The only acceptable multiple is zero, since eqn (8.3.9) implies that $\int \psi^{(1)} d\xi = 0$, while eqn (8.3.9) states that the corresponding integral for $\psi^{(0)}$ does not vanish.

The final result is that, to order μ^2, $n(x,t)$satisfies

$$\frac{\partial n}{\partial t} = \mu \int d\xi L' \psi^{(0)} + \mu^2 \int d\xi L'' \psi^{(0)} + \mu^2 \int d\xi L' \psi^{(1)} \tag{8.3.16}$$

where $\psi^{(0)}$ and $\psi^{(1)}$ are given by solutions of eqns (8.3.8) and (8.3.15) respectively.

This has been a long and complex derivation. The result is abstract and requires further computation to reduce it to an explicit equation for any particular case. Nevertheless, the effort of the derivation is justified because of the applicability of the result to many physical situations. We shall use the method several more times in this book. We now show how the Smoluchowski equation follows from the Kramers–Klein equation using these results.

8.4 The Smoluchowski Equation Continued

We return to eqn (8.3.1), divided through by β, and set $\beta t = \tau$, as dimensionless time variable.

$$\frac{\partial p}{\partial \tau} = -\frac{1}{\beta}(u\frac{\partial p}{\partial x} + \frac{K}{M}\frac{\partial p}{\partial u}) + \frac{\partial}{\partial u}(up + \frac{kT}{M}\frac{\partial p}{\partial u}) \tag{8.4.1}$$

In this equation, x is the slow variable, u is the fast variable. The operators L and Λ can be identified as follows

$$L = -(u\frac{\partial}{\partial x} + \frac{K}{M}\frac{\partial}{\partial u})$$

$$\Lambda = \frac{\partial}{\partial u}(u + \frac{kT}{M}\frac{\partial}{\partial u}) \tag{8.4.2}$$

To complete the story, we must compute the functions f and $\psi^{(1)}$. First let us look at f.

f satisfies

$$\frac{\partial}{\partial u}(u + \frac{kT}{M}\frac{\partial}{\partial u})f = 0 \tag{8.4.3}$$

One solution to this equation is the Maxwell distribution φ_M, which we have already met in eqn (7.7.4). The second linearly independent solution of this linear second order equation decays for large u as u^{-1}. It is unacceptable as a probability distribution function, since it is not integrable, decaying too slowly for large velocities. Consequently $\bar{f}$is

$$\bar{f} = \left(\frac{M}{2\pi kT}\right)^{1/2} e^{-Mu^2/2kT} \tag{8.4.4}$$

We are now in a position to evaluate the right-hand member of eqn (8.3.15). This equation becomes

$$\frac{\partial}{\partial u}(u\psi^{(1)} + \frac{kT}{M}\frac{\partial\psi^{(1)}}{\partial u}) = -\frac{kT}{M}\frac{d\bar{f}}{du}(\frac{\partial n}{\partial x} - \frac{Kn}{kT}) \tag{8.4.5}$$

The solution of this equation is

$$\psi^{(1)} = -\left(\frac{M}{2\pi kT}\right)^{1/2} ue^{-Mu^2/2kT}(\frac{\partial n}{\partial x} - \frac{Kn}{kT}) \tag{8.4.6}$$

Now that $\psi^{(0)}$ and $\psi^{(1)}$ have been determined, we can substitute them into eqn (8.3.16). Note that L'' vanishes for the case being considered; there is no term of order μ^2 in L. It is now merely a question of carrying out the required differentiations and finally multiplying through by ζ to return from the variable τ to t. The result is precisely eqn (8.2.7), the Smoluchowski equation. A similar technique has been used by Samoletov (1999) for this same purpose.

Thus we have reached the Smoluchowski equation by two routes, the first quick and simple, the second long and complicated. The easy way, unfortunately, had some flaws, so we presented it with due caution because of its intuitive nature. The complex way was done by a technique which has many applications to other problems in which several time scales exist. We presented it here in a form more general than needed for the present application because of its value in other areas. The method is easily generalized to situations where x and ξ are sets of variables, not single variables.

8.5 Passage over Potential Barriers

We shall illustrate the use of the Kramers–Klein equation by applying it to the problem originally envisaged by Kramers (1940), the passage of particles over potential barriers. This problem arises in many areas, chemical kinetics, nuclear reactions, diffusion in solids are a few of them. It can be stated as follows. Particles are moving in a potential with two metastable wells, A and B in Fig. 8.1.

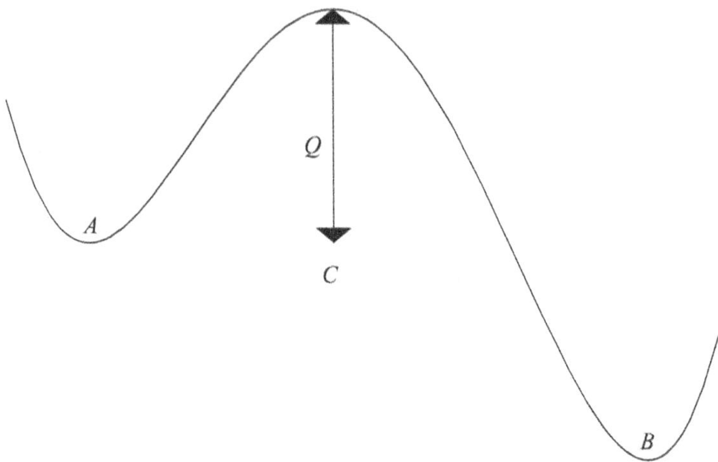

FIG. 8.1. Potential energy curve for the passage from the potential well at x_A to the potential well at x_B over the barrier at x_C

In addition to the force given by the negative gradient of the potential, the particles are subject to Brownian motion due to random interactions with their environment. They are therefore subject to friction and a random force. Under these conditions, what is the rate at which particles surmount the barrier between A and B? Alternatively, what is the lifetime of particles in well A?

The system is in a steady state; particles are injected into the well A and removed from the well B at such a rate so as to compensate, on the average, for the passage of particles over the barrier at C separating A from B. The question is, what is this rate? This problem can model the passage of reacting molecules through the transition state in a chemical reaction, the process of fission in the liquid drop model of the nucleus, the motion of an impurity particle from one cell to another in a solid, and many other physical systems.

The character of the problem depends on whether the temperature is high or low relative to the height, Q, of the potential barrier, and whether the friction is high or low relative to the typical undamped periods of motion of particles near the minimum of the well A. Since this chapter is devoted to the Smoluchowski equation, we shall only consider here the case of high friction, which is the case for which the Smoluchowski equation is valid. When kT is comparable to or greater than Q, there is no separation of time scales between thermalization and barrier passage. The full phase space description is needed, not just the Smoluchowski equation.

We shall approach the problem by Kramers' original method, known nowadays as the 'flux over population' method. The flux, w, of particles over the barrier C and the population of particles in well A are calculated separately; their ratio is the desired rate, the inverse of the lifetime. The flux of particles over the barrier from well A to well B is determined by writing eqn (8.2.7) in the form

$$\frac{\partial n}{\partial t} = \frac{\partial}{\partial x}\left(D\frac{\partial n}{\partial x} - \frac{Kn}{kT}\right) \tag{8.5.1}$$

This has the structure of a continuity equation with the current density

$$j = -D\frac{\partial n}{\partial x} + \frac{Kn}{\zeta} = De^{-U/kT}\frac{\partial}{\partial x}(e^{U/kT}n) \tag{8.5.2}$$

where $U$is the potential for the force K, $K = -\partial U/\partial x$. j is a constant in both space and time. It is constant in time because we are considering a steady state; it is constant in space because the present model is one-dimensional and time constancy implies space constancy by the equation of continuity. In higher dimensions, a stationary current vector would only have to satisfy $\nabla \cdot \mathbf{j} = 0$; this is not sufficient to imply constancy of $\mathbf{j}$ in space.

Equation (8.5.2) is integrated over x from x_A to x_B (which we abbreviate hereafter as A to B), yielding

$$j = \frac{-Dne^{U/kT}\,|_A^B}{\int_A^B e^{U(x)/kT}dx} \tag{8.5.3}$$

We have assumed that the density of particles in well B is very small and is kept small by a sink. Furthermore, $exp(U/kT)$ is small at B (see Fig. 8.1). Thus eqn (8.5.3) can be written as

$$j = \frac{-Dn(x_A)e^{U_A/kT}}{\int_A^B e^{U/kT}dx} \tag{8.5.4}$$

Having the flux, we next turn to the population in A. This population is

$$c_A = n(x_A)\int e^{-U/kT}dx \tag{8.5.5}$$

where the integral extends over the well A. This characterization of the domain of integration is vague and apparently not well defined. However, we are considering the case when $Q \gg kT$, so that the integrand is negligible before the ambiguity becomes worrisome. In the neighborhood of its minimum, the potential U may be approximated as a quadratic function, $U(x) \approx \frac{1}{2}m\omega^2(x - x_A)^2$. Because the integrand is negligible away from the minimum of the well A, the limits of the integral may be taken to be infinite without sensible error, yielding

$$c_A = n(x_A)\left(\frac{2\pi kT}{M\omega^2}\right)^{1/2} \tag{8.5.6}$$

Consequently, the lifetime, the inverse of the rate, is given by

$$\tau = \frac{1}{r} = \frac{c_A}{j} = \int_A^B e^{U/kT}dx \times \left(\frac{2\pi kT}{M}\right)^{1/2}\frac{1}{\omega D} \tag{8.5.7}$$

The last step is to evaluate the integral in this expression. The integrand is largest near C, and the region near C therefore makes the dominant contribution to the integral. Writing

$$U(x) \approx Q - \frac{M}{2}\omega'^2(x - x_C)^2$$

and again extending the limits of the integral to $\pm\infty$, yields

$$\frac{1}{r} = \left(\frac{2\pi kT}{M}\right)\frac{1}{\omega\omega'}e^{-Q/kT} \tag{8.5.8}$$

This is the final result for the high friction limit. A remarkable feature of this derivation is that we did not have to solve the Smoluchowski equation to obtain it; by following the clever route originated by Kramers, we were able to avoid an explicit solution.

As stated above, Kramers also treated the low friction limit. This is the case most closely related to gas phase chemical kinetics and which can be compared to other popular theories of barrier passage which neglect friction, for example transition state theory. Since the theory of the low friction region is not connected with the Smoluchowski equation, the subject of this chapter, we shall not treat it here. Of course, we have presented here only the case of a structureless particle. In any application to chemical reactions, the internal structures of the reacting particles play an essential role, for this structure is what changes in a reaction. Thus, quite aside from removing the restriction to large friction, the Kramers' theory must be generalized to a multidimensional phase space before most chemical kinetic applications are envisaged. See the review article Hänngi *et al.* (1990) for details. There is also an entire issue of the *Berichte der Bunsen Gesellschaft* (1991) devoted to modern developments of Kramers theory.

8.6 Concluding Remarks

The Kramers–Klein equation and the Smoluchowski equation are second order linear partial differential equations in $2n + 1$ and $n + 1$ variables respectively, where n is the dimension of the configuration space of the system. It is too much to expect that we will be able to write down solutions in explicit analytic form for either of them except in very special circumstances. Indeed, Chapter 7 was devoted to special cases where explicit solutions are possible. As is commonly the case in theoretical work, in more general situations we must content ourselves with approximate analytical or numerical methods.

There are a wide variety of such methods available. They are discussed in some detail in the monograph (Risken 1989), where many examples are given. The general idea is that many of the approximation methods that one learns when studying quantum mechanics are also applicable to the equations we have been discussing. The Schrödinger equation is also a second order linear partial differential equation, first order in the time. This is the fundamental mathematical fact underlying the well known approximation methods of quantum mechanics. It is true that the Fokker–Planck and Smoluchowski operators are not necessarily Hermitian, but they can be made so by a simple device. Hence the tool chest of available approximations for quantum mechanics and for the statistical dynamics we have been studying are essentially identical.

When we have a differential equation, boundary conditions are needed to specify the solution. So far, we have been working in infinite space, both position space and momentum space. The proper boundary condition is that the probability density should be integrable when integrated over all space. this usually entails that the distribution function vanish at infinity at least as rapidly as $r^{-(d+\epsilon)}$ in both position and velocity as $r \to \infty$ for any $\epsilon > 0$; r is $\mid x \mid$ or $\mid u \mid$. d is the dimensionality of the system.

On the other hand, physical situations often arise where boundary conditions must be applied in a finite portion of space. for example, the system could be enclosed in a box, forcing the flux of probability through the surface of the box to vanish. Another example, this time in momentum space, is if particles with energy greater than some given amount are removed from the system. For example they could decompose. If the decomposition products are of no interest, they are most simply treated as having merely been removed from the system. Thus we have an absorbing barrier in energy space, or alternatively, in velocity space. We shall discuss boundary conditions appropriate to specific applications when we come to treat these applications in later chapters.

We have derived the Smoluchowski equation by an expansion in powers of β^{-1} and have stopped with terms of order β^{-2}. We should ask if this derivation can be extended to higher orders in β^{-1}. This has been done by Wilemski (1976), correcting several previous attempts which contained errors, and by Titulaer (1978), who used essentially the same method exposed in Section 8.3. Including the β^{-2} term, Wilemski's result is

$$\begin{aligned}\frac{\partial n}{\partial t} &= -\nabla \cdot \mathbf{j}_0 + \frac{1}{M\beta^2}\nabla \cdot [\frac{1}{2}\nabla \cdot (\mathbf{K}\mathbf{j}_0 + \mathbf{j}_0\mathbf{K}) - \mathbf{K}\nabla \cdot \mathbf{j}_0] \\ \mathbf{j}_0 &= -D\nabla n + \frac{\mathbf{K}n}{\zeta}\end{aligned} \tag{8.6.1}$$

Here we have written the equation in three-dimensional form. $\mathbf{j}_0$ is the current vector for the Smoluchowski equation up to terms in β^{-1}. The complete current vector, up to terms of order β^{-2}, is everything under the divergence sign on the right-hand side of the first of eqns (8.6.1).

We have seen, eqn (7.7.7), that the exact kinetic equation for the probability density of a free particle is just like the Smoluchowski equation except that the diffusion 'constant' is time dependent. This does not contradict the result of Wilemski, which we have just quoted. The correction terms found by Wilemski vanish in the absence of an external force, which is consistent with our result. Wilemski also notes that there will be corrections depending on initial conditions varying as $\exp(-\beta t)$; this is also consistent with our results for the free particle.

Note that the correction term exhibited in eqn (8.6.1) contains third derivatives of the unknown function, n. The equation will be harder to solve than the second order Smoluchowski equation, and furthermore will require additional boundary data to specify a unique solution. These corrections to the Smoluchowski equation have rarely been used in modeling actual physical problems, so we shall not carry their discussion any further.

9

RANDOM WALK

9.1 The Random Walk

In 1905, the statistician Karl Pearson published a short letter in *Nature*, entitled 'The problem of the random walk' (Pearson 1905). It read in part

> A man starts from a point O and walks l yards in a straight line. He then turns through any angle whatever and walks another l yards in a second straight line. He repeats this process n times. I require the probability that after n of these stretches he is at a distance between r and $r + dr$ from his starting point, O.

In the very next issue of the journal Lord Rayleigh replied with an even shorter letter (Rayleigh 1905) pointing out that

> This problem ... is the same as that of the composition of n iso-periodic vibrations of unit amplitude and of phases distributed at random.

the solution to which he had published six years earlier. Rayleigh's solution was an asymptotic one, for the case of large n.

Actually, Pearson was not the first to pose, nor Rayleigh to solve, a problem in random walk. Christian Huygens (1657) posed a problem in the theory of games of chance which can be interpreted as a random walk with boundary. The problem was solved by Jacob Bernoulli many years later (Bernoulli 1713). A treatment of Huygens' problem accessible to someone who does not read Latin can be found in Bartlett (1955). These problems and their generalizations have since generated an enormous literature.

From the mathematical point of view, random walk is the study of the statistical properties of the sums of arbitrary numbers of random variables. In the simpler cases, such as those alluded to above, the random variables are independent, but they need not be. From the physical point of view, random walk is the study of the sum of independent vectors. In Pearson's case, the vectors were ordinary two-dimensional vectors in the plane. In Bernoulli's case the vector was a one-dimensional object representing the amount of money available to a gambler.

The subject of random walk is a fascinating facet of the theory of stochastic processes. However, in this book we shall only be interested in it as a model for Brownian motion. The reader who is interested in pursuing more general aspects of random walk theory may wish to consult the excellent monographs by Hughes (Hughes 1995, 1996) and Weiss (Weiss 1994), as well as the review articles by Chandrasekhar (1943), Weiss and Rubin (1983) and Montroll and Schlesinger (1984). Purely mathematical aspects are treated by Spitzer (1964).

To see the relevance of random walk to Brownian motion we need only recall Smoluchowski's model for Brownian motion in a gas; see Section 4.2. There the path of a Brownian particle was taken to be a broken line in space; the angle between two succeeding segments of the broken line was given (and small, for the particular case considered there). This is a random walk in three dimensions, but not of the precise type envisaged by Pearson. In Pearson's case the angle between successive segments of the walk is to be chosen at random. In Smoluchowski's case the polar angle between successive segments is fixed, although the azimuthal angle is to be taken at random. The former case is known as an *unrestricted random walk* or a *Pearson walk.* The latter case is called a *persistent random walk* or a *correlated walk.*

In Section 4.2 we determined the mean-square end-to-end distance of the walk; this is the same as the variance of the sum of the vectors making up the walk. For the Pearson walk, the variance is trivial to determine; if $\mathbf{l}_j$ is the vector corresponding to the jth segment then $< \mathbf{l}_j \cdot \mathbf{l}_k >= l^2 \delta_{j,k}$. It is immediately clear that the mean-square end-to-end distance for N steps is Nl^2. But there is much more to the theory of random walk than the second moment of the walk. We present some of those aspects that help in the understanding of Brownian motion.

9.2 The One-Dimensional Pearson Walk

In this section we treat the simplest case of a random walk. The states of the system are the lattice points on a line, jl, $j = 0, \pm 1, \pm 2, \ldots$, and l is the lattice spacing. At times $k\tau$, $k = 0, 1, 2, \ldots$, a particle on site j moves to site $j+1$ or to site $j-1$, each with probability $\frac{1}{2}$. Assume that the particle starts off at the origin, $k = 0$, at time 0; what is the probability, $p_k(j)$, that it is on site j at time k? This is obviously Pearson's problem reduced to one dimension.

The solution may be obtained by a combinatorial argument. Suppose that the particle has taken n steps. n_+ of these will have been to the right (i.e. in the positive direction) and n_- of them to the left (in the negative direction). Of course, $n_+ + n_- = n$. At the end of this realization of the walk, the particle will find itself at site $k = n_+ - n_-$. The n_+ and n_- steps to the right and left can be taken in any order, so there are $n!/n_+!n_-!$ possible realizations of the walk compatible with n_+ and n_-. Therefore the desired probability is

$$p_n(k) = \frac{1}{2^n}\frac{n!}{n_+!n_-!} = \frac{1}{2^n}\frac{n!}{(\frac{n+k}{2})!(\frac{n-k}{2})!} \tag{9.2.1}$$

In other words, the probability distribution for the position of the particle after n steps is a binomial distribution. Using the information on the binomial distribution given in Chapter 2, it is easy to write down the mean distance and its variance after n steps. Furthermore, from the de Moivre Laplace limit theorem, we know that the probability distribution approaches a Gaussian as the number of steps increases.

Even though we now know the solution to the problem, it is useful to try to solve the problem by other methods also. The combinatorial method used above is not useful in more than one dimension. To this end we search for other methods that may be generalizable. We start by writing down a balance of probabilities for the particle's position. A particle on site j at time $n+1$ must have been on site $j+1$ or $j-1$ at time n. Hence

$$p_{n+1}(j) = \frac{1}{2}p_n(j+1) + \frac{1}{2}p_n(j-1) \tag{9.2.2}$$

This difference equation may be solved by the method of generating functions.

Suppose we are given a discrete probability distribution $\{p_i\}$. The generating function of the sequence $\{p_i\}$ is defined by

$$Q(z,j) = \sum_i z^i p_i(j) \tag{9.2.3}$$

Multiplying eqn (9.2.2) by z^n and summing from $n=0$ to ∞ yields

$$\frac{1}{z}(Q(z,j) - \delta_{j,0}) = \frac{1}{2}Q(z,j+1) + \frac{1}{2}Q(z,j-1) \tag{9.2.4}$$

Here the function Q has two arguments because the original probabilities, $p_j(k)$ depended on two variables and we have only introduced the generating function with respect to one of them. Equation (9.2.4) is simpler than eqn (9.2.3), but not yet simple enough.

We now introduce a generating function with respect to the variable k by multiplying (9.2.4) by $\exp(ik\alpha)$ and summing over k from $-\infty$ to ∞. The reason for using a complex exponential instead of a power lies in the translational invariance of the set of eqns (9.2.2). The set of complex exponentials preserves this property. Introducing the symbol S for the sum just defined, we obtain

$$S(z,\alpha) = \frac{1}{1 - z\cos\alpha} \tag{9.2.5}$$

$p_j(k)$ can now be obtained by expanding our closed form expression for S in powers of z and $\exp(i\alpha)$. The expansion coefficients just give the required probabilities; of course, these are exactly the ones which we obtained by the combinatorial method earlier.

The generating function method is a very powerful one for studying random walks on lattices in arbitrary dimension even though the generating function can only be evaluated in closed form in one or two dimensions. It is discussed in detail in the book by Hughes (1995).

To make a connection to Brownian motion, we are particularly interested in the behavior of the probability for long times compared to τ (large k) and distances large compared to l (large j). Although this can be studied using the generating function method, it is better to go back to the fundamental balance equation (9.2.2), which we write in the form

$$\frac{p_{n+1}(j) - p_n(j)}{\tau} = \frac{1}{2}\frac{p_n(j+1) - 2p_n(j) + p_n(j-1)}{l^2} \times \frac{l^2}{\tau} \tag{9.2.6}$$

If the function p were in fact a smooth function, instead of only being defined on the integers, then the left-hand side of eqn (9.2.6) would be recognized as the finite difference approximation to $\partial p/\partial t$, while the difference quotient on the right-hand side would be recognized as the finite difference approximation to $\partial^2 p/\partial x^2$. This suggests that we take the limit $l \to 0$, $\tau \to 0$, in such a way that $x = jl$ and $t = k\tau$ remain fixed. $p_j(k)$ approaches a smooth function $p(x,t)$ in this limit. But l and τ cannot go to 0 in an arbitrary way, for then the coefficient l^2/τ that appears in eqn (9.2.6) would not necessarily have a limit.

To have a sensible limit, we must require that l and τ approach 0 in such a way that $l^2/2\tau \to D$, a constant. In this fairly complicated limit, eqn (9.2.6) becomes

$$\frac{\partial p(x,t)}{\partial t} = D\frac{\partial^2 p}{\partial x^2} \tag{9.2.7}$$

Equation (9.2.7) will immediately be recognized as the diffusion equation, the basic evolution equation of the Einstein theory and of the Wiener process. We have already discussed its fundamental solution in Chapter 3.

Equations analogous to eqn (9.2.2) can be formulated in any number of dimensions; there are just more terms on the right hand side corresponding to the possibility of taking steps in various directions. For example, on a square lattice in two dimensions, the probability of being at a lattice point (k, m) after n steps, $p_n(j,k)$, satisfies

$$\begin{aligned} &p_{n+1}(k,m) - p_n(k,m) = \\ &\tfrac{1}{4}[p_n(k+1,m) + p_n(k,m+1) + p_n(k-1,m) + \\ &p_n(k,m-1) - 4p_n(k,m)] \end{aligned} \tag{9.2.8}$$

In the same way as in one dimension, in the asymptotic limit

$$\frac{\partial p}{\partial t} = D\left[\frac{\partial^2 p}{\partial x^2} + \frac{\partial^2 p}{\partial y^2}\right] \tag{9.2.9}$$

Similar results hold in any dimension.

9.3 The Biased Random Walk

Let us go on to discuss a slightly more complicated case. Again we restrict ourselves to a one- dimensional walk, but now we do not require that the probabilities for a step to the right and one to the left are the same. Let the probability of a step to the right be p and that to the left be q, $p + q = 1$. The probability can be obtained by the same combinatorial argument as in the previous section. The only difference is that the factor $(\frac{1}{2})^n$ is replaced by $p^{n_+}q^{n_-}$.

It is more instructive to solve the problem by the method of balance equations. The balance equation is

$$p_{j+1}(k) = pp_j(k-1) + qp_j(k+1) \tag{9.3.1}$$

In the same way as in the previous section, this can be expressed in the form

$$\begin{aligned} p_{j-1}(k) - p_j(k) = p[p_j(k-1) - p_j(k)] + q[p_j(k+1) - p_j(k)] = \\ \frac{1}{2}[p_j(k+1) - 2p_j(k) + p_j(k-1)] - \frac{s}{2}[p_j(k+1) - p_j(k)] \end{aligned} \tag{9.3.2}$$

where we have written $p = (1+s)/2$, $q = (1-s)/2$.

The first term in parentheses on the right-hand side of eqn (9.3.2) is a finite difference approximation to the second space derivative of p. The second parentheses give a finite difference approximation to the first space derivative. Again, we let l and τ approach zero while j and k approach infinity so as to keep x and t fixed, and $l^2/2\tau \to D$. This time, however, we must also let $s \to 0$ in such a way that sl/τ approaches the finite value v. Without this last stipulation, the coefficient of the first space derivative becomes infinite in the limit. This entails that both p and q approach 1/2 in the limit being considered. Under these conditions, eqn (9.3.2) becomes

$$\frac{\partial p}{\partial t} = D\frac{\partial^2 p}{\partial x^2} - v\frac{\partial p}{\partial x} \tag{9.3.3}$$

What is the physical meaning of the conditions that l^2/τ and sl/τ should approach finite limits? We are looking for asymptotic approximations to the probability distribution in the region where it is reasonably large. From Einstein's theory or from the explicit expression derived in Section 9.2, we know that this is the region $(\delta x)^2 \leq t$ (or a few multiples thereof). Hence we insist that our parameters scale so as to treat this region as accurately as possible.

Similarly, in the case of a biased walk the mean displacement of the particle is $nl(p-q)$; this is a simple computation from eqn (2.4.5) expressed in the notation of this section. This does not have a finite limit unless p and q approach $\frac{1}{2}$ as l and $\tau \to 0$. Intuitively this might seem to eliminate the bias in the walk, but it does not. It is a question of respecting the proper order of limits. To be sure, if p and q are sent to 1/2 before the space and time step limits are taken, the bias is certainly eliminated. But all of these limits are to be taken simultaneously, with prescribed ratios of the rate of approach to the limits. The effect of the bias gets magnified by taking many time steps, while the magnitude of the bias for a given time step decreases. If these relative rates are scaled as described above, a net effect of the bias is left.

Notice that eqn (9.3.3) has the same structure as the Smoluchowski equation in the presence of a constant field of force. In fact, if we introduce a force K

and a friction constant ζ by the relation $K = \zeta v$, it is precisely Smoluchowski's equation. Here v is the velocity introduced two paragraphs above. Since the transition probabilities p and q were constant, we are bound to get a constant field when translated into force language. If we had chosen p and q to vary from site to site, we would have obtained, by additional algebra, the general Smoluchowski equation. Again, this derivation can easily be generalized to higher dimensions.

The model of this section and that of the last were restricted to nearest neighbor walks; that is, walks on a lattice in which the elementary steps are to nearest neighbors only. An obvious generalization of these models is to walks in which the transition probabilities may be non zero for sites more distant than nearest neighbors. If $\pi(k,k')$ is the transition probability from site k' to site k, the balance equation reads

$$p_{j+1}(k) = \sum_{k'} \pi(k,k')p_j(k') \tag{9.3.4}$$

Calculations similar to those we have carried out can be performed yielding the same Smoluchowski equations that we have already derived, provided that the first and second moments of the transition probabilities are finite, both for finite values of l and τ and in the limit as these quantities approach zero.

$$\begin{aligned}\sum_{k'} (k-k')\pi(k,k') &< \infty \\ \sum_{k'} (k-k')^2\pi(k,k') &< \infty\end{aligned} \tag{9.3.5}$$

A derivation of this fact by a method different from the one we have used can be found in Weiss (1994).

Walks whose transition probabilities have infinite second moments are often called *Lévy walks* or *Lévy flights.*[22] In the continuum limit, they do not yield a Smoluchowski equation. Their sample paths are rather different from Brownian sample paths. It is even possible to envisage random walks with infinite first moments, or means. Consider, for examples, a walk in which $\pi(k,k') = 2^{-n}$ when $\mid k-k' \mid = 2^n$, and zero otherwise. This walk has infinite mean. This is a translation into random walk language of the problem of the St. Petersburg paradox in probability theory; see Feller (1950).

It is also possible to divorce these considerations from the concept of an underlying lattice in space or time, or both. Consider a random walk in an n dimensional space, the individual steps of which have a probability distribution $w(\mathbf{r},\mathbf{r}';dt)$ of moving from $\mathbf{r}'$ to $\mathbf{r}$ in time dt. Here $\mathbf{r}$ and $\mathbf{r}'$ are vectors in n-space. The basic balance equations for probabilities are

[22]Lévy flights serve as a model for *anomalous diffusion*, where $< \Delta x^2 > \propto t^\alpha, \alpha \neq 1$. Since the variance of the individual step lengths is infinite, so will be the variance of the total displacement of the walk. It is necessary to introduce a distribution of waiting times between steps, correlated with the step size, before a finite variance (as a function of time) results.

$$p(\mathbf{r}, t + dt) = \int [w(\mathbf{r}, \mathbf{r}'; dt)p(\mathbf{r}', t) + w(\mathbf{r}', \mathbf{r}; dt))p(\mathbf{r}, t)]\, d^n\mathbf{r}' \qquad (9.3.6)$$

This expresses the probability of being at $\mathbf{r}$ as that of jumping into $\mathbf{r}$ from $\mathbf{r}'$ plus the probability of leaving $\mathbf{r}$, having been there already at t, in the interval dt. As long as the first two moments of w are finite, this model too leads to the Smoluchowski equation. A detailed discussion ofthis is given in (Chandrasekhar 1943) for three dimensions; The method is very similar to the one we have already used in the lattice case, so we shall not go into detail.

9.4 The Persistent Walk

It is quite satisfying to see how the equations of Brownian motion, in particular the Smoluchowski equation, can arise from the simple model of an unrestricted random walk. However, the original walk envisaged by Smoluchowski was not an unrestricted walk. Because the Brownian particle is supposed so much heavier than the particles of the medium, it does not change its direction of motion at random on every collision, but retains a large component of velocity in the direction of motion before collision. Walks with this property are called *persistent*.

Unrestricted random walks are Markov processes. The position of the walker at time t depends only on the position at the immediately previous time and on the increment of position; the distribution of the increment is specified by the transition probability. Persistent walks, however, are non-Markovian. The position at time t depends, in addition to the factors above, on the direction of motion of the walker as it entered the volume element around $\mathbf{r}'$.

It was relatively easy to compute the mean-square displacement of the walk introduced in Section 4.2. We shall call this the Smoluchowski walk. It is not so easy to find the evolution equation for the distribution function. In fact, as far as I am aware, this has never been done for the case of the Smoluchowski walk in spite of its occurrence in polymer science. We shall consider here only a much simpler case, a one-dimensional walk in the limit of large times and large length scales. In view of what has been said, we shall not be surprised to find that the governing equation is not the Smoluchowski equation. The model is specified more precisely as follows.

The walk takes place in one dimension and in discrete time, exactly as described at the beginning of Section 9.2. However, the rules for movement are different. If the particle finds itself at position j at time k, *having arrived there from site* $j-1$, i.e. having moved from left to right on the last move, it moves from j to $j+1$ with probability a, or to $j-1$ with probability b, on the next move. If it has arrived at j from the left, i.e. from site $j+1$, then it moves to $j-1$ with probability a or to $j+1$ with probability b. That is, the particle moves in the direction of its last step with probability a and in the opposite direction with probability b.

Let $p_k(j,+)$ be the probability that the particle is at j at time k having arrived there by a move from $j-1$. Let $p_k(j,-)$ be the probability that the particle is at j at time k having gotten there from $j+1$. Then the two functions $p_k(j,\pm)$ satisfy the simultaneous equations

$$\begin{aligned} p_{k+1}(j,+) &= ap_k(j-1,+) + bp_k(j-1,-) \\ p_{k+1}(j,-) &= bp_k(j+1,+) + ap_k(j+1,-) \end{aligned} \tag{9.4.1}$$

This set of equations can be solved by the generating function method illustrated in Section 9.2. Since we are not interested here in the detailed solution, but only in its asymptotic properties for large space and time scales, we shall not pursue the exact solution.

In fact, we are not interested in the separate functions $p_k(j,\pm)$ but only in their sum, which is the required probability $p_k(j)$ for being at j at time k. This suggests that we introduce new variables

$$\begin{aligned} p_k(j) &= p_k(j,+) + p_k(j,-) \\ q_k(j) &= p_k(j,+) - p_k(j,-) \end{aligned} \tag{9.4.2}$$

The reason for introducing q is merely that the starting equations, eqn (9.4.1), contained two unknown functions. If we introduce new dependent variables we must introduce two of them. p is clearly one new variable to be chosen, since it is what we want in the end. The choice of the second is not uniquely determined, but q is a convenient choice. For these new variables, eqn (9.4.1) becomes

$$\begin{aligned} p_{k+1}(j) &= \frac{(a-b)}{2}(q_k(j-1) - q_k(j+1)) + \frac{1}{2}(p_k(j+1) + p_k(j-1)) \\ q_{k+1}(j) &= \frac{(a-b)}{2}(q_k(j+1) + q_k(j-1)) + \frac{1}{2}(p_k(j-1) - p_k(j+1)) \end{aligned} \tag{9.4.3}$$

To go to the continuum limit we want to send l, the unit of length, and τ, the unit of time, to zero in some appropriate ratio; we shall have to scale the transition probabilities a and b, also. We could go to the limit by writing eqn (9.4.3) in terms of difference quotients, as we did in going from eqn (9.3.2) to eqn (9.3.3), but let us take a slightly different route, in order to see an alternative method.

Assume that the functions $p_k(j)$ and $q_k(j)$ are the values taken at integral multiples of l and τ of smooth functions $p(x,t)$ and $q(x,t)$. Develop these functions in a Taylor's series around $x = jl$ and $t = k\tau$.

$$p_{k+1}(j) = p_k(j) + \tau\frac{\partial p}{\partial t} + \cdots$$

$$p_k(j+1) = p_k(j) + l\frac{\partial p}{\partial x} + \cdots \tag{9.4.4}$$

with similar expansions for q. Putting these expansions in eqn (9.4.3), and retaining terms up to order l and τ only yields

$$\begin{aligned} \frac{\partial p}{\partial t} &= \frac{(b-a)l}{\tau}\frac{\partial q}{\partial x} \\ \frac{\partial q}{\partial t} &= \frac{l}{\tau}\frac{\partial p}{\partial x} - \frac{2b}{\tau}q \end{aligned} \tag{9.4.5}$$

The coefficients in this equation will be finite only if we send l and τ to zero in such a way that $v = l/\tau$ remains fixed. Furthermore, b/τ must approach some fixed constant which we call $\frac{1}{2}T$ (The factor of 2 is convention; it removes an unsightly factor of 2 from the final equation.). Consequently a must approach unity since $a + b = 1$.

With these requirements, eqns (9.4.5) become

$$\begin{aligned} \frac{\partial p}{\partial t} &= -v\frac{\partial q}{\partial x} \\ \frac{\partial q}{\partial t} &= -v\frac{\partial p}{\partial x} - \frac{1}{T}q \end{aligned} \tag{9.4.6}$$

Finally, we can eliminate q by differentiating the first of eqns (9.4.6) with respect to t and the second with respect to x. Algebraic manipulation then yields

$$\frac{\partial^2 p}{\partial t^2} + \frac{1}{T}\frac{\partial p}{\partial t} = v^2\frac{\partial^2 p}{\partial x^2} \tag{9.4.7}$$

This equation is known as the *telegrapher's equation* for historical reasons. It was originally derived by W. Thomson (Lord Kelvin) to describe the propagation of telegraph signals in underwater cables, a far cry from its use here.

Equation (9.4.7) is a hyperbolic equation. It is quite different in character from the parabolic eqn (9.2.7). Properties of this equation are given in several textbooks (Morse and Feshbach 1953; Sommerfeld 1952); here we shall just mention several properties of the fundamental solution. The first is that there is a finite velocity of propagation of the probability density. Recall that the Gaussian solution of eqn (9.2.7) entails that, for any time greater than zero, no matter how small, there is some probability at any distance from a delta function source. For the telegrapher's equation this does not hold. The solution corresponding to a delta function source drops abruptly to zero and remains zero for distances from the source greater than vt.

Also, since the equation is second order in the time, it does not suffice to give merely the initial value of the solution. One must also specify the initial rate of change of the solution. This can be seen from the observation that the original set of equations, eqns (9.4.6), contained two unknown functions and hence required two initial conditions. Eliminating one unknown does not change the number of needed conditions because if it did, different qs would give rise to the same p.

Finally, for large values of t/T, the fundamental solution of the telegrapher's equation approaches a Gaussian. Consequently, for long times the walk will look like a Pearson walk, except, of course, for large displacements. This fact is shown directly in (Weiss 1994; Section 5.3d). The reader may guess from this that there may be a generalization of the central limit theorem to at least a class of dependent random variables. Indeed, there are such generalizations, but the hypotheses under which they are valid are much more delicate and may be harder to verify than in the case of independent random variables.

The one-dimensional model just treated is merely an extension of the over-simplified model of Section 4.1 to take account of persistence. For Brownian motion, we are more interested in the Smoluchowski walk of Section 4.2. In that section we computed the variance (mean-square end-to-end distance); what can be said about the distribution function?

Explicit calculations showing the approach to a Gaussian distribution for large values of the number of steps have been given by Moran (1948), Tchen (1952), and Kac (1959; Section II.3). These are not based on the reduction of the problem to a Smoluchowski equation in the continuum limit. In fact, the problem does not appear to have been treated in the literature from this point of view. It would be useful to have a functional equation satisfied by the distribution function, for then there would be a convenient route for computing deviations from Gaussian behavior. This is not important for Brownian motion but would be interesting for problems of polymer conformational statistics. Nevertheless, the fact that the Smoluchowski walk is asymptotically Gaussian makes the final connection between the kinetic approach of Smoluchowski and the more abstract approach of Einstein.

9.5 Boundaries and First Passage Times

So far we have been considering walks in infinite space, with no boundaries. We now want to investigate what happens when the walk is forbidden from entering certain regions of space. This is the problem of walks with boundaries.

Also, we have been asking for the probability of the walker being at a certain position at a given time. We can turn this question around; what is the probability that the walker reached a given position *for the first time* at a certain time. In addition to reversing the roles of time and space, so to speak, a new condition has been added, that of the first arrival at the specified spatial position. This is called the problem of *first passage times.* Indeed, the original random walk problem of Huygens and Bernoulli was a first passage time problem. Roughly speaking it consisted of two players, each with fixed initial capital, playing a certain game of chance against each other. The game ended when one of the players lost all of his money. What is the expected duration of the game? Clearly this can be immediately translated into the language of the mean first passage time of a random walk.

We shall see in this section that the problem of random walk with boundaries and the problem of first passage times are related. Consider, for simplicity, a symmetrical lattice walk in one dimension. The walker starts from the origin at time 0; we ask for the distribution of the time, k, that it first reaches the point a. Call this distribution $f_a(k)$. Here as before, a and k are integers; to get real lengths and times, they must be multiplied by l and τ respectively. For any time, s, the walker may have reached a by passing through a at least one time previously or by arriving there for the first time at s. Adding up the probability of these independent events

$$p_a(s) = \sum_{h=0}^{s} f_a(h) p_0(s-h) \tag{9.5.1}$$

Here p_0 is the probability of going from a to a in the indicated time. Since the lattice is homogeneous, this probability is the same as that for returning to the origin, 0, from the origin in the same time interval; this explains the choice of the zero subscript.

$p_a(s)$ is known; we have computed it in Section 9.2. So we only need to solve eqn (9.5.1) for f. This is easily done by the method of generating functions. Multiplying eqn (9.5.1) by z^s and summing from $s = 0$ to infinity, we find

$$Q(z, a) = F_a(z) Q(z, 0) \tag{9.5.2}$$

where Q and F are the generating functions for p and f respectively, defined as in eqn (9.2.3).

Equation (9.5.2) can be used to obtain the mean first passage time. In fact, this quantity is infinite for the model currently being considered. To see this, note that $< s >= dF(a, z)/dz \mid_{z=1}$. From eqn (9.2.4), $p_a(z)$ is given by

$$p_a(z) = \frac{1}{2\pi} \int_0^{2\pi} \frac{e^{ia\alpha}}{1 - z\cos\alpha} d\alpha \tag{9.5.3}$$

This integral can be found in standard integral tables (Gradshteyn and Ryzhik 1980; 3.613,1) or evaluated by contour integration; its value is

$$p_a(z) = (1 - z^2)^{-\frac{1}{2}} \left(\frac{1 - \sqrt{1 - z^2}}{z} \right)^a \tag{9.5.4}$$

It is now easy to form $p_a(z)/p_0(z)$ and take its derivative; as $z \to 1$ the derivative is seen to approach infinity.

Symmetric random walks on a lattice in one and two dimensions are known to return to their starting point with probability one; we shall not prove this here. The fact that the mean first passage time is infinite in one dimension means that although the walker will, in general, return, he will most usually take a very long walk before doing so.

It is, of course, possible to ask for the mean first passage time to a set of points, not just a single point. Indeed, the Huygens–Bernoulli gambling problem mentioned above was of this type. The game ends when the original capital of either of the two players is exhausted. Such problems are more difficult to treat than the simple case of a single point treated above. First passage time problems can also be studied without introducing the distribution function f. Instead, one imposes a special condition on p. Let us return to the probability balance equations of eqn (9.2.2). If we are interested only in the time of first passage to a, we may as well ignore what happens to the walker after a has been reached. That is, we may as well remove the walker from the system at its first passage time. Thus we treat the site a as an *absorbing site* or a *trap*; the walker cannot get to site $a+1$ or $a-1$ from site a. For the sake of definiteness, we suppose that a is positive. For $j = a-1$ we must have

$$p_{k+1}(a-1) = \frac{1}{2} p_k(a-2) \tag{9.5.5}$$

This relation is compatible with eqn (9.2.2) only if $p_k(a) = 0$ for all k. This is the boundary condition for an absorbing boundary.

To calculate the first passage time from this point of view, one first must solve the balance equations (9.2.2) with the absorbing boundary condition. The solution is not so easy to find, since the translational invariance of the original problem has been broken. Nevertheless, assuming that the solution is in hand, form

$$R(j,a) = \sum_{j=-\infty}^{a-1} p_k(j) \tag{9.5.6}$$

which is just the probability that the walker has not reached a up to time j. This quantity is not unity because of the absorbing boundary condition; walkers are removed from the system when they reach a. The probability that the walker reaches a for the first time at time j is then

$$f_a(j) = R(a,j) - R(a,j+1) \tag{9.5.7}$$

This form of the first passage time distribution is useful in the passage to the continuum limit and in higher dimensions, when the phenomenon of interest is the time of first passage out of regions.

Our considerations have been only for the simple symmetric random walk with steps to nearest neighbor points only. Walks with steps of varying length (governed by a step size distribution), and walks in continuous time are also of interest. These are not very different conceptually from the simple walks we have discussed but the technical tools needed to study their properties can become much more complicated. We refer the interested reader to the references given in Section 9.1.

We have discussed absorbing boundary conditions and shown how concept of absorbing boundary can be related to that of first passage time. There are other boundary conditions that are also of interest in applications. The first of these is the *reflecting* boundary condition. There are several ways of defining a reflecting barrier. Sticking to the one dimensional case, our definition will be that there is a reflecting boundary at point a if a walker at $a-1$ at time $j+1$ either has arrived there from $a-2$ or was at $a-1$ at time j. The interpretation of the latter eventuality is that it tried to go to a and was reflected back to $a-1$ instantaneously. We have again assumed for definiteness that a is positive. In terms of the balance equation at a, this means

$$\begin{aligned} p_{k+1}(a-1) &= \frac{1}{2}p_k(a-1) + \frac{1}{2}p_k(a-2) \\ &= \frac{1}{2}p_k(a) + \frac{1}{2}p_k(a-2) + \frac{1}{2}(p_k(a-1) - p_k(a)) \end{aligned} \tag{9.5.8}$$

This is consistent with eqn (9.2.2) if the last term in brackets vanishes. In the case of a biased walk, eqn (9.3.1), one can make an analogous computation, and finds

$$pp_k(a-1) = qp_k(a) \tag{9.5.9}$$

which can be written in the form

$$p(p_k(a-1) - p_k(a)) + sp_k(a) = 0 \tag{9.5.10}$$

As in Section 9.3, $s = p - q$. The usefulness of this form will become apparent below. It should be interpreted as stating that the flux of probability into the point a vanishes.

To get a better understanding of the physical meaning of these boundary conditions, let us go to the continuum limit. In the case of absorbing boundaries, $p_j(a) = 0$ translates directly into

$$p(c,t) = 0 \quad \text{absorbing} \tag{9.5.11}$$

where c is the x coordinate of the boundary; i.e. $a \to 0$, $l \to \infty$ with $q = al$ fixed. For a reflecting boundary, eqn (9.5.10) becomes

$$-D\frac{\partial p}{\partial x} + vp\,|_a = 0 \quad \text{reflecting} \tag{9.5.12}$$

This can be seen by the methods used in Section 9.3 to pass to the continuum limit for the differential equation itself.

The quantity on the left side of eqn (9.5.12), which vanishes at a reflecting boundary, is the probability current density of the system. Such an interpretation is called for by the governing differential equation eqn (9.3.3); this equation states that the rate of change of the probability density is the negative divergence of the quantity in question. Since the probability is a conserved object, the left-hand side of eqn (9.5.12) must be the probability current. In d dimensions this should be written as

$$-D\nabla p + \mathbf{v}p\,|_{\text{bndry}} = 0 \tag{9.5.13}$$

where ∇ is the d-dimensional gradient and $\mathbf{v}$ is a d-dimensional vector. The probability current is made up of two parts. One part is the diffusive part, $-D\nabla p$, which expresses the effect of the randomness of the random walk. The second part is $\mathbf{v}p$, which is the convective part of the walk. It expresses the part of the probability current due to the walker being carried along with the velocity $\mathbf{v}$; it is due to the bias of the underlying walk.

We can illustrate the use of these considerations to compute the first passage time distribution for a one-dimensional random walk in the continuum. What is the first passage time distribution for a particle starting at $x = 0$ at time $t = 0$ to arrive at $x = \xi$? We have to solve the differential equation (9.2.6) with the boundary condition (9.5.10) at ξ and with the initial condition

$$p(x, 0) = \delta(x) \tag{9.5.14}$$

This is a textbook exercise in books on partial differential equations; the solution is

$$p(x, t) = \frac{1}{\sqrt{4\pi Dt}}\left[e^{-x^2/4Dt} - e^{-(2\xi - x)^2/4Dt}\right] \tag{9.5.15}$$

Consequently the probability that the walker has not yet reached ξ up to time t is

$$R(\xi, t) = \int_{-\infty}^{\xi} p(x, t)dx \tag{9.5.16}$$

This can be written as

$$R(\xi, t) = \frac{2}{\sqrt{\pi}}\int_0^{\xi/\sqrt{4Dt}} e^{-u^2}du \tag{9.5.17}$$

As we have argued in eqns (9.5.6) and (9.5.7), the first passage time density is the negative derivative of R with respect to t. Taking this derivative yields

$$f(\xi, t) = \frac{\xi^2}{\sqrt{4\pi Dt^3}}e^{-\xi^2/4Dt} \tag{9.5.18}$$

It is readily seen that the mean first passage time is infinite since $f \sim t^{-3/2}$ at infinity. Of course this is not surprising, since we have already found the same behavior for the discrete case.

In addition to the absorbing and reflecting boundary conditions that we have just discussed, mixed boundary conditions corresponding to partially reflecting and partially absorbing boundaries may also be considered. In one dimension such boundary conditions may be written as

$$\alpha\frac{\partial p}{\partial x} + \beta p \mid_a = 0 \tag{9.5.19}$$

where α and β are given numbers. We shall not discuss these boundary conditions here.

Also, in the case of a persistent walk, the boundary conditions are more complicated than those that we have discussed. This is because the telegrapher's equation for the persistent walk comes from eliminating one dependent variable from a pair of partial differential equations. Although the function is eliminated from the equation, its effect still remains in the boundary conditions. The interested reader is referred to Masoliver and Weiss (1992).

9.6 Random Remarks on Random Walks

The subject of random walks is a vast one, and we have barely scratched the surface. Our object in this chapter was to show the usefulness of random walks as a model for Brownian motion. The realization that there is a relationship goes back to the beginnings of the subject, the paper by Smoluchowski, even before the random walk was named. Nevertheless we cannot leave the subject of random walks without mentioning, however briefly, some of the further ramifications of the subject.

There is an extensive literature on random walk models in financial and commodity markets (Granger and Morgenstern 1970; Osborne 1959). Such markets seem to have little in common with a heavy particle moving in an ambient medium. It is graphic testimony to the seeming universality of randomness that similar models describe, even approximately, such different phenomena.

There is also an enormous mathematical literature on random walks (Spitzer 1964). This literature is mostly concerned with questions such as limit theorems, convergence, relations of random walks to other problems in probability and statistics, etc.

Returning to aspects closer to the physical and biological sciences, we note the book by Berg (1983) on random walks in biology; the theory presented in this work does not go much beyond what has been discussed in the current chapter, but the applications are very interesting. The references given in Section 9.1 take the theory in different directions. We have considered only random walks in discrete time and space, and then gone to the continuum limit. Walks in continuous time still keeping space discrete, have attracted much interest; they enable one to study processes with a random amount of time between steps.

We have discussed walks with one trap (absorbing site). Walks with a distribution of absorbing sites in d-dimensions ($d > 1$) are also of great interest. They are used to describe electronic processes in crystals that trap electrons and diffusion controlled chemical reactions.

The walks we have considered all have finite second moments (cf. eqn (9.3.5)). The case when the second moment is infinite is also of interest. These Lévy walks are made up of clusters of short steps, the clusters being separated by long steps. Clusters of clusters are separated by still longer steps, and so on. The trails of the walk are self-similar as the length scale increases.

Another set of problems deals with the mean number of sites visited in a random walk of n steps. This number will be less than n since a site may be visited more than once, but the multiplicity of the visit is not counted. Call the mean number of sites visited S_n; for a nearest neighbor walk on a (hyper-)cubic lattice, for large values of n it is given by (Dvoretsky and Erdos 1950)

$$S_n = \begin{cases} \sqrt{8n/\pi} & d = 1 \\ \pi n/\ln n & d = 2 \\ c_d n & d \geq 3 \end{cases} \tag{9.6.1}$$

where c_d is a constant depending on the dimension, d.

We must also mention random walks on random structures. The simplest of these is a random walk on a random walk (Kehr and Kuttner 1982). A walker is constrained to move on the sample path of a random walk. The mean-square displacement of such a walk is proportional to $n^{\frac{1}{2}}$ instead of to n as in a normal walk. This can be seen intuitively as follows. After n steps the walker will find itself at point s along the chain of the underlying walk, and s will be close to $n^{\frac{1}{2}}$ on the average. The mean-square displacement in the ambient space should then be proportional to s, which leads to the quoted conclusion. Of course this must be shown more precisely, as has been done by Kehr and Kuttner along with a study of many other properties of the model.

The exponent relating the mean-square displacement to the number of steps is not what one would naively expect. This is the first example we have seen in this book of anomalous exponents. They show up also in the Lévy walks mentioned above (Hughes 1995). We shall see other anomalous exponents in Chapter 18 when we discuss the fractal nature of Brownian motion. The subject of random walks on random environments is one on which entire books have been written, (Hughes 1996).

Lastly we want to mention the self-avoiding walk. This is a random walk that is forbidden to revisit a point it has already visited; it is not allowed to cross itself. Such a process is highly non-Markovian, since the entire history of the walk must be known before it is possible to tell if the next step is legal. Much of what is known about this class of walks comes from computer simulations. The subject is not germane to Brownian motion, and we refer the interested reader to the references of Section 9.1 for further information.

We have by no means mentioned all the ramifications of the simple questions posed by Huygens and Pearson. We have tried to whet the reader's curiosity about this fascinating subject, and have given references to enable the motivated reader to begin a more serious and complete study.

10

STATISTICAL MECHANICS

10.1 Molecular Distribution Functions

So far we have studied Brownian motion as a physical realization of a random process. Our model was a particle immersed in a medium subject to a statistical law of evolution; the nature of the medium entered our considerations only through one parameter, the friction constant. We know however, that the medium is made up of atoms and molecules ultimately subject to deterministic, not statistical, evolution. The statistical hypotheses we have used must be justified on the basis of the molecular structure and molecular dynamics of matter. This physical justification lies, as we have previously discussed, in the enormous number of collisions of the molecules of the ambient medium with the Brownian particle. Even though the individual particles obey deterministic dynamical laws, the effect of a single collision is very small, and the number of collisions is very great. Consequently only relatively large scale fluctuations about the mean behavior are observable.

The motion of these atoms and molecules is governed by deterministic dynamics, not statistical dynamics. In principle the applicable dynamical law is the Schrödinger equation. In practice for atoms and molecules, Newton's laws or Hamilton's equations furnish an entirely satisfactory substitute. In this book we shall study only those situations in which the classical laws describe atomic motions well.

The statistical hypotheses must be related quantitatively to the deterministic dynamics. Furthermore, the friction constant, the only parameter in the stochastic theory bearing on the physical nature of the ambient medium, ought to be derivable from the basic atomic dynamics. We shall carry out these two tasks in Chapter 11. The current chapter is devoted to background material on non-equilibrium statistical mechanics, the theoretical tool for realizing these two aims. We shall assume that the reader has an understanding of equilibrium statistical mechanics, and some familiarity with quantum mechanics. Although no specifically quantum systems will be treated, some of the mathematics that we will use is very similar to that used in quantum mechanics. Experience with the mathematics of operator algebra will be helpful. Helpful review articles covering the necessary operator algebra in the statistical mechanics context have been written by Berne (1971, 1977) and Hynes and Deutch (1975).

We shall be dealing with functions of many variables. It is therefore useful to introduce some simplifying notation. The functions will be functions of the positions and momenta of N particles. A function, g, of the $6N$ variables

$x_1, \ldots, z_N, p_{x,1}, \ldots, p_{z,N}$, will be written as a function of the $2N$ vector variables $\mathbf{r}_1, \ldots, \mathbf{r}_N, \mathbf{p}_1, \ldots, \mathbf{p}_N$. In turn, this will be shortened to $\mathbf{r}^N, \mathbf{p}^N$: thus $g(\mathbf{r}^N, \mathbf{p}^N)$. The volume element in the $6N$ dimensional $(\mathbf{r}, \mathbf{p})$ space will be written $d^N\mathbf{r}\, d^N\mathbf{p}$. The set $\{\mathbf{r}^N, \mathbf{p}^N\}$ is called the *phase point* or more simply just the *phase* of the system. In general we shall deal with N medium particles and one Brownian particle. The Brownian particle will always be labeled with particle number zero. Thus there will be $N + 1$ particles in the system. In a later chapter, when we discuss the concentration dependence of Brownian motion, we shall have to deal with more than one Brownian particle, and additional notation will be introduced. For the present, the notation sketched here will suffice.

Even though the state of a molecular system evolves according to deterministic laws, probability enters into its description. First, except for a very few simple systems, the equations of motion are too complicated for us to solve exactly. Secondly, even if we could solve the equations of motion to any accuracy desired, this is not enough to describe the evolution of the system. The initial conditions, the positions and momenta of all of the $N + 1$ particles at the initial time, must also be known. The only way we could handle such a mass of information, assuming that we could ever obtain it, is to treat it statistically. The positions and momenta of all the molecules become stochastic processes, albeit of a very special kind. The initial coordinates and momenta of the molecules are taken to be random variables. The subsequent deterministic evolution of the system does not eradicate this initial randomness so the positions and momenta remain random variables at later times. Thus they form a $6N$ dimensional stochastic process. This process is even a Markov process, for given the coordinates and momenta at any time, the future values are uniquely determined by the equations of motion, and the values before that time are irrelevant for the future evolution. On the other hand, this Markovian property is of little help in deriving Brownian motion equations from molecular considerations.

Recall that the distribution function of $\{\mathbf{r}^N, \mathbf{p}^N\}$ is the probability density of finding the particles of the system at that phase. We shall call the distribution function of the coordinates and momenta of the particles $f_{N+1}(\mathbf{r}^{N+1}, \mathbf{p}^{N+1}, t)$; when no ambiguity arises, we may omit writing the time variable in order to lighten the notation. This function contains an enormous amount of information, much more than we could handle, and much more than we need. We are, after all, interested in particle zero, the Brownian particle, not the medium particles. Consequently we discard information we do not want by averaging over it, or integrating it out. This leads us to define the distribution functions for $n + 1$ particles by

$$f_{N+1}^{(n+1)}(\mathbf{r}^{n+1}, \mathbf{p}^{n+1}) = \frac{N!}{(N-n)!} \int f_{N+1}(\mathbf{r}^{N+1}, \mathbf{p}^{N+1}) d^{N-n}\mathbf{r} d^{N-n}\mathbf{p} \quad (10.1.1)$$

This is the probability of finding the $n+1$ particles with the specified coordinates and momenta, regardless of the coordinates and momenta of the other $N - n$

particles. The prefactor of the integral is conventionally included in the definition because it is possible to choose n medium particles out of N in $N!/(N-n)!$ ways. Thus the distribution function for the Brownian particle itself is

$$f_B(\mathbf{r}_0, \mathbf{p}_0, t) = \int f_{N+1}(\mathbf{r}^{N+1}, \mathbf{p}^{N+1}, t) d^N \mathbf{r} d^N \mathbf{p} \tag{10.1.2}$$

Molecular distribution functions are non-negative and f_{N+1} is normalized to unity. The other $f_N^{(n)}$ are normalized to $N!/(N-n)!$.

$$\int f_{N+1}(\mathbf{r}^{N+1}, \mathbf{p}^{N+1}) d^{N+1}\mathbf{r} d^{N+1}\mathbf{p} = 1$$
$$\int f_{N+1}^{(n+1)}(\mathbf{r}^{(n+1)}, \mathbf{p}^{(n+1)}) d^{n+1}\mathbf{r} d^{n+1}\mathbf{p} = \frac{N!}{(N-n)!} \tag{10.1.3}$$

We shall shortly be performing linear operations on distribution functions. It would be convenient if distribution functions formed a linear space (vector space). Unfortunately they do not. The positivity and normalization requirements are not preserved under addition of functions and multiplication by scalars. On the other hand, distribution functions do form a subset of the space of absolutely integrable functions over phase space, the space of coordinates and momenta; this space is often called L^1. Although the set of distribution functions is not a subspace of L^1 it is a subset, so we can work within a vector space framework by working within L^1, which includes all of the distribution functions. It is necessary to keep in mind that the result of applying linear operators to distribution functions are not necessarily distribution functions.

10.2 The Liouville Equation

We now inquire into the equation of evolution of the distribution function f_{N+1}. First we note that if a system is at a point $(\mathbf{r}^{N+1}, \mathbf{p}^{N+1})$ of phase space at time t, it will necessarily be at the point $(\mathbf{r}^{\mathbf{N+1}} + \delta\mathbf{r}^{N+1}, \mathbf{p}^{N+1} + \delta\mathbf{p}^{N+1})$ at time $t + \delta t$. Here $\delta\mathbf{r}^{N+1}$ and $\delta\mathbf{p}^{N+1}$ are the increments in $\mathbf{r}^{N+1}$ and $\mathbf{p}^{N+1}$ due to the natural motion of the system under Newton's laws or Hamilton's equations in the time interval δt. They are not arbitrary increments. Hence

$$f_{N+1}(\mathbf{r}^{N+1} + \delta\mathbf{r}^{N+1}, \mathbf{p}^{N+1} + \delta p^{N+1}, t + \delta t) = f_{N+1}(\mathbf{r}^{N+1}, \mathbf{p}^{N+1}, t) \tag{10.2.1}$$

Now expand $f_N(\mathbf{r}^{N+1}, \mathbf{p}^{N+1}, t + \delta t)$ about $(\mathbf{r}^{N+1}, \mathbf{p}^{N+1}, t)$

$$\begin{aligned} f_{N+1}(\mathbf{r}^{N+1} + \delta\mathbf{r}^{N+1}, \mathbf{p}^{N+1} + \delta\mathbf{p}^{N+1}, t + \delta t) = \\ f_{N+1}(\mathbf{r}^{N+1}, \mathbf{p}^{N+1}, t) + \textstyle\sum_{j=0}^{N} \delta\mathbf{p}_j \cdot \nabla_{\mathbf{p}_j} f_{N+1} + \\ \textstyle\sum_{j=0}^{N} \delta\mathbf{r}_j \cdot \nabla_{\mathbf{r}_j} f_{N+1} + \frac{\partial f_{N+1}}{\partial t} + o(\delta t) \end{aligned} \tag{10.2.2}$$

For sufficiently small δt

$$\begin{aligned}\delta \mathbf{r}_j &= \dot{\mathbf{r}}_j \delta t + o(\delta t) \\ \delta \mathbf{p}_j &= \dot{\mathbf{p}}_j \delta t + o(\delta t)\end{aligned} \tag{10.2.3}$$

Hence, using eqn (10.2.1) and taking the limit as $\delta t \to 0$, we obtain

$$\frac{\partial f_{N+1}}{\partial t} + \sum_{j=0}^{N} \dot{\mathbf{r}}_j \cdot \nabla_{\mathbf{r}_j} f_{N+1} + \sum_{j=0}^{N} \dot{\mathbf{p}}_j \cdot \nabla_{\mathbf{p}_j} f_{N+1} = 0 \tag{10.2.4}$$

This equation is called the *Liouville equation.* It has the structure of a continuity equation and expresses the conservation of probability in phase space.

The Liouville equation looks rather innocent because of the compact notation we have used, but it is really a very complicated object. It is equivalent to the set of Hamilton's equations from which we started, and its exact integration is no simpler than the solution of the original equations of motion.

Alternate ways of writing this equation are occasionally useful. Hamilton's equations of motion can be written as

$$\begin{aligned}\dot{\mathbf{r}}_j &= \nabla_{\mathbf{p}_j} H \\ \dot{\mathbf{p}}_j &= -\nabla_{\mathbf{r}_j} H\end{aligned} \tag{10.2.5}$$

where H is the Hamiltonian function of the system, $H = K + V$. K is the kinetic energy of the system, and V is the potential energy. The potential energy includes the potential of external sources, including the confining potentials of the walls of the system, and the potential of intermolecular interactions. Using eqn (10.2.5), eqn (10.2.4) can be written

$$\frac{\partial f_{N+1}}{\partial t} + \sum_{j=0}^{N} (\nabla_{\mathbf{p}_j} H \cdot \nabla_{\mathbf{r}_j} f_{N+1} - \nabla_{\mathbf{r}_j} H \cdot \nabla_{\mathbf{p}_j} f_{N+1}) \tag{10.2.6}$$

This can, in turn, be written as

$$\frac{\partial f_{N+1}}{\partial t} + \{f_{N+1}, H\} = 0 \tag{10.2.7}$$

The symbol $\{A, B\}$ is the *Poisson bracket* of the quantities within the brackets, and is defined for a set of generalized coordinates, q_i and momenta p_i, by

$$\{A, B\} = \sum_j \left(\frac{\partial A}{\partial q_j} \frac{\partial B}{\partial p_j} - \frac{\partial A}{\partial p_j} \frac{\partial B}{\partial q_j} \right) \tag{10.2.8}$$

Another common notation is to define the operator L_{N+1} by

$$L_{N+1}\phi(\mathbf{r}^{N+1}, \mathbf{p}^{N+1}) = -i\{\phi, H\} \tag{10.2.9}$$

The operator, L, is called the *Liouville operator*. The imaginary unit, i, is inserted into the definition by convention, for with it the Liouville equation becomes

$$i\frac{\partial f_{N+1}}{\partial t} = L_{N+1} f_{N+1} \tag{10.2.10}$$

Of course, this is purely formal; it does not get us any closer to a solution, although a formal solution can immediately be produced:

$$f_{N+1}(t) = e^{-iLt} f_{N+1}(0) \tag{10.2.11}$$

To shorten the writing we have omitted the phase space arguments of the function f_{N+1}. The advantage of writing the Liouville equation in the form (10.2.10) is that it has a formal resemblance to the Schrödinger equation of quantum mechanics. Since a great deal of work has been done devising approximate solutions for the many body Schrödinger equation, because of the similarity of form of the equations we may hope to adapt some of these techniques to the present problem.

The analogy between the Liouville equation and the Schrödinger equation should not be pushed too far, however. Planck's constant, $\hbar$, does not appear in the Liouville equation, and this emphasizes that the physics described by the two equations is quite different. Furthermore, the Schrödinger equation is a second order differential equation, while the Liouville equation is first order. The solution of the Liouville equation is a probability *density*, while the solution of the Schrödinger equation is a probability *amplitude*, which must be squared to obtain physical quantities. The formal analogy is all that we shall need, however.

10.3 Projection Operators—The Zwanzig Equation

As stated above, there is much more information in the phase space distribution function, f_{N+1} than is needed. In this section we present a method due to R. W. Zwanzig (1960) for extracting the relevant information without computing the entire f_{N+1}. The method is based on the idea of a projection operator.

A *projection operator* is an operator, $\wp$, with the property $\wp^2 = \wp$. i.e. $\wp$ is idempotent. Some authors require that $\wp$ also be Hermitian to be a projection operator, but since we have not introduced a scalar product on our function space, we cannot make such a requirement. Suppose that we have a projection operator, $\wp$, that projects out of f_{N+1} the function f

$$\begin{aligned} \wp f_{N+1} &= f \\ (1-\wp) f_{N+1} &= g \end{aligned} \tag{10.3.1}$$

f is called the *relevant part* of f_{N+1}, and g is called the *irrelevant part*. Of course, the meaning of 'relevant' and 'irrelevant' depend on the problem being treated, and are completely dependent on the choice of the projection operator. At this stage we can proceed purely formally and search for the evolution equations satisfied by f and g without being specific about which projector we are using.

Operating on eqn (10.2.10) by $\wp$ and $1 - \wp$ respectively, we get coupled equations for f and g.

$$\begin{aligned}\frac{\partial f}{\partial t} &= \wp L(f+g) \\ \frac{\partial g}{\partial t} &= (1-\wp)L(f+g)\end{aligned} \tag{10.3.2}$$

In the second of eqns (10.3.2), pretend that f is a known function, and solve the equation as though it were a first order inhomogeneous linear differential equation for the unknown function g. The result is

$$g(t) = e^{-i(1-\wp)Lt}g(0) - i\int_0^t e^{-i(1-\wp)Ls}(1-\wp)Lf(t-s)ds \tag{10.3.3}$$

Here we have omitted the phase space arguments of f and g and the subscript $N+1$ on L for the sake of economy of notation.

It is now a simple matter to substitute eqn (10.3.3) in the first of eqns (10.3.2) to obtain an equation for the relevant part, f.

$$i\frac{\partial f}{\partial t} = \wp Lf - i\int_0^t \wp Le^{-i(1-\wp)Ls}(1-\wp)Lf(t-s)ds + \wp Le^{-i(1-\wp)Lt}g(0) \tag{10.3.4}$$

This is the Zwanzig equation. It is a closed equation for f. To be sure, it does contain the irrelevant part, g, but only the initial value of this function, the value at $t = 0$.

The projection operator is usually chosen so that the number of effective degrees of freedom in f is small. Nevertheless, the equation is a very complicated one. It contains projected Liouville operators in exponentials. Notice also the presence of the time integral term in eqn (10.3.4); the equation is not a differential equation, but an integro-differential equation, an added source of complication. There is no more hope of finding explicit exact solutions to the Zwanzig equation than there was to the Liouville equation from which we started. However, exact solutions are not our aim. While exact solutions are useful when available, they are usually not available except in simple special cases. It is necessary to search for physically motivated simplifications and approximations that will reduce the equation to tractable form while retaining the essential features of the description of the phenomena. The Zwanzig equation has proved itself to be useful for this purpose; we shall see in the next chapter how it can be used to derive the Fokker–Planck equation from a molecular point of view.

10.4 Projection Operators—The Mori Equation

The distribution functions which we have been studying are not the only quantities of interest to us. There are also dynamical variables, positions and momenta of the particles, and simple functions of these quantities. Recall that in the phenomenological theory there was the Fokker–Planck equation for the distribution function and the Langevin equation for the velocity, a dynamical variable. Of course, fundamentally Hamilton's equations or Newton's laws govern the evolution of the dynamical variables. But here also there is a over-abundance of information. Suppose, for example, that we are interested in the velocity of a Brownian particle. This velocity will depend, in the course of time, on the positions and momenta of all of the medium particles because of the forces exerted on the Brownian particle by the medium particles (collisions). Can we find a way of determining the average velocity of the Brownian particle without having to determine those of the medium particles in detail? We can, indeed; the method of doing so was originated by H. Mori (1965a). It also utilizes projection operators.

To derive the Mori equation, we need to use some operator algebra on the set of dynamical variables. Since linear combinations of dynamical variables are also dynamical variables, it is natural to regard dynamical variables as elements of a vector space (linear space). We are usually interested in variables with finite mean-square equilibrium averages. This suggests that we define a norm on the vector space. We do this by defining the norm of the dynamical variable A to be

$$\| A \|^2 = \int f_{N+1}^{(eq)} \mid A(\mathbf{r},\mathbf{p}) \mid^2 d^{N+1}\mathbf{r} d^{N+1}\mathbf{p} \tag{10.4.1}$$

Here $f_{N+1}^{(eq)}$ is $\exp(-H_{N+1}/kT)/Z_{N+1}$, the equilibrium $N+1$ body distribution function. H_{N+1} is the Hamiltonian of the system, and Z_{N+1} is the partition function. The linear space becomes a normed linear space. We can also define a scalar product on the space by

$$(A, B) = \int f_{N+1}^{(eq)} A^*(\mathbf{r},\mathbf{p}) B(\mathbf{r},\mathbf{p}) d^{N+1}\mathbf{r} d^{N+1}\mathbf{p} \tag{10.4.2}$$

Thus the norm is the one given by the scalar product. The space is just the space of square integrable functions with weight $f_{N+1}^{(eq)}$; this space is well known to be a Hilbert space.

Integration by parts shows that the Liouville operator is formally self-adjoint with respect to the scalar product we have chosen. That was one reason for introducing the factor of i into its definition. Whether the operator is actually self-adjoint involves subtle questions of whether its domain is dense in the space. As far as I know, this question has not been addressed; workers have assumed that formal self-adjointness sufficed for their application, and we shall assume this also.

Suppose we are interested in some dynamical variable $A(\mathbf{r},\mathbf{p})$. A need not depend on the full set of coordinates and momenta; it could be the velocity of a Brownian particle, for example. Therefore we omit the superscripts on $\mathbf{r}$ and $\mathbf{p}$ in our general presentation. We consider only As that are not explicitly time dependent, though they depend on time through the time dependence of their phase space arguments, the various coordinates and momenta.

Since

$$\frac{dA}{dt} = \sum_j (\dot{\mathbf{r}}_j \cdot \nabla_{\mathbf{r}_j} A + \dot{\mathbf{p}}_j \cdot \nabla_{\mathbf{p}_j} A) \tag{10.4.3}$$

using eqn (10.2.5), A obeys the evolution equation

$$\frac{dA}{dt} = iLA \tag{10.4.4}$$

Note the difference in sign between eqns (10.2.10) and (10.4.4). This sign difference is characteristic of the difference in evolution between dynamical variables and probability distributions. We can write

$$A(\mathbf{r}, \mathbf{p}) = e^{iLt} A(\mathbf{r}(0), \mathbf{p}(0)) \tag{10.4.5}$$

where $\mathbf{r}(0)$ and $\mathbf{p}(0)$ are the coordinates and momenta at time zero that will evolve into $\mathbf{r}$ and $\mathbf{p}$ in time t. We shall abbreviate $A(\mathbf{r}, \mathbf{p})$ by $A(t)$ and $A(\mathbf{r}(0), \mathbf{p}(0))$ by $A(0)$.

Now let us introduce a projection operator, $\wp$, which projects an arbitrary dynamical variable onto the variable A

$$\wp B = \frac{(B, A)}{(A, A)} A \tag{10.4.6}$$

Then

$$\frac{dA}{dt} = e^{iLt}(\wp + (1 - \wp))iLA(0) \tag{10.4.7}$$

Now

$$\begin{aligned} e^{iLt} i\wp LA(0) &= e^{iLt} \frac{(iLA(0), A(0))}{(A(0), A(0))} A(0) = i\Omega A(t) \\ i\Omega &= \frac{(iLA(0), A(0))}{(A(0), A(0))} \end{aligned} \tag{10.4.8}$$

so that

$$\frac{dA}{dt} = i\Omega A + e^{iLt}(1 - \wp)iLA(0) \tag{10.4.9}$$

Now we use an operator identity

$$e^{iLt} = e^{i(1-\wp)Lt} + \int_0^t ds e^{iL(t-s)} i\wp L e^{i(1-\wp)Ls} \tag{10.4.10}$$

This identity can be verified by differentiating both sides with respect to t and noting that the results are the same. A derivation is given in Appendix C.

Using eqn (10.4.10), the second term of the right member of eqn (10.4.9) can be written

$$\begin{aligned} e^{iLt}(1-\wp)iLA(0) = e^{i(1-\wp)Lt}(1-\wp)iLA(0) + \\ \textstyle\int_0^t e^{iL(t-s)}\wp iLe^{i(1-\wp)Ls}(1-\wp)iLA(0) \end{aligned} \tag{10.4.11}$$

The first term on the right of eqn (10.4.11), $\exp(i(1-\wp)Lt)(1-\wp)iLA(0)$, is called the random force, $F(t)$. Because $\wp(1-\wp) = 0, (\wp^2 = \wp), F(t) = (1-\wp)F(t)$, so that

$$(F(t), A(0)) = ((1-\wp)F(t), A(0)) = (F(t), (1-\wp)A(0)) \tag{10.4.12}$$

Since $(1-\wp)A(0)$ vanishes

$$(F(t), A(0)) = 0 \tag{10.4.13}$$

This is indeed a characteristic of the random force that we postulated in the phenomenological theory; there is no correlation between the random force and the variable A at an earlier time. We also postulated that the average of the random force vanish. This translates, in the notation of this section, to $(F(t), 1) = 0$. However, this is not necessarily true for $F(t)$ as defined above nor should we necessarily expect it to be true. After all, the development so far has been for a perfectly general Hamiltonian system with no special or simplifying properties having been assumed.

Now we analyze the last term on the right of eqn (10.4.11). We may write

$$\begin{aligned} i\wp Le^{i(1-\wp)Ls}i(1-\wp)LA = i\wp LF(t) = \\ \frac{((1-\wp)iLF(s), A)}{(A, A)}A = \frac{(F(s), F(0))}{(A, A)}A \end{aligned} \tag{10.4.14}$$

where we have used the fact that L and $1-\wp$ are formally self-adjoint operators. The quantity $(F(t), F(0))/(A, A^*)$ is called the *memory function* , $K(t)$. It is a numerically valued function, not an operator.

Putting all of these results together, eqn (10.4.9) finally becomes

$$\frac{dA}{dt} = i\Omega A + \int_0^t K(s)A(t-s)ds + F(t) \tag{10.4.15}$$

This is the Mori equation. It is often also called the *generalized Langevin equation.*

The quantity $i\Omega$ (eqn (10.4.8)) will vanish unless A is a sum of terms with different time reversal symmetries. This is because we deal with Hamiltonians that are quadratic in momenta, consequently L is odd under time reversal and the integrand defining Ω will be odd under time reversal unless A has parts which are odd and other parts which are even under this symmetry. This is a very unlikely circumstance. However, if we consider a set of dynamical variables $A_1, A_2, \ldots, A_m$ we can go through a similar analysis to derive coupled Mori equations for the members of the set of the form

$$\frac{dA_j}{dt} = i\sum_k \Omega_{jk} A_k + \sum_k \int_0^t K_{jk}(s) A_k(t-s) ds + F_j(t) \qquad (10.4.16)$$

where $i\Omega_{jk}$ and K_{jk} are scalar products similar to those defined in the one variable case. Here the symmetry rule will be: $\Omega_{jk} = 0$ if A_j and A_k have the same time reversal symmetry and may be non-zero if they have different time reversal symmetries.

It is also possible to introduce projection operators different from that of eqn (10.4.6). This is sometimes advantageous in order to exploit special properties of the function $A(r, \mathbf{p})$. We shall see an example of this in Chapter 11, where we shall take $A = \mathbf{p}_0$, the momentum of the Brownian particle (Mazur and Oppenheim 1970).

10.5 Concluding Remarks

In this chapter we have studied some formal aspects of the statistical mechanics of non- equilibrium processes. We have not given a general treatment of the subject; this would take a treatise in itself. Instead we have centered on those aspects that we shall need in the subsequent study of the molecular theory of Brownian motion.

Since the development has been so formal, it is worthwhile to indicate why it is useful. After all, we have merely rewritten the Liouville equation (Section 10.3) and Hamilton's equations (Section 10.4) in an equivalent form. The form may look simpler and suggestive, but in reality this apparent simplicity is misleading. The operators appearing in the Zwanzig and Mori equations are complicated objects. The advantage of the Zwanzig and Mori equations is that they lead more easily to the introduction of physically motivated approximations. In particular, when there is a good separation of time scales for the motions of the components of the system, we may hope to use the projection operator technique to remove the fast modes from explicit consideration, concentrating attention on the slow modes of interest. There is a natural small parameter, the ratio of the short to the long time scales. A systematic development in terms of this small parameter may be possible.

Without the $i\Omega$ term, eqn (10.4.14) is very similar to the Langevin equation. The friction constant, ζ, has become a friction kernel, $K(t)$. However, if $K(t)$ decays to zero sufficiently rapidly, we may write

$$\int_0^t K(s)A(t-s)ds \approx A(t)\int_0^t K(s)ds \approx A(t)\int_0^\infty K(s)ds \tag{10.5.1}$$

Thus the friction term in the Mori equation may be approximated by the friction term in the Langevin equation, provided that the correlation time of the random force is short compared to the time in which A changes appreciably. Although the random force does not necessarily have a non-vanishing average, it is reasonable to suppose that this average will decay to zero in a time of the order of the correlation time. It will be the task of he next chapter to make these ideas more quantitative.

With respect to the Zwanzig equation, it is harder to be specific at the present stage because we have not specified the projection operator. Nevertheless, the following general remark can be made. The $\wp Lf$ term in eqn (10.3.4) contains one factor of L and hence is linear in the space and momentum derivatives of f. The second term, on the other hand, contains two factors of L, and is thus apparently of second order in these derivatives. We used the word 'apparently' because it is not clear from the formal expression how the operator $\exp(-i(1-\wp)Ls)$ might modify this. At any rate it looks not unreasonable that a Fokker–Planck equation for f may arise under suitable circumstances. Again, this will be made more quantitative in the following chapter.

11

STOCHASTIC EQUATIONS FROM A STATISTICAL MECHANICAL VIEWPOINT

11.1 The Langevin Equation A Heuristic View

We have seen in the previous chapter that formal equations for a dynamical variable or for a distribution function can be derived using the technique of projection operators. The purpose of this chapter is to apply the lessons of the last one to the case of Brownian motion. The dynamical variable we will study is the momentum of the Brownian particle, $\mathbf{p}_0$. The distribution function we shall study is that of the Brownian particle, $f_{N+1}^{(1)}(\mathbf{r}_0, \mathbf{p}_0)$. The reason we entitled this section 'a heuristic view' is that we shall have to make some assumptions in going from the formally exact eqns (10.3.4) and (10.4.15) to the Langevin equation and the Fokker-Planck equation respectively. While we shall point out these assumptions, we shall not justify them until later sections. Even then, we shall not be able to justify them with complete rigor.

The system we wish to study consists of one particle, particle number zero, of mass M, momentum $\mathbf{p}_0$, and position $\mathbf{r}_0$, together with N particles each of mass m, with coordinates and momenta $\mathbf{r}^N$, $\mathbf{p}^N$ respectively. Particle zero is the Brownian particle; the remaining N particles are the bath particles. The Hamiltonian for the system is

$$
\begin{aligned}
H &= \sum_{i=1}^{N} \frac{p_i^2}{2m} + \sum_{0 \leq i < j} V_{i,j}(\mathbf{r}_i, \mathbf{r}_j) + \frac{p_0^2}{2M} \\
&= H_0 + H_B
\end{aligned}
\tag{11.1.1}
$$

H_0 is the Hamiltonian of the bath particles in the presence of the *stationary* Brownian particle. H_B is the kinetic energy of the Brownian particle, $\mathbf{p}_0^2/2M$. The Liouville operator of the system is

$$
\begin{aligned}
iL &= \sum_{i=0}^{N} \left(\frac{\mathbf{p}_i}{m} \cdot \nabla_{r_i} + \mathbf{F}_i \cdot \nabla_{p_i} \right) + \frac{\mathbf{p}_0}{M} \cdot \nabla_{r_0} + \mathbf{F}_0 \cdot \nabla_{p_0} \\
&= iL_0 + iL_B
\end{aligned}
\tag{11.1.2}
$$

Here L_0 consists of the summation over the bath particles in eqn (11.1.2), while L_B is the Liouville operator of the Brownian particle. Note that although

$L = -i\{\cdots, H\}$ (cf. eqn (10.2.9)) it is not true that $L_B = -i\{\cdots, H_B\}$. This is because there are potential energy terms in H_0 which depend on both the Brownian and bath particle coordinates. These terms give contributions to both L_0 and L_B.

We write again, for convenience, the Mori equation, taking for the dynamical variable A, the momentum of the Brownian particle, $\mathbf{p}_0$.

$$\frac{d\mathbf{p}_0}{dt} = \int_0^t K(t-s)\mathbf{p}_0(s)ds + F(t) \tag{11.1.3}$$

$$F(t) = \exp[i(1-\wp)Lt](1-\wp)iL\mathbf{p}_0(0) \tag{11.1.4}$$

We have not written the term in Ω, since it vanishes for the single variable $\mathbf{p}_0$ which has a definite time reversal parity (compare Section 10.4). Equation (11.1.3) already looks quite a bit like the Langevin equation. However, the first term on the right, ostensibly linear in $\mathbf{p}_0$, is an integral over the history of $\mathbf{p}_0$ and does not depend on its value only at time t, as does the Langevin equation. Furthermore there is no a priori assurance that the memory function, K, does not depend on $\mathbf{p}_0$; the linearity may be only apparent, an artifact of the notation. It also must be shown that the 'random force', F, has the properties assumed in the Langevin equation.

To investigate these matters, we first look at the random force (11.1.4). We first note that

$$iL_0\mathbf{p}_0 = 0 \tag{11.1.5}$$

since there are no derivatives with respect to $\mathbf{p}_0$ in L_0. This implies that

$$iL\mathbf{p}_0 = iL_B\mathbf{p}_0 \tag{11.1.6}$$

where $\mathbf{F}_0$ is the intermolecular force exerted on particle zero by all of the bath particles. Therefore $\wp iL\mathbf{p}_0$ is zero; the projection operator contains a factor $(\mathbf{F}_0, \mathbf{p}_0)$ which vanishes since the scalar product is calculated with the equilibrium distribution function (compare eqn (10.4.2)). At equilibrium, the force on a particle is uncorrelated with its momentum, and both averages vanish.

The result, then, is that

$$\mathbf{F}(t) = \exp[i(1-\wp)Lt]\mathbf{F}_0 \tag{11.1.7}$$

Note that the Liouville operator does not appear alone in the exponent, but in combination with the projection operator $\wp$; this makes the exponential operator a very difficult one to analyze. So far we have not made any approximations; all of the manipulations have been formally exact. We must now introduce the fact that we are dealing with Brownian motion, i.e. that the mass ratio

$$\lambda^2 = \frac{m}{M} \tag{11.1.8}$$

is small, $\lambda \ll 1$.

The momenta of the bath particles are of the order of $(mkT)^{\frac{1}{2}}$, while that of the Brownian particle is of the order of $(MkT)^{\frac{1}{2}}$. Of course, these momenta are not fixed, they are distributed according to a momentum distribution function. What this statement means is that the bulk of the momentum distributions are centered about the two values given above. Consequently we scale the momentum of the Brownian particle to make it comparable to those of the bath particles. Define $\mathbf{P}$ by

$$\mathbf{P} = \lambda \mathbf{p}_0 \tag{11.1.9}$$

Thus $\mathbf{P}$ has the desired order of magnitude. We will also write $\mathbf{r}_0 = \mathbf{R}$; this is only a notational change , to keep the position and momentum of the Brownian particle in the same size type.

Expressing the Liouville operator in these new variables, we note that L_0 is unchanged , but L_B becomes

$$iL_B = \lambda \left(\frac{\mathbf{P}}{M} \cdot \nabla_R + \mathbf{F}_0 \cdot \nabla_P \right) \tag{11.1.10}$$

In other words, L_B is of order λ relative to the individual terms of L_0. Now let us recall the operator identity eqn (10.4.10), which we write in the form

$$e^{i(A+B)t} = e^{iAt} + \int_0^t e^{i(A+B)(t-s)} iBe^{iAs} ds \tag{11.1.11}$$

If we take A as iL and B as $\wp iL$, then the integral term in eqn (11.1.11) is of order λ relative to the first term. In fact, iterating eqn (11.1.11) yields a formal power series in λ. Therefore, to lowest order in λ we may replace iL in the exponent by iL_0. This is our first heuristic assumption.

Having done this, we may now replace $(1 - \wp)iL_0$ by iL_0. This is because $(1 - \wp)iL_0\mathbf{F}_0$ is independent of $\mathbf{P}$ and is equal to $iL_0\mathbf{F}_0$. The same is true for every power $(1 - \wp iL_0)^n$. Since the exponential function of an operator is defined by its power series expansion, the replacement is justified.

We finally have

$$\mathbf{F}(t) = e^{iL_0 t}\mathbf{F}_0(0) \tag{11.1.12}$$

That is, the random force is to be interpreted as the intermolecular force on the Brownian particle exerted by the bath particles when they move in the field of the *fixed* Brownian particle. In other words, in calculating $\mathbf{F}$ the Brownian particle should be considered as infinitely heavy to this order of approximation. Notice that there is nothing intrinsically random in the random force. It is to be computed from the standard Hamiltonian equations of motion of an N body system in the field of a fixed center of force. Of course N is enormous, of the order of Avogadro's number. Consequently we are not able to carry out the computation of $\mathbf{F}(t)$ in practice; We must make some assumptions about the properties of $\mathbf{F}$. The usual assumption is that $\mathbf{F}(t)$ has a short correlation time with respect to the time scale on which $\mathbf{P}$ varies.

The only remaining part of eqn (11.1.3) to be analyzed is the memory function $K(t)$. From (10.4.14) that the memory function is

$$K(s) = \frac{< \mathbf{F}(0)\mathbf{F}(s) >}{MkT} \tag{11.1.13}$$

where we have used the fact that $(\mathbf{p}_0, \mathbf{p}_0) = MkT$. We make the same approximations made in the discussion of the random force, replacing $\mathbf{F}$ as defined by eqn (11.1.7) by the approximation (11.1.12). Thus the friction kernel becomes the correlation function of the random force.

If we assume that the random force has a very short correlation time, as stated above, then we may replace

$$\int_0^t K(t-s)\mathbf{p}_0(s)ds \approx \mathbf{p}_0(t) \int_0^t K(s)ds = \zeta \mathbf{p}_0 \tag{11.1.14}$$

We have made two approximations at the same time here. First, we have taken $\mathbf{p}_0$ outside the integral sign, by our assumption that it varies slowly compared to K; second, we have extended the limit of the integral to infinity since we are assuming that the integrand is negligible after some short correlation time. Finally, subject to all of the assumptions and approximations we have made in this section, we arrive at the equation

$$\frac{d\mathbf{p}_0}{dt} = -\zeta \mathbf{p}_0 + \mathbf{F} \tag{11.1.15}$$

This is precisely the Langevin equation that we introduced and discussed in Sections 4.6 and 5.1. Although it is illuminating to see how this equation arises, there were a number of assumptions made that must be justified more fully before the result can be accepted without reservation.

11.2 The Fokker–Planck Equation—A Heuristic View

Just as we went from the formally exact eqn (10.4.15) to the Langevin equation, we can also go from eqn (10.3.4) to the Fokker–Planck equation. Again, we shall have to make some assumptions, very similar to the assumptions made in the preceding section. The starting point is eqn (10.3.4), which we write again for ease of reference. Here we denote the projection operator by $\wp_Z$ to distinguish it from the Mori projection operator used in Section 11.1.

$$i\frac{\partial f}{\partial t} = \wp_Z iLf - \int_0^t \wp_Z iLe^{i(1-\wp_Z)Lt}(1-\wp_Z)iLf(t-s)ds + \wp_Z iLe^{i(1-\wp_Z)Lt}g(0) \tag{11.2.1}$$

It is now necessary to choose the projection operator. We shall set

$$\wp_Z\phi(\mathbf{r}^{N+1},\mathbf{p}^{N+1}) = f_N^{(eq)}(\mathbf{r}^N,\mathbf{p}^N)\int \phi(\mathbf{r}^{N+1},\mathbf{p}^{N+1})d^N\mathbf{r}d^N\mathbf{p} \qquad (11.2.2)$$

Here, ϕ is an arbitrary operand. $f_N^{(eq)}$ is, as in Section 11.1, the equilibrium distribution function of the bath particle in the presence of the fixed Brownian particle. The presence of some function of the bath particles is required to insure that $\wp_Z\phi$ lies in the same $N+1$ variable space as does ϕ itself.

Two important properties of this projection operator are

$$i\wp_Z L_0 = 0$$
$$\wp_Z\left(\frac{\mathbf{p}_0}{M}\cdot\nabla_{\mathbf{r}_0}\right)(1-\wp_Z) = 0 \qquad (11.2.3)$$

The first of these follows immediately upon integration by parts. The second arises from computing $\wp_Z(\mathbf{p}_0/M)\cdot\nabla_{\mathbf{r}_0}\phi$:

$$\wp_Z\frac{\mathbf{p}_0}{M}\cdot\nabla_{\mathbf{r}_0}\phi = \frac{\mathbf{p}_0}{M}\cdot\nabla_{\mathbf{r}_0}\phi - \frac{\mathbf{p}_0}{M}\cdot\nabla_{\mathbf{r}_0}\ln f_N^{(eq)}\cdot\wp_Z\phi \qquad (11.2.4)$$

Taking $\phi = (1-\wp_Z)\psi$, and using the fact that $\wp_Z(1-\wp_Z) = 0$, the second of eqns (11.2.3) follows.

We eliminate the last term of eqn (11.2.1) by choosing $g(0) = 0$. The last term then vanished for all time. Since $g = f_{N+1} - f_N^{(eq)}f_{N+1}^{(1)}$ this initial condition is equivalent to saying that, at the initial time, the Brownian particle finds itself in a bath at thermal and mechanical equilibrium. Of course, such a distribution will not stay invariant in time; it is not a solution of the Liouville equation. The motion of the Brownian particle will throw the neighboring bath slightly out of equilibrium, but equilibrium will again be established after long times when the Brownian particle thermalizes. It is believed, though not proved, that solutions of eqn (11.2.1) are insensitive to initial conditions as long as they describe states reasonably close to the actual physical state. On the basis of this belief, we have chosen the simplest such initial condition.

Using eqns (11.2.3) it is straightforward to verify that eqn (11.2.1) can be written

$$\frac{\partial f_{N+1}^{(1)}}{\partial t} + \frac{\mathbf{p}_0}{M}\cdot\nabla_{\mathbf{r}_0}f_{N+1}^{(1)} =$$
$$\nabla_{\mathbf{p}_0}\cdot\int_0^\infty ds < \mathbf{F}_0 e^{i(1-\wp_Z)L(t-s))}\mathbf{F}_0 >_b (\nabla_{\mathbf{p}_0} + \frac{\mathbf{p}_0}{MkT})f_{N+1}^{(1)}(s) \qquad (11.2.5)$$

Here, the symbol $<\cdots>_b$ denotes an average over the bath distribution function $f_N^{(eq)}$. This average could still depend on $\mathbf{p}_0$. Even more, it could be an operator containing derivatives with respect to $\mathbf{p}_0$ and $\mathbf{r}_0$. Nevertheless, we can scale the momentum of the Brownian particle exactly as we did in eqn (11.1.9) and use the same argument as was used to obtain (11.1.12) to replace $\exp(-i(1-\wp_Z)Lt)$ by $\exp(-iL_0t)$. We then get

$$< \mathbf{F}_0 e^{i(1-\wp_Z)L_0(t-s))} \mathbf{F}_0 > = < \mathbf{F}(0)\mathbf{F}(t-s) >_0 \tag{11.2.6}$$

This is the same as the expression for $K(s)$ given by eqn (11.1.13) (except for a numerical factor).

We have assumed that $K(s)$ decays rapidly compared to the decay rate of the momentum of the Brownian particle. We shall assume that this is equivalent to asserting that the decay rate of the force correlation function is rapid with respect to the rate of change of the distribution function of the Brownian particle. Once this assumption is made, we can take $(\nabla_{p0} + \mathbf{p}_0/MkT)f_N^{(1)}(t)$ outside the integral over s, and extend the upper limit of that integral to ∞. The argument is the same as that made in eqn (11.1.14). The final result is that eqn (11.2.1) has been put into the form

$$\frac{\partial f_{N+1}^{(1)}}{\partial t} + \frac{\mathbf{p}_0}{M} \cdot \nabla_{\mathbf{r}_0} f_{N+1}^{(1)} = \zeta \nabla_{\mathbf{p}_0} \cdot \left(\nabla_{\mathbf{p}_0} + \frac{\mathbf{p}_0}{MkT} \right) f_{N+1}^{(1)} \tag{11.2.7}$$

This is precisely the Fokker–Planck equation which we studied in detail in Chapter 7.

So, we have here given molecular derivations of the Langevin equation (Section 11.1) and the Fokker–Planck equation (this section) on the basis of precisely the same hypotheses and assumptions. The two derivations are then mutually consistent. Furthermore, we now have a molecular expression for the friction constant, ζ. It is the time integral of the autocorrelation function of the intermolecular force exerted by the bath particles on a *stationary* Brownian particle.

$$\zeta = \frac{1}{MkT} \int_0^\infty < \mathbf{F}(0)\mathbf{F}(t) >_0 dt \tag{11.2.8}$$

Before the introduction of statistical mechanical considerations, the friction constant was a purely phenomenological factor; the theory gave no indication of its molecular meaning or how it was to be determined.

Even now, with our newly acquired precise definition of how ζ is to be calculated, we cannot actually compute it except in some simple cases. The detailed computation is much too complicated. Nevertheless, it is a definite advantage to have a specific algorithm, albeit an exceedingly difficult one to implement, for the friction constant.

On the other hand, we have derived the Langevin and Fokker–Planck equations only on the basis of assumptions on the behavior of certain functions and operators in the limit of small λ^2. These assumptions must be verified before the results we have so far obtained can be accepted. That is why we have called both this and the preceding section 'heuristic.'

11.3 What is Wrong with these Derivations?

It seems as though we have succeeded in deriving the Langevin and Fokker–Planck equations starting from basic molecular theory, augmented by several seemingly innocuous approximations. After the original publication of the derivations, workers were satisfied with them but it was soon realized that the validity of the approximations used was not as obvious as first thought. One of the observations that led to this realization was the discovery of 'long time tail' in molecular correlation functions; the decay of certain molecular correlation functions had an asymptotic slow inverse power decay, not the rapid exponential decay that had been assumed.

The two basic assumptions that were made and that must be investigated more carefully are first, that the operator L_B can be neglected in the operators $\exp[-i(1-\wp)Lt]$ because it is small compared to L_0, and second, that the correlation function $K(t)$ so obtained has a short lifetime relative to the relaxation time of the Brownian particle's momentum. Why might these assumptions be suspect?

The first might be in error because the propagator in question occurs under an integral sign; the integral is over the time variable. A small cause that is able to act over a long time may have an appreciable effect.

The second assumption may be in error because the basic assumption that the fluid variables move on a much faster time scale than the Brownian particle is not correct. It is true that *most* of the fluid variable move on the more rapid time scale. This is evidenced by our argument that the mass ratio m/M determines the ratio of time scales. But there are a few slow fluid variables. These are the long wave length components of the densities of conserved collective quantities.

A conserved quantity is one that has a density, $\rho(\mathbf{r},t)$, and a current density, $\mathbf{j}(\mathbf{r},t)$, that satisfy an equation of the form

$$\frac{\partial \rho}{\partial t} + \nabla \cdot \mathbf{j} = 0 \tag{11.3.1}$$

This equation says that the quantity in question accumulates in a given volume only because of transport across the boundary of that volume. Expressing this equation in Fourier space, where $\rho_{\mathbf{k}}$ and $\mathbf{j}_k$ are the Fourier transforms of ρ and $\mathbf{j}$ respectively, one has

$$\frac{\partial \rho_k}{\partial t} + i\mathbf{k} \cdot \mathbf{j}_k = 0 \tag{11.3.2}$$

For small $\mathbf{k}$, that is for long wave lengths, $\rho_{\mathbf{k}}$varies slowly.

There are few of these conserved densities. For a simple fluid they are the mass density, momentum density, and energy density. Their rarity means that they do not have much statistical weight. Yet they couple strongly to the degrees of freedom of suspended bodies. Therefore the existence of these fluid variables provides a candidate for a mechanism leading to a long lifetime for fluid correlation functions.

There is a qualitative argument that this may be the case. We have seen that the ratio of characteristic velocities of Brownian and fluid particles is $(m/M)^{\frac{1}{2}}$. Therefore the ratio of characteristic times for single particle fluid motion to Brownian particle motion is the inverse of this, $(M/m)^{\frac{1}{2}}$. But there are other characteristic times relating to this problem. The characteristic time for relaxation of the velocity of the Brownian particle is $M/\zeta \sim M/\eta a$. Here we have used Stokes' law to estimate ζ; η is the viscosity of the fluid, and a is the radius of the Brownian particle. Then there are the characteristic relaxation times of the slow fluid variables that we discussed in the previous paragraph.

To estimate these, we shall assume that we have an incompressible fluid, and that thermal conduction plays no appreciable role in Brownian motion. (Both of these assumptions can be lifted at the expense of much algebra and with little profit.) The only slow conserved quantity of interest is then the momentum density, $\rho\mathbf{v}$. The linearized equation of motion for the velocity of a continuum incompressible fluid is

$$\begin{aligned} \frac{\partial \mathbf{v}}{\partial t} = -\frac{1}{\rho}\nabla p + \frac{\eta}{\rho}\nabla^2 \mathbf{v} = 0 \\ \nabla \cdot \mathbf{v} = 0 \end{aligned} \tag{11.3.3}$$

Expressed in Fourier space, this is

$$\begin{aligned} \frac{\partial \mathbf{v}_k}{\partial t} &= -\frac{i\mathbf{k}}{\rho}p_{\mathbf{k}} - \frac{\eta k^2}{\rho}\mathbf{v}_k \\ \mathbf{k}\cdot\mathbf{v}_k &= 0 \end{aligned} \tag{11.3.4}$$

The second of these equations says that the motion is purely transverse.[23] The longitudinal modes of a real fluid are sound waves, which do not exist in an incompressible fluid. Therefore we restrict ourselves to the transverse modes. Again, this restriction can be lifted but it is not worth the trouble to do so here.

For the transverse modes, eqn (11.3.4) becomes

$$\frac{\partial \mathbf{v}_{k\,tr}}{\partial t} = -\frac{\eta k^2}{\rho}\mathbf{v}_{k\,tr} \tag{11.3.5}$$

This clearly has as solution an exponential relaxation with time constant $\rho/\eta k^2$. Let us consider modes with wave length of the order of the Brownian particle's radius, a. The appropriate time constant for this mode is then $\rho a^2/\eta$. The ratio of this relaxation time to the relaxation time of the Brownian particle's velocity is $\rho a^2/\eta \div M/\eta = \rho a^3/M \approx \rho/\rho_B$. ρ is the mass density of the fluid and ρ_B is

[23] A continuous vector field $\mathbf{V}$ vanishing at infinity can be represented as a sum of two terms, $\mathbf{V}_T$ and $\mathbf{V}_L$, where $\nabla\cdot\mathbf{V}_T = 0$ and $\nabla\times\mathbf{V}_L = 0$. In Fourier space this translates to $\mathbf{k}\cdot\mathbf{V}_{k,T} = 0$ and $\mathbf{k}\times\mathbf{V}_{k,L} = 0$. $\mathbf{V}_L$ is called the longitudinal part of $\mathbf{V}$, $\mathbf{V}_T$ the transverse part.

the mass density of the Brownian particle. Thus the fluid will relax rapidly on the Brownian particle time scale if the density of the Brownian particle is much greater than the density of the fluid. If this is not the case, then the separation of time scales that was assumed as the basis for the approximations made in Sections 11.1 and 11.2 cannot be accepted. Compare the observation of Lorentz, Section 4.3 and Appendix A.

For liquid molecular systems, the density of the Brownian particles is of the same order of magnitude as the density of the ambient fluid although for particles suspended in dilute gases ρ/ρ_B may indeed be small. But while bath effects with a long time scale exist, they might have a small amplitude and thus not be important. It is not at all obvious from the type of arguments we have been making what the magnitude of the effects of the slowly varying bath modes are. We shall have to reexamine the derivations of the Langevin and Fokker–Planck equations taking the slow modes of the fluid into account.

There is a large literature on this problem (Mazur and Oppenheim 1970; Cukier and Deutsch 1969; Van Kampen and Oppenheim 1986; Bocquet 1998) among others. These works have taken the limit $m/M \to 0$ without changing the intermolecular forces. Since it is the intermolecular forces that determine the 'sizes' of molecules, this means that a stays fixed and consequently M/a^3 becomes very large as M becomes large. Thus ρ/ρ_B becomes small with m/M. Therefore, when these authors obtain the Fokker–Planck or Langevin equations, they do so in the limit that the slow processes in the bath have disappeared.

11.4 Eliminating Fast Processes

The derivations discussed in the previous sections of this chapter essentially assumed that the velocity of the Brownian particle was the only slow variable that in the system. Of course, if the velocity is small, as we have argued, and changes slowly, so will the position of the particle, but the important assumption was that the dynamical variables describing the *bath* varied on a rapid time scale. In the last section, we have argued that this was unlikely to be correct, but the magnitude of the effect of slow bath variables must still be estimated.

We used the Zwanzig–Mori projection operator formalism to put the equations of motion of the system in a form in which it was easy to eliminate the presumably fast bath variables, leaving only the slow Brownian particle variables. Fortunately, the Zwanzig–Mori formalism is robust enough so that it can also be applied advantageously when there are other slow variables.

Suppose that there are many slow variables; let us label them by A_k. Then we can define a projection operator to project onto the subspace spanned by all of these slow variables in the obvious way

$$\wp B = \sum_{\mathbf{k}} A_k \frac{(A_{-k}, B)}{(A_{-k}, A_k)} \tag{11.4.1}$$

We then derive the coupled Mori equations for the variables A_k as indicated in eqn (10.4.16). Presumably all of the fast variables have now been projected out, and no important slow variables have been lost in the process.

The question remains, what variables, in addition to the Brownian particle's momentum, should be included in the set $\{A_k\}$? Clearly the long wave length (low wave number) components of the conserved hydrodynamic densities should be included, as we have argued in the previous section. But there are more, for if $A_{k'}$ is slowly varying, so is $A_{k-k'}$. This means that $A_{k'}A_{k-k'}$ should be included in the set of slow variables if k and k' are both small. Furthermore, triple and higher order products should also be included. In actual computations, bilinear products are the only ones considered; this is a matter of practicality, not of principle.

We mention in passing that there is another possible source of slow variables. It may happen that a system has a broken symmetry; that is, the state of the system has less symmetry than the underlying Hamiltonian. There will usually be many such states. Then there are slow modes which move the system between the various broken symmetry states, tending to restore the symmetry. A case in point is a ferromagnet (say the Heisenberg model). The basic Hamiltonian is spherically symmetrical, yet in the ferromagnetic phase the magnetic moment is pointed in a given direction. Any direction will do; it depends on the preparation of the system. The symmetry restoring slow variables are the spin waves which rotate the direction of the magnetization vector.

As far as I know, except for liquid crystals there are no broken symmetry phenomena in classical liquids. Consequently we do not pursue the problem of broken symmetry in this book; it is mentioned here only to keep the record complete. If broken symmetry slow variables should arise in some problem different from those considered here, they could be handled by techniques similar to those we are going to discuss.

Returning now to the conserved densities, for a one component classical liquid these are the mass density $\rho(\mathbf{r})$, the flow velocity $\mathbf{v}(\mathbf{r})$, and the energy density $e(\mathbf{r})$

$$\begin{aligned}\rho(\mathbf{r}) &= m\sum_i \delta(\mathbf{r}_i - \mathbf{r})\\ \mathbf{v}(\mathbf{r}) &= \frac{1}{n}\sum_i \mathbf{v}_i\delta(\mathbf{r}_i - \mathbf{r})\\ e(\mathbf{R}) &= \frac{1}{2}\sum_i [mv_i^2\delta(\mathbf{r}_i - \mathbf{r}) + \sum_{j\neq i} V(\mathbf{r}_i, \mathbf{r}_j)\delta(\mathbf{r}_i - \mathbf{r})]\end{aligned} \tag{11.4.2}$$

n is the average fluid density. For our purposes it is sufficient to assume that the fluid is incompressible; this means $\nabla\cdot\mathbf{v} = 0$, or $\mathbf{k}\cdot\mathbf{v}_k = 0$ in Fourier space. There is no longitudinal component to the velocity fluctuations. The mass density is also constant in time due to the incompressibility condition. Moreover, we shall neglect energy fluctuations since Brownian motion is an essentially isothermal phenomenon. Thus the only slow fluid variables that we shall take into account are the transverse velocity components and their bilinear products. In order to try to eliminate all of the fast fluid variables without eliminating the slow ones, we first express the memory function of the exact Mori equation in terms of a transport coefficient of the Brownian particle, the self-diffusion coefficient. Then we study the time dependent self-diffusion coefficient by mean of what is called *mode–mode coupling* theory. This is a technique originally devised for the study of dynamical critical phenomena, where long wave length slow modes are very important. The method is equally applicable to the current problem.

If we multiply the exact Mori equation, eqn (11.1.3) by $\mathbf{p}_0(0)$, we get an equation for the autocorrelation function of the momentum of the Brownian particle, $\phi(t) =< \mathbf{p}_0(0) \cdot \mathbf{p}_0(t) >$

$$\begin{aligned} \frac{d\phi}{dt} &= -\int_0^t K(t-s)\phi(s)ds \\ \phi(0) &= 3MkT \end{aligned} \tag{11.4.3}$$

If we let superscript carets denote Laplace transforms and z denote the Laplace transform variable, this equation can be written as

$$z\hat{\phi}(z) - \phi(0) = -\hat{K}(z)\hat{\phi}(z) \tag{11.4.4}$$

whose solution is

$$\hat{\phi}(z) = \frac{\phi(0)}{z + \hat{K}(z)} \tag{11.4.5}$$

We now have to interpret $\hat{K}$. To do this, we look at the mean square displacement, $< (\Delta \mathbf{r}_0(t))^2) >$, of the Brownian particle. This is

$$\begin{aligned} < \Delta\mathbf{r}_0^2(t) > &= \tfrac{2}{M^2} \textstyle\int_0^t ds \int_0^s ds' < \mathbf{p}_0(s)\mathbf{p}_0(s') > = \\ & \tfrac{2}{M^2} \textstyle\int_0^t ds \int_0^s ds' \phi(s-s') \end{aligned} \tag{11.4.6}$$

Here we have made use of the fact that the momentum correlation in equilibrium is even in the time, and only depends on the time difference of its velocity arguments. Equation (11.4.6) can be written in the form

$$\frac{2}{M^2}\int_0^t ds \int_0^s ds' \phi(s-s') = \frac{2}{M^2}\int_0^t (t-s)\phi(s)ds \tag{11.4.7}$$

The Laplace transform of this is

$$< \Delta \mathbf{r}_0^2 > (z) = \frac{2\hat{\phi}(z)}{M^2 z^2} \tag{11.4.8}$$

Now if $< (\Delta \mathbf{r}_0(t))^2) >$ were described by the classical diffusion equation (the Smoluchowski equation in the absence of external forces), it would be given by $6Dt$, and have the Laplace transform $6D/z^2$. Consequently, we may regard $2\hat{\phi}(z)/M^2$as a generalization of the self-diffusion coefficient, D. We call it $D(z)$. As we have seen in Section 7.3, even when the motion of the Brownian particle is described by the relatively simple Ornstein–Uhlenbeck process, $D(z)$ is not constant.

Putting together eqns (11.4.3) and (11.4.6), we end up with an expression for $\hat{K}$

$$\hat{K}(z) = \frac{6MkT - M^2 z D(z)}{M^2 D(z)} \tag{11.4.9}$$

So, finally we have $\hat{K}$ expressed in terms of a generalized transport coefficient. We must now calculate this transport coefficient, $D(z)$.

To do this, we relate it to a different correlation function. Suppose that the density of Brownian particles obeyed the ordinary diffusion equation

$$\frac{\partial n}{\partial t} = D\nabla^2 n \tag{11.4.10}$$

Taking the space-time Fourier transform of this equation, we easily find that

$$\hat{n}_{\mathbf{k}}(z) = \frac{n_{\mathbf{k}}(0)}{z + k^2 D} \tag{11.4.11}$$

Thus the correlation function of the Fourier components of the density are

$$< \hat{n}_{\mathbf{k}}^*(z)\hat{n}_{-\mathbf{k}}(0) >= \frac{< n_{\mathbf{k}}(0) n_{-\mathbf{k}}^*(0) >}{z + k^2 D} \tag{11.4.12}$$

Since the density–density correlation function will not necessarily obey the diffusion equation on all space-time scales, we cannot use this equation to calculate the correlation function. Our reason for introducing it is to motivate the *definition* of the generalized frequency and wave number dependent diffusion coefficient, $D(\mathbf{k}, z)$, by

$$< \hat{n}_{\mathbf{k}}(z)\hat{n}_{-\mathbf{k}}^*(0) >= \frac{< n_{\mathbf{k}}(0) n_{-\mathbf{k}}^*(0) >}{z + k^2 D(k, z)} \tag{11.4.13}$$

We now assert that $D(z)$ is the limit, as $\mathbf{k}$ approaches zero of $D(\mathbf{k}, z)$. To see this, note that

$$< n_{\mathbf{k}}(t) n_{-\mathbf{k}}^*(0) >=< e^{i\mathbf{k}\cdot(\mathbf{r}(t)-\mathbf{r}(0))} > \tag{11.4.14}$$

Thus

$$\nabla_{\mathbf{k}}^2 < n_{\mathbf{k}}(t)n_{-\mathbf{k}}^*(0) >= - < (\Delta\mathbf{r}(t))^2 e^{ik\cdot(\mathbf{r}(t)-\mathbf{r}(0))} > \tag{11.4.15}$$

and hence

$$< \Delta\mathbf{r}^2(t) >= - \lim_{k\to 0} \nabla_{\mathbf{k}}^2 < n_{\mathbf{k}}(t)n_{-\mathbf{k}}^*(0) > \tag{11.4.16}$$

Switching to Laplace transform language, and keeping only the lowest order terms in $\mathbf{k}$ on the right-hand side (since we are sending $\mathbf{k}$ to zero), we obtain

$$\begin{aligned} \frac{6D(z)}{z^2} &= - \lim_{\mathbf{k}\to 0} \nabla_{\mathbf{k}}^2 \frac{< n_{\mathbf{k}} n_{-\mathbf{k}}^* >}{z + k^2 D(k,z)} \\ &= - \lim_{\mathbf{k}\to 0} \frac{\nabla_{\mathbf{k}}^2 < n_{\mathbf{k}} n_{-\mathbf{k}}^* >}{z}(1 - \frac{k^2}{z} D(k,z)) \end{aligned} \tag{11.4.17}$$

Only the $k^2 D$ term contributes to the Laplacian, and we obtain $D(z) = \lim D(\mathbf{k}, z)$ as $\mathbf{k} \to 0$. Consequently the quantity we want to study is the density–density correlation function.

To calculate $< n_{\mathbf{k}}(t)n_{-\mathbf{k}}^*(0) >$ we shall write a Mori equation for $n_{\mathbf{k}}$. We need to use the mode–mode coupling technique, for we have just argued that it is important to project out all of the slow variables. Thus it is necessary to identify the other slow variable that couple to the Brownian particle density. Oppenheim and coworkers have studied this problem (Keyes and Oppenheim 1973; Michaels and Oppenheim 1975; Tokuyama and Oppenheim 1978). The calculations are quite complicated and technical. We shall only give an outline of them here. Naturally, in the course of such a difficult calculation a number of approximations have been made, and we shall try to point them out.

Oppenheim *et al.* first assert that the only slow variables that need to be considered, in addition to the Brownian particle momentum, are the Brownian particle density and the fluid velocity (Fourier components). This is not superficially obvious; it is a conclusion from an analysis of the dependence of the various bilinear products of conserved variables on the number of particles in the system.

Now we assemble all of the slow variables in a vector V_α, where the subscript α stands for both the label specifying which variable is in question and for the wave number of that variable. In general, the set $\{V_\alpha\}$ is infinite, because of the infinitely many wave vectors that couple in the limit of large numbers of particles. One can now define a projection operator $\wp_V$ that projects onto the subspace spanned by the set $\{V_\alpha\}$

$$\wp_V A_\alpha = \sum_{\beta,\gamma} (A_\alpha, V_\beta)(G^{-1})_{\beta,\gamma} V_\gamma \tag{11.4.18}$$

where the matrix $\mathbf{G}$ has matrix elements $G_{\mu\nu} = (V_\mu, V_\nu)$. It is the generalization of the projection operator of eqn (10.4.6) to the case when the space being projected onto is higher than one-dimensional. One then gets eqn (10.4.15) with

A replaced by the column vector $\{V_\alpha\}$, F replaced by a corresponding vector, and the memory function, K, replaced by a matrix of memory functions.

Mori showed that a consequence of the Mori equation is

$$\mathcal{L}[(\dot{V}(t)V^*(0))] = \frac{1}{\tau(z)}\mathcal{L}[(V(t)V^*(0))]$$
$$\tau^{-1}(z) = (-(\dot{V}(0)V^*(0)) + \mathcal{L}[(\dot{V}(0)\dot{V}^*(0))^\dagger])G^{-1} \tag{11.4.19}$$

$\mathcal{L}$ denotes Laplace transform; the transform variable is z. Oppenheim *et al.* call this 'the Mori identity.' Here (CD) is the matrix whose elements are (C_μ, D_ν) and the symbol † means

$$(C_\mu, D_\nu)^\dagger = ((1-\wp_V)C_\mu[e^{i(1-\wp_V)Lt}(1-\wp_V)D_\nu]) \tag{11.4.20}$$

This matrix is the matrix of memory functions when the arguments are the V.

Now we make an important approximation. We assume that the set $\{V_\alpha\}$ contains only the linear variables, Brownian particle density, momentum, and fluid velocity, together with their bilinear products. Trilinear or higher order products are not taken into account. Of course, higher order products in the set $\{V_\alpha\}$ could be included if desired, but then the subsequent analysis would get even more complex than is the case here. The conjecture is that the bilinear products contain the bulk of the coupling effects, but this has never been proven.

Since we assume that all of the important slow variation is contained in the variables of the set $\{V_\alpha\}$, the matrix K should contain only the rapidly varying variables, and consequently should have a short lifetime. That, at any rate, is the hypothesis made to substitute for the hypothesis that the memory function for $\mathbf{p}_0$ alone decayed rapidly. If this is the case, then $\tau^{-1}(z)$ may be replaced by $\tau^{-1}(0)$ for small z. The range of small z is the range of interest, for the small z behavior tells us the large t behavior of the memory function itself, and this is what we are trying to learn.

With this approximation, eqn (11.4.16) is the Laplace transform of a set of coupled linear differential equations with constant coefficients, the unknowns of which are $< n_k(z)n_{-k}(0) >$ and correlation functions of the other slow variables. We are not interested in these latter functions. Finding them is part of the solution of the coupled equations, but after the solution has been found, they may be discarded. We shall not go into the complex computations involved in evaluating $\tau^{-1}(0)$ and G^{-1}. The formal solution to the equations for the correlation functions is

$$(V(t)V^*(0)) = e^{-\tau^{-1}(0)t}(V(0)V^*(0))) \tag{11.4.21}$$

Of course, $< n_k(z)n_{-k}(0) >$ is one element of (VV), and must be extracted from the formal result.

Michaels and Oppenheim (1975) carry out this program, a number of approximations being made in the course of the computation. Comparing the result with that of eqn (11.4.11) in the $\mathbf{k} \to 0$ limit yields $D(z)$. For large Brownian particles in a dense fluid, $D(z)$ can be written

$$D(z) = D_{SE}(\frac{6}{5} - a(\frac{\rho}{\eta})^{\frac{1}{2}} z^{\frac{1}{2}})$$

$$D_{SE} = \frac{kT}{6\pi\eta a} \tag{11.4.22}$$

D_{SE} is, of course, the Stokes–Einstein expression for the self-diffusion coefficient (compare Section 4.5).[24] We have not quoted here the full formula given by the original authors, but only that approximation to it appropriate for a large particle in a dense medium.

Having found $D(z)$ we can use eqn (11.4.9) to calculate the memory function $K(t)$. Recall that this memory function was the one defined with the aid of the projection operator onto $\mathbf{p}_0$ alone. We want to see if it decays rapidly or not.

The Laplace transform of K is given by

$$\hat{K} = \mathcal{L}K = \frac{kT}{MD_{SE}}(1 + (\frac{a\rho}{\eta})^{\frac{1}{2}} z^{\frac{1}{2}}) \tag{11.4.23}$$

This is not the complete form of $\mathcal{L}K$; since the calculation of the density–density correlation function was only carried out to order $z^{\frac{1}{2}}$, terms of higher order in (11.4.9) have been neglected.

We can use a Tauberian theorem[25] to determine how $K(t)$ depends on t for long times. From such a theorem, it follows that

$$K(t) \sim -\frac{3(kT)^2 a}{2D_{SE}(\pi\eta/\rho)^{\frac{1}{2}}} t^{-3/2} \tag{11.4.24}$$

as $t \to \infty$.

The conclusion is then that $K(t)$ decays relatively slowly at long times. Furthermore, this slow asymptotic decay does not have a small coefficient that decreases as the mass ratio λ^2 decreases. Therefore the long time tail of K cannot be ignored. Said another way, the diffusion coefficient is not a diffusion constant. It varies with time. Note also from eqn (11.4.23) that$\mathcal{L}K(t)$ has the same form as that proposed by Lorentz (Appendix A) on macroscopic hydrodynamic grounds.[26]

We next inquire about the range of times that we expect these considerations to be valid. We would like them to be valid for time greater than microscopic

[24]Those consulting the original paper should note that what the authors write as η_0 is not the viscosity of the solvent, but the kinematic viscosity.

[25]Tauberian theorems tell how the asymptotic (long time) behavior of a function is characterized by the small z behavior of its Laplace transform (Feller 1966, pp 418 ff.).

[26]The minor difference is that Lorentz gives the Fourier transform while we give the Laplace transform.

collision times, or about 10^{-12} s, in other words for z less than about $10^{12}\mathrm{s}^{-1}$. Recall that we neglected terms of order higher than $z^{\frac{1}{2}}$ in $\hat{K}$. For this to be justified, we must have, at least $z < 6kT/MD)$. Replacing $D(z)$ by D_{SE} for an estimate, we get

$$\frac{6kT}{MD_{SE}} > z \tag{11.4.25}$$

or

$$\frac{36\pi\eta a}{M} < 10^{12}s^{-1} \tag{11.4.26}$$

In case $a(\rho z/\eta)^{\frac{1}{2}} \ll 1$ for such values of z, $\hat{K}$ may be approximated by $(z + (kT/MD_{SE}))^{-1}$. The approximation yields the friction description originally proposed by Langevin. This will hold when $\rho_{med}/\rho_B \ll 1$, as can be deduced from eqn (11.4.22). Again, this is the criterion for Langevin-like behavior found by Lorentz.

To summarize, we have found that the Langevin equation in its original form is not valid for Brownian particles in a dense medium. The Generalized Langevin equation, or Mori equation, can be reduced to a Langevin-like equation with a memory function that is not short ranged in time. In the limit that the Brownian particle is much denser than the ambient medium, the original form of the Langevin equation holds. This will be the case for solid particles suspended in dilute gases, for example.

We have not investigated the random force, $F(t)$, appropriate to the new projection operator. The reason is that there is very little known about it. The correlation function of the random force decays in the same way as the memory function, $K(t)$. This is a consequence of the fluctuation–dissipation theorem. Because the random force is the resultant of a large number of molecular collisions, the central limit theorem suggests that it is safe to assume that it may be treated as a Gaussian random variable. But it is certainly not Markovian; its correlation function is not delta correlated.

11.5 The Distribution Function

Having studied the evolution equation of the dynamical variables via the Mori equation, we now proceed to find the distribution function for the momentum. We shall not call it a Fokker–Planck equation, because that name is reserved by custom for the equation for the transition probability of a Markov process, and we have seen that the momentum is not a Markov process. It does take the place of the Fokker–Planck equation when the generalized Langevin equation must be used. If we start with the generalized Langevin equation, eqn (11.1.3), Adelman (1976) and Mazo (1978) have shown how to get a distribution function equation by purely statistical reasoning. They both assume that the random force is a Gaussian random variable. If we denote the Laplace transform of $\mathbf{p}_0$ by $\pi(z)$, that of $K(t)$ by $\hat{K}$, that of $\mathbf{r}_0$ by $\rho(z)$, and that of $\mathbf{F}$ by $\xi(z)$, eqn (11.1.3) can be rewritten in transform space as

$$z\pi(z) - \mathbf{p}_0(0) = -\hat{K}(z)\pi(z) + \xi(z)$$
$$z\rho(z) - \mathbf{r}_0(0) = \frac{\pi(z)}{M} \tag{11.5.1}$$

We have appended the Laplace transform of $M\dot{\mathbf{r}}_0 = \mathbf{p}_0$. The solution of eqn (11.5.1) is, of course,

$$\pi(z) = \frac{\mathbf{p}_0(0) + \xi(z)}{z + \hat{K}(z)}$$
$$\rho(z) = \frac{\mathbf{p}_0(0)}{z} + \frac{\mathbf{p}_0(0) + \xi(z)}{Mz(z + \hat{K}(z))} \tag{11.5.2}$$

where we have written

$$\chi(z) = \frac{1}{z + \hat{K}(z)}$$
$$\psi(z) = \frac{1}{z(z + \hat{K}(z))} \tag{11.5.3}$$

Notice that $\pi - \mathbf{p}_0(0)\chi$ and $\rho - \mathbf{r}_0(0)\psi$ are linear functions of ξ. This means that their inverse transforms are linear functionals of F. Since F is assumed to be Gaussian, it follows that $\mathbf{p}_0(t) - \mathbf{p}_0(0)w_1(t)$ and $\mathbf{r}_0(t) - \mathbf{r}_0(0) - \mathbf{p}_0(0)w_2(t)$ is a bivariate Gaussian random process. Herew_1 and w_2 are those functions whose Laplace transforms are $\chi$and ξ respectively. This process has zero mean. We denote its covariance matrix by $\mathbf{Q}$. Consequently, if for the sake of brevity we write y_1 and y_2 for the components of the bivariate process just defined, we have

$$P(y_1, y_2) = \pi^{-1} \mid \det \mathbf{Q} \mid^{-\frac{1}{2}} e^{-\mathbf{y}^T\mathbf{Q}^{-1}\mathbf{y}} \tag{11.5.4}$$

The covariance matrix $\mathbf{Q}(t)$ is given by

$$\mathbf{Q} = \frac{3kT}{M}\begin{bmatrix} 1 - \chi^2(t) & \psi(t)(1 - \chi(t)) \\ \psi(t)(1 - \chi(t)) & 2\int_0^t \psi(s)ds - \psi^2(t) \end{bmatrix} \tag{11.5.5}$$

To see where this comes from, we derive the expression for Q_{11}. Q_{11} is $< (\mathbf{p}_0 - \mathbf{p}_0(0)\chi(t))^2 >$, or

$$Q_{11} = \tfrac{1}{M^2}\int_0^t ds \int_0^t ds' \chi(s)\chi(s') < F(s)F(s') >=$$
$$\tfrac{1}{M^2}\int_0^t ds \int_0^t ds' \chi(s)\chi(s')\xi(s - s') \tag{11.5.6}$$

Now we have to put in a physical assumption. If $\mathbf{p}_0(0)$ has a Maxwell, or equilibrium distribution, then $\mathbf{p}_0$ will have a Maxwell distribution for all time. That is, thermal equilibrium is stable. Under this assumption, $< \mathbf{p}_0^2 >=< \mathbf{p}_0(0)^2 >= 3MkT$. It then follows that

$$Q_{11} = \frac{1}{M^2}(< p_0^2 > \chi^2 + < p^2 > -2 < pp_0 > \chi) \tag{11.5.7}$$

from which the Q_{11} element of eqn (11.5.6) follows immediately. The other matrix elements of the covariance matrix can be calculated by similar methods.

Now that we have the two point probability function, it is reasonable to ask, what equation does this function satisfy? It may seem redundant to look for the equation after we have already found its solution, but this is really not the case. We have found only a particular solution, the solution for infinite space that approaches a δ function as $t \to 0$. If want to discuss other initial conditions, or systems with conditions imposed at spatial boundaries, then it is necessary to have the governing equation.

It is most convenient to pass to the characteristic function. The characteristic function of eqn (11.5.4), which we denote by $C(\lambda, \mu)$, is

$$\begin{aligned} C(\lambda, \mu) = \exp[i\lambda \cdot \tfrac{\mathbf{p}_0(0)}{M}\chi + i\mu \cdot (r_0(0) - \tfrac{\mathbf{p}_0(0)}{M}\psi) - \\ \tfrac{1}{2}(Q_{11}\lambda^2 + 2Q_{12}\lambda \cdot \mu + Q_{22}\mu^2] \end{aligned} \tag{11.5.8}$$

It obeys the differential equation

$$\frac{1}{C}\frac{\partial C}{\partial t} = i\frac{\mathbf{p}_0}{M} \cdot (\lambda\dot{\chi} + \mu\dot{\psi}) - \frac{1}{2}(\lambda^2\dot{Q}_{11} + 2\lambda \cdot \mu\dot{Q}_{12} + \mu^2\dot{Q}_{22}) \tag{11.5.9}$$

We should like to eliminate the terms depending on initial conditions, i.e. those depending on $\mathbf{p}_0(0)$. This can be done by noting that

$$\begin{aligned} \frac{\lambda}{C}\frac{\partial C}{\partial \lambda} &= i\lambda \cdot \tfrac{\mathbf{p}_0(0)}{M}\chi - Q_{11}\lambda^2 - Q_{12}\lambda \cdot \mu \\ \frac{\mu}{C}\frac{\partial C}{\partial \mu} &= i\mu \cdot \tfrac{\mathbf{p}_0(0)}{M} - Q_{12}\lambda \cdot \mu - Q_{22}\mu^2 \end{aligned} \tag{11.5.10}$$

From this, it is only a matter of algebra to eliminate $\mathbf{p}_0(0)$ from eqn (11.5.9) obtaining

$$\frac{\partial C}{\partial t} - \mu \cdot \frac{\partial C}{\partial \lambda} = -\frac{\dot{\chi}}{\chi}(-\lambda\frac{\partial C}{\partial \lambda} = \frac{kT}{M}\lambda^2 C) + \lambda \cdot \mu(\frac{\dot{\chi}}{\chi}\psi - 1 - \chi)C \tag{11.5.11}$$

Reverting from the characteristic function back to the probability distribution itself (i.e. from Fourier transform space to phase space) we finally arrive at

$$\frac{\partial P}{\partial t} + \frac{\mathbf{p}}{M} \cdot \nabla_{\mathbf{r}} P = \zeta(t)\nabla_{\mathbf{p}} \cdot (\mathbf{p} + MkT\nabla_{\mathbf{p}})P + kT\sigma(t)\nabla_{\mathbf{p}} \cdot \nabla_{\mathbf{r}\mathbf{P}} \tag{11.5.12}$$

where

$$\zeta(t) = \frac{\dot{\chi}}{\chi}$$
$$\sigma(t) = \zeta\psi - 1 - \chi \qquad (11.5.13)$$

This is the evolution equation satisfied by the distribution function of the random variables described by the generalized Langevin equation. Note that we have derived this under the assumption that the random force was Gaussian though not necessarily Markovian. If the random force is not Gaussian, then the proper form for the evolution equation is not known at the present time.

There are several interesting features of eqn (11.5.12). The term describing the evolution in a spatially homogeneous system has precisely the same structure as in the Fokker–Planck equation for Markov processes. The only difference is that the friction 'constant' is not a constant but is a function of time. A second feature is the term involving the mixed derivatives with respect to space and momentum variables. This has no counterpart in the Markovian Fokker–Planck equation.

Still another feature is that $\zeta(t)$ can vanish and even become negative; there is no prohibition against the mean velocity, conditional on its initial value, having an extremum. It is even possible for χ to vanish, so that ζ jumps from $-\infty$ to ∞. At first glance this seems unphysical for it would require the Brownian particle to approach equilibrium in a damped oscillatory fashion. On the other hand, this is just what happens in the case of the harmonically bound Markovian Brownian particle. Although in this case the friction is assumed constant, the distribution function does approach a Maxwell–Boltzmann distribution in an oscillatory way. A similar phenomenon occurs in the present case for a harmonically bound particle. Adelman (1975) has worked out ζ for a Brownian particle in a lattice with a frequency spectrum given by the Debye continuum model. The explicit formulae worked out for this model show the oscillatory behavior.

For the case of a spatially homogeneous distribution, the term multiplied by $\sigma(t)$ vanishes. eqn (11.5.12) is no more difficult to solve than the ordinary Fokker–Planck equation. First let us suppose that $\zeta(t)$ does not vanish. Then we can introduce a new time

$$\tau = \int_0^t \zeta(s)ds \qquad (11.5.14)$$

and eqn (11.5.12) has exactly the form of the ordinary Fokker-Planck equation with t replaced by τ. If ζ has a zero, than τ is not a one-to-one function of t. The equation must be solved individually in each region in which ζ has constant sign, and the solutions pieced together at the zeros of ζ. Unfortunately, in the spatially inhomogeneous case this simplification is not available.

11.6 Discussion

In this chapter we have discussed to what extent the equations assumed in the stochastic treatment of Brownian motion can be justified in terms of our current knowledge of molecular theory. Our first general conclusion, conjectured first by Lorentz long ago, is that the splitting of the force on the Brownian particle into an *instantaneous* friction and a random force is not valid in dense media. It is necessary to take account of the coupling of the motion of the Brownian particle with the slowly varying (hydrodynamic) modes of the ambient medium. It is quite difficult to take this coupling into account very accurately, for we have to deal with a strongly coupled many body phenomenon. Nevertheless, using the methods of mode coupling theory, we have indicated the modifications that must be made, and the modified Langevin equation that is obtained.

With respect to the evolution equation for the distribution function, the analog of the Fokker–Planck equation, we have arrived at it from the generalized Langevin equation supplemented by the hypothesis that the random force is a Gaussian random variable. This hypothesis is certainly reasonable, because the random force is due to the fluctuations of an enormous number of molecular collisions. The central limit theorem should surely apply. The purist will object that, in a dense medium , the collisions are not all independent; therefore the hypotheses of the central limit theorem are not strictly obeyed. This observation is correct in principle. Nevertheless it is believed to be almost certain, even though not strictly proved, that a sufficient amount of independence remains to ensure the Gaussian nature of the random force.

Before eqn (11.5.12) was derived by Adelman, the fact that the generalized Langevin equation (11.1.3) contained the friction as a retarded kernel, caused some authors to conjecture that the analog of the Fokker-Planck equation would also contain a retarded friction (Mazo 1971, 1973). Indeed, such a retarded friction occurs in the developments of Lebowitz and Résibois (Lebowitz and Rubin 1963; Résibois and Davis 1964; Lebowitz and Résibois 1965). These works, however, emphasized the case when the force correlation on the Brownian particle was short lived, so that the friction kernel could be replaced by a δ function. Then the retardation is irrelevant and the conventional Fokker–Planck equation is obtained.

The equation in question is

$$\frac{\partial P}{\partial t} + \mathbf{p} \cdot \nabla_{\mathbf{r}} P = \int_0^t K(t-s) \nabla_{\mathbf{p}} \cdot (\nabla_{\mathbf{p}} + \frac{\mathbf{p}}{MkT}) P(s) ds \tag{11.6.1}$$

We shall not derive it here since it follows directly from the Zwanzig equation (10.3.4) together with the same assumptions that were made in going from eqn (11.1.5) to (11.1.15). As we have just discussed in much detail in Section 11.5, these assumptions are unjustified. Furthermore, it has been shown in at least one special case that eqn (11.6.1) leads to an incorrect result (Chang, Mazo, and Hynes 1974).

Bocquet and Piasecki (1997) have studied the evolution equation by the methods of modern kinetic theory for the case of a hard sphere Brownian particle in a dense gas of hard spheres. The methods used in the derivation are rather specialized, and we shall not discuss them here. They find a very complicated non-Markovian evolution for the initial stages of the process. If their general evolution equation is specialized to long times, the claim is that eqn (11.6.1) is obtained. However, although the derivation of the general result is detailed and presumably exact, the specialization leading to eqn (11.6.1) is based more on intuitive arguments. At the present time, the question of whether one gets eqn (11.6.1), eqn (11.5.12), or indeed, something still different from the Boquet–Piasecki equation is open. It is even conceivable that the proper equation is model dependent. we shall see in the next chapter that eqn (11.6.1) does not describe properly the motion of a heavy particle embedded in a lattice of light particles. On the other hand, the derivation of Bocquet and Piasecki is for a very anharmonic system—hard spheres.

12

TWO EXACTLY TREATABLE MODELS

12.1 Two Illustrative Examples

The developments presented in Chapter 11 were long, formal, and used many assumptions and approximations. It would be worthwhile to have concrete examples for which we could carry out the computations in detail. This could teach us to what extent the approximations are reasonable and the assumptions are valid, at least in special cases.

There are two such examples available. These are, a Brownian particle immersed in a dilute gas, and a heavy particle linearly coupled to a lattice of harmonically bound light particles. The first of these examples is a model for an aerosol particle in a gaseous medium, and thus has some practical importance. The second could be a model for a heavy substitutional impurity in a crystal, but is more important as a soluble example for a strongly coupled system.

Before doing any calculation, let us ask what should be expected in these two cases? For the first model, the heavy particle in a dilute gas, the parameter $\rho_B/\rho \gg 1$. The coupling of the Brownian motion to the slow hydrodynamic modes of the ambient medium will be minimal, and retrieval of the classical Fokker–Planck equation in the Brownian limit should be anticipated. In the case of the harmonic lattice, the coupling to slow modes of the lattice ought to be appreciable and a generalized Langevin equation for the velocity of the heavy particle should be expected. Both models will give explicit expressions for the friction kernels (friction constant in the former case).

In a very direct sense, the dilute gas example is a modern treatment of the model treated by Smoluchowski in his original paper, discussed in detail in Chapter 4. The case of the harmonic lattice is more recent, having first been treated by Rubin (1960).

12.2 Brownian Motion in a Dilute Gas

We consider a dilute gas of particles of mass m, together with Brownian particles of mass M. The Brownian particles are supposed to be much more dilute than the ambient gas, so that we may completely neglect collisions of the Brownian particles with each other. The dynamics of the system, gas plus Brownian particles, is supposed to be completely described by the Boltzmann equation (Chapman and Cowling 1952)

$$\frac{\partial f_g}{\partial t} + \mathbf{v} \cdot \nabla_{\mathbf{r}'} f_g + \frac{\mathbf{F}}{m} \cdot \nabla_{\mathbf{v}} f_g = n_B J_{gB}(f_g, f_B) + n_g J_{gg}(f_g, f_g)$$

$$\frac{\partial f_B}{\partial t} + \mathbf{c} \cdot \nabla_{\mathbf{r}} f_B + \frac{\mathbf{F}}{M} \cdot \nabla_{\mathbf{c}} f_B = n_g J_{gB}(f_g, f_B) + n_B J_{BB}(f_B, f_B) \tag{12.2.1}$$

Here J_{gB}, J_{BB} and J_{gg} are the collision integrals for gas–Brownian particle, Brownian–Brownian particle, and gas–gas particle collisions respectively. The Boltzmann equation is known to be a valid description of a gas in the limit $L \gg l \gg a$. Here L. is a typical dimension of the container, l the mean free path, and a the diameter of a molecule. Particle –particle collisions dominate particle – wall collisions, and the particles travel relatively long distances on free trajectories between collisions.

Our object is to take the set of coupled integrodifferential equations (12.2.1), eliminate f_g, and derive an equation for f_B alone. This was first done by Wang Chang and Uhlenbeck (1970). The model has since been studied by several other authors.

We have assumed that there are no Brownian–Brownian particle collisions or, alternatively, n_B is so small that the term in eqn (12.2.1) containing J_{BB} can be neglected. Similarly the J_{gB} term in the equation for f_g can be neglected. A given light particle collides with other light particles much more often than with Brownian particles; consequently the contribution of $n_B J_{gB}$ is negligible compared to that of $n_g J_{gg}$ in the equation for $\partial f_g/\partial t$.

Furthermore, the relatively frequent gas–gas particles collisions will drive the distribution function of the gas particles close to equilibrium before any appreciable change takes place in the Brownian particle distribution function. So eqn (12.2.1) can be further simplified by replacing f_g by the equilibrium Maxwell–Boltzmann distribution φ_M.

We assume elastic collisions between the gas molecules. The collision term takes the form

$$J_{gB}(f_B, \varphi_M) = -\int d\mathbf{v}d\mathbf{v}'d\Omega g\sigma(\theta, g)[f_B(\mathbf{c})\varphi_M(\mathbf{v}) - f_B(\mathbf{c}')\varphi_M(\mathbf{v}')] \tag{12.2.2}$$

Here $\mathbf{c}$ and $\mathbf{v}$ are the velocities of the Brownian and gas particle before collision; $\mathbf{c}'$ and $\mathbf{v}'$ are those after collision. $\sigma(\theta, g)$ is the differential cross section for scattering of the relative velocity $\mathbf{g} = \mathbf{v} - \mathbf{c}$ into $\mathbf{g}' = \mathbf{v}' - \mathbf{c}'$. $d\Omega$ is the increment of solid angle of scattering. In some texts $\sigma d\Omega$ will be found written as $bdbd\epsilon$, where b is the impact parameter of the scattering event, and ϵ is the azimuthal angle of the collision. Of course, $\mathbf{c}'$ and $\mathbf{v}'$ must be calculated from $\mathbf{c}$ and $\mathbf{v}$ by the dynamical laws of two body collisions.

As we discussed in Chapter 4, $\mathbf{c}' - \mathbf{c}$ is small when $\lambda^2 = m/M$ is small. Therefore we may expand $f_B(\mathbf{c}')$ around $f(\mathbf{c})$.

$$f_B(\mathbf{c}') = f_B(\mathbf{c}) + (\mathbf{c}' - \mathbf{c}) \cdot \nabla f_B + \frac{1}{2}(\mathbf{c}' - \mathbf{c})(\mathbf{c}' - \mathbf{c}) : \nabla\nabla f_B + \cdots \tag{12.2.3}$$

The laws of elastic collisions require that the velocity of the center of mass of the colliding pair, $\mathbf{V}$, does not change upon collision, and the magnitude of the relative velocity also does not change

$$\begin{aligned}\mathbf{V} &= \mathbf{V}' \\ g &= g'\end{aligned} \tag{12.2.4}$$

Consequently we may extend the integral in eqn (12.2.2) into an integral over $d\mathbf{v}d\mathbf{c}'d\mathbf{v}'$ by inserting δ functions, $\delta(\mathbf{V}-\mathbf{V}')\delta(g-g')$ in the integrand. The collision integral becomes, to order λ^2,

$$\begin{aligned}J_{Bg}(f_B,\varphi_M) = \textstyle\int g\sigma(g,\theta)\delta(\mathbf{V}'-\mathbf{V})\{(\varphi_M(\mathbf{v}')-\varphi_M(\mathbf{v}))f_B(\mathbf{c})+ \\ +\varphi_M(\mathbf{v}')(\mathbf{c}-\mathbf{c}')\cdot\nabla f_B+\tfrac{1}{2}\varphi_M(\mathbf{v}')(\mathbf{c}-\mathbf{c}')(\mathbf{c}-\mathbf{c}'):\nabla\nabla f_B(\mathbf{c})\}d\mathbf{v}d\mathbf{v}'d\mathbf{c}'\end{aligned} \tag{12.2.5}$$

This has terms linear in $f_B, \nabla f_B$, and $\nabla\nabla f_B$, just as does the Fokker–Planck equation. We need only check that the coefficients are correct and we do this term by term.

Consider the first term, proportional to f_B. First change variables to $\mathbf{g}, \mathbf{g}', V$, noting that the Jacobian of the transformation is unity.

$$d\mathbf{v}d\mathbf{v}'d\mathbf{c}' = d\mathbf{g}d\mathbf{g}'d\mathbf{V} \tag{12.2.6}$$

We also expand $\varphi_M(\mathbf{v})$ and $\varphi_M(\mathbf{v}')$ about $\varphi_M(\mathbf{g})$ and $\varphi_M(\mathbf{g}')$ respectively. It is only necessary to take this expansion to terms linear in $\mathbf{v}-\mathbf{g}=\mathbf{c}$. After some manipulation, the term in question becomes

$$\int g\sigma(\theta,g)\delta(\mathbf{V}-\mathbf{V}')\delta(\frac{g^2-g'^2}{2})(\mathbf{c}-\mathbf{c}')\cdot\nabla\varphi_M(g')d\mathbf{g}d\mathbf{g}' \tag{12.2.7}$$

But $\mathbf{c}-\mathbf{c}' = [\lambda^2/(1+\lambda^2)](\mathbf{g}-\mathbf{g}') \approx \lambda^2(\mathbf{g}-\mathbf{g}')$. This is why the expansion of φ only had to be taken to first order. Therefore, to order λ^2, the coefficient of $f_B(\mathbf{c})$ is

$$\lambda^2\int g\sigma(\theta,g)(\mathbf{g}-\mathbf{g}')\cdot\nabla_g\varphi_M(\mathbf{g}')\delta(\frac{g^2-g'^2}{2})d\mathbf{g}d\mathbf{g}' \tag{12.2.8}$$

The second term may be written to order λ^2 using the same technique as for the first term, as

$$\lambda^2\int g\sigma(\theta,g)\delta(\frac{g^2-g'^2}{2})\{\varphi_M(\mathbf{g})(\mathbf{g}-\mathbf{g}')+\mathbf{c}\nabla\varphi_M(\mathbf{g}')\}d\mathbf{g}d\mathbf{g}' \tag{12.2.9}$$

The first term of eqn (12.2.9) vanishes for elastic collisions with spherically symmetrical potentials. The last term in eqn (12.2.5) is already of order λ^2, so we may merely replace $\varphi_M(\mathbf{v}')$ by $\varphi_M(\mathbf{g}')$ in it.

Putting these results together, we finally get

$$\frac{\partial f_B}{\partial t} + \mathbf{c} \cdot \nabla_r f_B + \frac{\mathbf{F}}{M} \cdot \nabla_c f_B = B f_B + \mathbf{B} \cdot \nabla_c f_B + \mathbf{D} : \nabla_c \nabla_c f_B \qquad (12.2.10)$$

where

$$\begin{aligned} \mathbf{B} &= \lambda^2 n_g \int g\sigma(\theta, g)(\mathbf{g} - \mathbf{g}') \nabla_{g'} \varphi_M(\mathbf{g}') \delta(\frac{g^2 - g'^2}{2}) d\mathbf{g} d\mathbf{g}' \\ \mathbf{D} &= \lambda^4 n_g \int g\sigma(\theta, g)(\mathbf{g} - \mathbf{g}')(\mathbf{g} - \mathbf{g}') \varphi_M(g') \delta(\frac{g^2 - g'^2}{2}) d\mathbf{g} d\mathbf{g}' \\ B &= Tr\mathbf{B} \end{aligned} \qquad (12.2.11)$$

These expressions need to be simplified further.

We first note that $\nabla \varphi_M = -M\mathbf{g}'\varphi_M/kT$, because φ_M is the Maxwell distribution. Because σ depends only on the magnitude of the relative velocity, and because of the δ function in the integrand, it is easy to see that $\mathbf{B} = M\mathbf{D}/kT$. Furthermore, because the only vector entering into the final equation is $\mathbf{c}$, and $\mathbf{c}$ does not appear in $\mathbf{D}$, we expect $\mathbf{D}$ to be a multiple of the unit tensor. Thus it is sufficient to evaluate $Tr\mathbf{D}$. This is a 6-fold integral. It is convenient to carry out the integration in polar coordinates using $\mathbf{g}$ as the polar axis for $\mathbf{g}'$. First integrate over the magnitude of $\mathbf{g}'$, and then do the angular integrations over the polar angles of $\mathbf{g}'$. The computation is long, but elementary, and one finds

$$\begin{aligned} \mathbf{D} &= D\mathbf{1} \\ D &= \lambda^4 n_g \frac{8\pi^2}{3} (\frac{m}{(2\pi kT)})^{3/2} \int_0^\infty g^5 e^{-mg^2/2kT} dg \times \\ &\quad \int_0^\pi \sigma(\theta, g)(1 - \cos\theta) \sin\theta d\theta \end{aligned} \qquad (12.2.12)$$

12.3 Discussion

We have obtained a Fokker–Planck equation for the model of a Brownian particle in a dilute gas that is precisely of the form discussed in chapters prior to Chapter 11. There is no trace of the objection raised by Lorentz; the friction coefficient is a constant. This is easy to understand. The model treated is the case $l \gg a$, mean free path much larger than the gas particle size. This can be written in the form $na^3 \ll 1$. Multiplication by M/m puts this in the form $\rho_B/\rho \ll \lambda^2$. But we are also working in the limit $\lambda^2 \ll 1$. Hence ρ_B/ρ is indeed quite small in this regime, and Lorentz's objection does not apply.

It is interesting that in the dilute gas limit, the friction constant, proportional to D, varies as a^2, while in the Stokes' law limit it varies as a. To understand the reason for the difference we note that a hydrodynamic description is valid on a scale large compared to the mean free path. In the dilute gas case, therefore, hydrodynamics does not describe the gas well in the region close to the Brownian particle. We have, in fact, treated the dynamics in this region by kinetic theory rather than hydrodynamics. The resistance to motion should be more or less uniformly distributed over the Brownian particles surface and therefore proportional to its surface area, or a^2.

In the case of a dense medium, where the mean free path (insofar as it is a meaningful concept in a dense medium) is small compared to the Brownian particle size, the distribution of stress over the particle's surface is not uniform. The flow at the poles (with respect to the particle's direction of motion as axis) is highly perturbed, while the flow near the equator is much less affected. Consequently not all elements of the Brownian particles surface contribute equally to the frictional resistance. This is not a convincing argument that the resistance should vary as a but does make it plausible that it need not vary as a^2. Additionally, if the Navier–Stokes equations of hydrodynamics are to be adopted as the description of the system, dimensional arguments show that the resistance must be linear in the radius, a, if it is linear in the velocity.

The derivation of Section 12.1 can be generalized to cover a situation when the Brownian particle has internal degrees of freedom that may exchange energy with the gas at a different rate than do the translational degrees of freedom (Mazo 1975). They may be pumped by an outside energy source. For example, on might think of a soot particle in air with light shining on it. It's temperature will then be different from the temperature of the surroundings. The generalization is based on a phenomenological model for the collision integral, whereas this chapter is devoted to examples that can be studied by basic dynamical principles.

12.4 The Particle Bound to a Lattice

The second model that we shall treat is that of a heavy particle harmonically bound to a harmonic lattice of lighter particles. In this case, the equations of motion are linear and can be solved exactly. We shall not search for the equations satisfied by the distribution function of the model, but shall compute the distribution function directly from the particle equations of motion.

To specify the model precisely, consider a simple hypercubic lattice of particles of mass m, except that one particle, which we denote by the subscript zero, has mass M. Nearest neighboring atoms in the lattice interact with a force that is a linear function of their relative displacement, with force constant k. In addition , each particle is bound to it's lattice point by a similar linear force with force constant k'. This latter force, tying each particle to its lattice point, is a purely formal device, introduced to make the force constant matrix positive definite. Without it, the translations and rotations of the lattice as a whole would complicate the computations without adding anything important to the problem. One may take the limit $k' \to 0$ at the end of the calculation.

The particles are each labeled by an n-tuple of integers $\mathbf{R} = (j_1, \ldots, j_\mathbf{n})$,$-N \leq j \leq N$ for an n-dimensional system. The heavy particle, of mass M has the label

$\mathbf{R} = (0, ..., 0)$. The equations of motion for the x components of the particle motions in the model are

$$(m + (M - m)\delta_{\mathbf{R},\mathbf{0}})\frac{d^2x(\mathbf{R},t)}{dt^2} = -k'x(\mathbf{R},t)$$
$$+\sum_{j=1}^{n} k[x(\mathbf{R}+\mathbf{l}_j,t) - 2x(\mathbf{R},t) + x(\mathbf{R}-\mathbf{l}_j,t)] \tag{12.4.1}$$

The other components are similar. The various components do not affect each other. To solve these equations and determine the velocity correlation function of the heavy particle, we follow the method of Rubin (1960).

We shall compute the velocity autocorrelation function and the velocity distribution function by direct solution of the equations of motion. The fact that makes this feasible is that the equations of motion are linear with constant coefficients. Let us denote by $\mathbf{Z}(t)$ the column vector whose components are $x(\mathbf{R},t)$ and $\dot{x}(\mathbf{R},t)$. Then the solution of equations (12.4.1) is

$$\mathbf{Z}(t) = \mathbf{Y}(t)\mathbf{Z}(0) \tag{12.4.2}$$

where $\mathbf{Y}$ is a time dependent matrix. We shall determine it below from the equations of motion. $\mathbf{Y}$ has two sets of indices; one set specifies the lattice sites $\mathbf{R}$ and $\mathbf{R}'$ connected by $\mathbf{Y}$. The other set specifies whether the matrix element in question connects two position variables (xs), velocity variables ($\dot{x}$s) or a position and a velocity variable.

We shall assume that the distribution of particles at the initial time, 0, is canonical

$$W(x,\dot{x}) = \frac{1}{N}e^{-\frac{1}{2kT}(\dot{\mathbf{x}}\mathbf{M}\dot{\mathbf{x}}+\mathbf{x}\mathbf{V}\mathbf{x})} \tag{12.4.3}$$

where N is a normalization factor, $\mathbf{M}$ is the mass matrix, a diagonal matrix with $M_{00} = M$ and all other diagonal elements equal to m. $\mathbf{V}$ is twice the force constant matrix. Then the velocity correlation function is

$$\langle \dot{x}(\mathbf{0},t)\dot{x}(\mathbf{0},0)\rangle = \int Y_{\mathbf{00},tt}\dot{x}^2(\mathbf{0},0)W(\mathbf{x}(0),\dot{\mathbf{x}}(0))d\mathbf{x}d\dot{\mathbf{x}}$$
$$= \frac{kT}{M}Y_{\mathbf{00},tt}(t) \tag{12.4.4}$$

Because $< \dot{x}(\mathbf{0},0)\dot{x}(\mathbf{R},0) >= 0$ for $\mathbf{R} \neq 0$, and $< \dot{x}(\mathbf{0},0)x(\mathbf{R},0) >= 0$, $Y_{00,tt}$ is the only element of the $\mathbf{Y}$ matrix that enters into the velocity correlation function. The two ts in the subscripts indicate that the matrix element connects $\dot{x}$ at times 0 and t respectively.

The velocity distribution function can be written

$$p(v,t \mid v_0,0) = \frac{\int \delta(v - \dot{x}(\mathbf{0},t))\delta(v_0 - \dot{x}(\mathbf{0},0)W(\dot{\mathbf{x}}(0),\mathbf{x}(0))d\mathbf{x}(0)d\dot{\mathbf{x}}(0)}{\int \delta(v_0 - \dot{x}(\mathbf{0},0)W(\mathbf{x}(0),\dot{\mathbf{x}}(0))d\mathbf{x}(0)d\dot{\mathbf{x}}(0)} \qquad (12.4.5)$$

Of course, $\dot{x}(\mathbf{0},t)$ must be expressed in terms of coordinates and momenta at time 0 through eqn (12.4.2). We shall see how eqn (12.4.5) can be evaluated after we have solved the equations of motion.

Equations (12.4.1) can be solved by the method of Fourier transforms. Multiply each of eqns (12.4.1) by $(2N+1)^{-\frac{1}{2}} \exp[(2\pi i/2N+1)\mathbf{S}\cdot\mathbf{R}]$ and sum over $\mathbf{R}$. $\mathbf{S}$ is also a vector with integral components. Setting

$$G(\mathbf{S},t) = \frac{1}{(2N+1)^{\frac{1}{2}}} \sum_{\mathbf{R}} x(\mathbf{R},t)e^{2\pi i\mathbf{S}\cdot\mathbf{R}/2N+1} \qquad (12.4.6)$$

we obtain for G

$$G_{tt}(\mathbf{S},t) + \frac{Q}{(2N+1)^{n/2}}\ddot{x}(\mathbf{0},t) = -\gamma' G(\mathbf{S},t) - 2\sum_{j=1}^{n} \gamma[1 - \cos\frac{\pi S_j}{2N+1}]G(\mathbf{S},t) \qquad (12.4.7)$$

The new notation is $Q = (M-m)/m, \gamma = k/m, \gamma' = k'/m$. This equation contains, not only G, but also $\ddot{x}(\mathbf{0},t)$, so we must analyze it further.

It is useful to Laplace transform the equation for G. Denoting the Laplace transform of G by $\Gamma(\mathbf{S},p)$ we obtain

$$\begin{gathered}\Gamma(\mathbf{S},p) = G_{tt}(\mathbf{S},0) + pG(\mathbf{S},0) + \tfrac{1}{(2N+1)^{n/2}}Q\times \\ [\dot{x}(\mathbf{0},0) + px(\mathbf{0},0) - p^2\xi(\mathbf{0},p)](p^2 + \gamma' + 2\textstyle\sum_{j=1}^{n}\gamma[1-\cos\frac{2\pi S_j}{2N+1}])^{-1}\end{gathered} \qquad (12.4.8)$$

where $\xi(\mathbf{0},p)$ is the Laplace transform of $x(\mathbf{0},t)$. The term in $\xi(\mathbf{0},p)$ would not appear in a normal lattice vibration problem, where all the masses are the same. It enters here because of the special mass of particle $\mathbf{0}$.

$\xi(\mathbf{0},p)$ can now be extracted from Γ by inverse Fourier transformation. Take the inverse Laplace transform of the result. The algebra is rather long and tedious, and the reader interested in the details is referred to the original paper of Rubin (1960). The important final result is that

$$Y_{\mathbf{00},tt} = \frac{Q+1}{2\pi i}\int_L \frac{p\zeta(\mathbf{0},p)e^{pt}}{1+Qp^2\zeta(\mathbf{0},p)}dp \qquad (12.4.9)$$

where

$$\zeta(\mathbf{0}, p) = \frac{1}{(2N+1)^n} \sum_{\mathbf{S}} (p^2 + \gamma' + 2\sum_{j=1}^{n} [1 - \cos \frac{2\pi S_j}{2N+1}])^{-1} \qquad (12.4.10)$$

L is a vertical line in the complex plane to the right of all singularities of the integrand, a Bromwich contour for the inversion of a Laplace transform.

In the limit of large N, the sum over $\mathbf{S}$ can be turned into an integral of a type familiar in the theory of lattice Green's functions. This integral, and consequently $Y_{\mathbf{00},tt}$, depend on n, the dimension of the system.

We have already shown that the velocity autocorrelation function is proportional to $Y_{\mathbf{00},tt}$. We now want to show that the distribution function for the x-component of the velocity is

$$p(v, t \mid v_0, 0) = [\frac{M}{2\pi kT(1 - Y_{\mathbf{00},tt}^2)}]^{\frac{1}{2}} e^{-\frac{M}{2kT}\frac{(v - v_0 Y_{\mathbf{00},tt})^2}{1 - Y_{\mathbf{00},tt}^2}} \qquad (12.4.11)$$

a Gaussian whose parameters depend only on the velocity autocorrelation function.

To obtain (12.4.11) we start with the definition, eqn (12.4.5) and introduce the integral representation for the delta functions

$$\delta(x - x') = \int_{-\infty}^{\infty} \frac{dk}{2\pi} e^{ik(x - x')} \qquad (12.4.12)$$

Then $\dot{x}(\mathbf{0}, t)$ is expressed as

$$\dot{x}(\mathbf{0}, t) = \sum_{\mathbf{R}} (Y_{\mathbf{0R},tt} x(\mathbf{R}, 0) + Y_{\mathbf{0R},tt} \dot{x}(\mathbf{R}, 0)) \qquad (12.4.13)$$

and substituted into the resulting integral. There are $4N + 4$ integrals left to perform in the numerator of eqn (12.4.5) and $4N + 3$ in the denominator. $4N + 2$ of these integrals in each case are over the x and x_t variables, and the remaining ones are over the auxiliary variables introduced by (12.4.12). The integrals are all simple Gaussian integrals, and there is a great deal of cancellation. A useful formula for carrying out the integration is

$$\int dx_1 \ldots dx_n e^{i\mathbf{t}\cdot\mathbf{x}} e^{-\frac{1}{2}\mathbf{x}\mathbf{A}\mathbf{x}} = \frac{(2\pi)^{n/2}}{|\,\mathbf{A}\,|^{\frac{1}{2}}} e^{-\frac{1}{2}\mathbf{t}\mathbf{A}^{-1}\mathbf{t}} \qquad (12.4.14)$$

where $\mathbf{A}$ is any positive definite matrix. The force constants k' were introduced just to insure that the potential energy matrix and consequently $\mathbf{Y}$ were positive definite ; it may safely be set to zero now that the integration has been performed.

The result for the distribution function is just (12.4.11). This is precisely what one would get by solving the Adelman equation (11.5.11), although the path we have taken to get to it, following Rubin, has been completely different.

12.5 The One-Dimensional Case

The result eqn (12.4.11) is rather formal. In order to see what it really means, one must actually evaluate the velocity correlation function from the contour integral expression (12.4.9). The integrand of the contour integral will depend on the dimension of the system, n, since $\zeta(\mathbf{0}, p)$ depends on n. We consider here only the one-dimensional case. The general idea is the same in all dimensions, but the analysis of the singularities of the integrand in the complex plane becomes more complicated as the dimension increases.

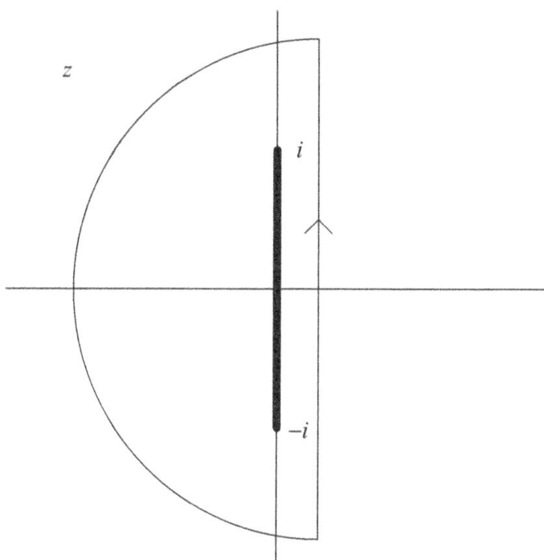

FIG. 12.1. The contour in the complex plane used in the evaluation of eqn (12.5.1). The branch cut has been chosen in a different but equivalent way to that of Rubin (1960).

For one dimension, ζ is given by $\zeta(\mathbf{0}, t) = p^{-1}(p^2 + 4\gamma)^{-\frac{1}{2}}$. Making the change of variable $p = 2\gamma^{\frac{1}{2}} s$, eqn (12.4.9) takes the form

$$Y_{\mathbf{00},tt} = \frac{Q+1}{2\pi i} \int_L \frac{e^{2\gamma^{\frac{1}{2}} st}}{Qs + (1+s^2)^{\frac{1}{2}}} ds \tag{12.5.1}$$

The integrand of eqn (12.5.1) has a pole at $s = -1/(Q^2 - 1)^{\frac{1}{2}}$ and branch cuts at $s = \pm i$. We are here assuming that Q is large and positive, as is appropriate for Brownian motion. A branch cut can be drawn between $-i$ and i as shown in Fig. 12.1. (This is a different, but equally valid, way of choosing the branch cut as that originally chosen by Rubin.) We then obtain

$$Y_{\mathbf{00},tt} = \frac{Q(Q+1)}{Q^2-1} e^{-2\gamma^{\frac{1}{2}}t/Q^2-1} + \frac{Q+1}{\pi} \int_{-1}^{1} e^{2i\gamma^{\frac{1}{2}}yt} \frac{(1-y^2)^{\frac{1}{2}}}{(Q^2+1)y^2+1} dy \qquad (12.5.2)$$

The first term is the contribution from the pole; the second term comes from the integration around the cut.

At short times, the first term dominates. To see this, note that the first term is $(Q^2+Q)/(Q^2-1) \approx 1 + 1/Q$ for large Q. But $Y_{\mathbf{00},tt}$ must be unity at $t = 0$. Hence, at $t = 0$, and by continuity at short times, the first term alone is a reasonable description of the decay of velocity correlations. For long times, however, the integral along the cut yields the dominant contribution. Indeed, an asymptotic analysis of the second term in eqn (12.5.2) shows that this term varies as an oscillating function times $t^{-\frac{1}{2}}$. Because the integrand is not analytic at the endpoints of integration, the analysis cannot be carried out by a straightforward integration by parts, but theorems covering the case of eqn (12.5.2) can be found in the literature (Erdélyi 1956, p. 48; Sirovich 1971, p. 79). Thus the velocity autocorrelation function has a 'long time tail' instead of an exponential decay for long times. The phrase 'long time tail' is jargon for power law decay. This is consistent with our general conclusions in Chapter 11.

We have obtained the velocity autocorrelation function and the velocity distribution function by direct solution of the equations of motion followed by averaging over initial conditions. We have not had to use the Langevin equation or the Fokker–Planck equation. The Langevin equation for this model has been studied by Deutch and Silbey (1971). Since the model is Gaussian but does not have exponential decay, it is not Markovian, and so does not satisfy a Fokker–Planck equation in the strict sense. Chang, Mazo, and Hynes (1974) have shown that an equation of the type of eqn (11.6.1) is inconsistent with the exact results of Rubin. On the other hand, an Adelman equation is consistent with the Rubin results. Rubin (1968) has also studied the autocorrelation in the case of a finite crystal (where the sum in eqn (12.4.10) is not approximated by an integral) and the error caused by such replacement.

Equation (12.5.2) is only valid for the one-dimensional case. In two dimensions, $\zeta(\mathbf{0}, p)$ is proportional to a hypergeometric function of ${}_2F_1$ type. The poles of the integrand of eqn (12.4.9) yield exponentially damped terms that dominate at short times. The contribution from the cut structure of the integrand can be written as a definite real integral (Rubin 1960), but the asymptotic behavior of this contribution has not been studied. A careful study would doubtless show the existence of a long time tail. Rubin (1961) has also studied the three- dimensional case. Again, there is an exponentially damped oscillation in the velocity correlation function that dominates for short times. The asymptotic form of the remainder has not been studied, but again, a long time tail should be expected.

12.6 Discussion

In this chapter we have treated two models simple enough to be treated exactly. The results were unsurprising. They illustrated just what we had been led to expect from the considerations of Chapter 11. For the case of a dilute gas as the ambient medium, the density ratio of Brownian particle to medium is very large, and we expect the classical theory of Brownian motion to describe the model very well. In fact, it does. And there is the added bonus that the theory yields automatically a value for the friction constant.

To be sure, the analysis of the model started from the Boltzmann equation, not the Liouville equation. The Boltzmann equation is well known to be a very accurate description of the dynamics of dilute gases. Its relation to the more fundamental Liouville equation has been well studied, and the approximations and assumptions necessary to pass from the Liouville equation to the Boltzmann equation have nothing to do with the mass ratio or density ratio that characterize the Brownian limit. It is true that long time tail effects appear when one goes beyond the lowest order in density in trying to derive the Boltzmann equation. From the discussion of Chapter 11, one also expects long time tail effects in the description of Brownian motion at higher densities. But, to lowest order in the density the Boltzmann description is valid.

The model of a heavy particle coupled to a harmonic solid is simple enough to be soluble, but yet realistic enough to approximate a real system. Again, the results are what one would expect from the considerations of Chapter 11. The velocity of the heavy particle, considered as a random process, is not Markovian. It is an interesting fact, which we do not derive here, that the two-dimensional random variable $(x(\mathbf{0},t), \dot{x}(\mathbf{0},t))$ for one-dimensional and three-dimensional lattices are Markovian, while for a two-dimensional lattice, it is non-markovian (Rubin, 1961). The physical reason for this difference remains obscure.

The two models discussed in this chapter are the only two cases known to the author where the Brownian motion of a physical system, albeit idealized, can be deduced from fundamental statistical mechanics without essential approximation. Their concordance with the results of the approximate general statistical mechanical theory give us confidence in the latter. This is fortunate, because the applications that we shall discuss in future chapters lend themselves to only approximate or phenomenological treatments.

13

BROWNIAN MOTION AND NOISE

13.1 Limits on Measurement

Every measurement has a certain error associated with it. In some cases the possibility of error is so small that it can be neglected altogether. This occurs mostly in situations involving counting small numbers of objects. But in most cases the possibility of error is ever present. The sources of error must be understood in order to guard against, or minimize the damage done by, the error.

Errors of measurement are usually divided into two categories, systematic and random. Systematic errors include miscalibration of instruments, misalignment due to mechanical wear, poor design, etc. We shall not discuss systematic error in this book. Random error is the error caused by uncontrollable factors whose effects are not reproducible from one repetition of the measurement to another. Random error occurs in many different contexts, including social science surveys and economic data. Here we shall only be interested in random error in measurements made by physical instruments. Random error in physical instruments is often called *noise*; this name is taken from the special case of audio electrical circuits where random currents or voltages manifest themselves as acoustic noise in the literal sense.

Random noise can come from a wide variety of sources. Mechanical vibrations caused by passing moving objects (including the experimenter) are a frequent cause. In electric circuits, the fact that charge is carried by discrete objects, the electrons, gives rise to a random component of current. In electronic amplifiers this is known as *shot noise* because, when converted to an audio signal, it sounds like shot being dropped on a reverberating surface. Electrical disturbances in the atmosphere are also a source of noise in receiving circuits. These sources of random noise are all very important, and must be taken into account in designing accurate measurements. Nevertheless, they do not fall within the subject matter of this book.

Measuring instruments are made of matter. There is one type of random motion, and hence of noise, that is common to all matter. That is Brownian motion. In this chapter we want to study Brownian motion as a fundamental limit to the precision of physical measurements. Fluctuations in general, and Brownian motion in particular, can affect measurements through random effects on the measuring instrument being used or through random variations in the quantity being measured. We shall see examples of both kinds of effects.

13.2 Oscillations of a Fiber

A very common ingredient of measuring instruments is an arrangement in which the quantity being measured is transformed into a torque on a suspended fiber (often quartz). This torque causes the fiber to rotate through an angle, say ϕ, about its axis. Measurement of the angle of twist then gives a measure of the torque, and hence of the original quantity. The angle of twist is very small, but can be magnified many times by mounting a small mirror on the fiber. A beam of light reflected from the mirror can undergo an appreciable linear deflection on a measuring scale resulting from a small rotation of the mirror if the light path between the mirror and the measuring scale is long enough. The setup is sketched in Fig. 13.1.

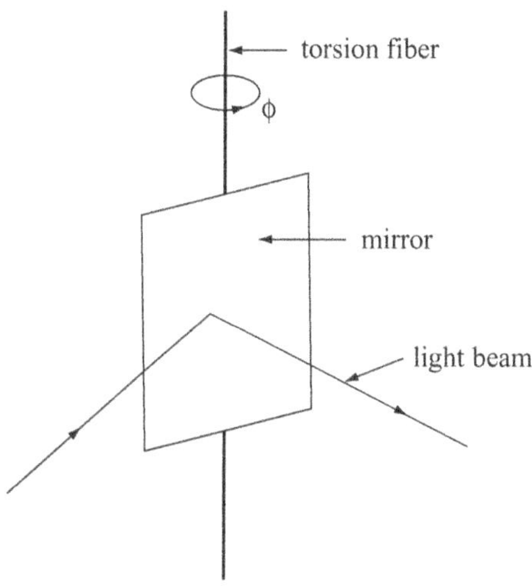

FIG. 13.1. The reflection of a light beam from a mirror mounted on a torsion fiber

The fiber and its associated mirror is, of course, suspended in a gas; this may be the ambient air or it may be the residual gas in an evacuated chamber. Consequently the mirror will be constantly bombarded by gas molecules, and undergo a rotational Brownian motion. For small angles of twist, the potential energy of the twist is assumed to be

$$V(\phi) = \frac{1}{2}b\phi^2 \tag{13.2.1}$$

where b is the torsional constant of the fiber. The system has one degree of freedom.

The fluctuation of the angle ϕ is a measure of the precision of the measurement. From the equipartition theorem alone, without anything more complicated, the mean-square fluctuation in the angle ϕ can immediately be written as

$$\frac{1}{2}b\langle\phi^2\rangle = \tfrac{1}{2}kT$$
$$\langle\phi^2\rangle = kT/b \tag{13.2.2}$$

This conclusion is subject to experimental test. Such a test was carried out by Gerlach and Lehrer (1927). To give an idea of the magnitudes involved, the deflection on the measuring scale was of the order of several centimeters, and $< \phi^2 >$ was measured with a precision of 7 per cent. The result agreed closely with the prediction of eqn (13.2.2).

Kappler (1932) repeated these measurements to try to obtain Avogadro's number, which is proportional to Boltzmann's constant, k, appearing in eqn (13.2.2). In a series of 101 hours of registrations of the light beam deflected from the mirror, a value of Avogadro's number, N, that is about 0.6 per cent different from the currently accepted value was determined.

Measurements were taken at several pressures, ranging from 10^{-4} mm of mercury to 1 atmosphere. The mean-square torsion angle was independent of the pressure, as is predicted by eqn (13.2.1). However, the qualitative appearance of the oscillating fluctuations was considerably affected by the pressure of the experimental run. Examples are shown in Fig. 13.2. This effect is quite understandable qualitatively. If collisions of the fiber plus mirror with gas molecules are very rare (extremely dilute gas), the natural harmonic motion of the mirror will be only seldom perurbed and the damping of the motion small. On the other hand, if the collisions are very frequent, the damping will be large, and the motion will be dominated by the collisions.

This effect can be treated quantitatively by the methods we have discussed in Chapters 3 and 7. The power spectrum of the fluctuations, $G(\omega)$, contains the information of interest. It can be calculated in several ways. One way is to note that the oscillating mirror is mathematically equivalent to a harmonically bound particle since we have assumed the restoring torque to be quadratic in the angular displacement. Therefore the correlation function of the angular displacement can be read off immediately from the second of eqns (7.4.20), merely substituting ϕ for x. From the correlation function $< \phi(t)\phi(0) >$, $G(\omega)$ can be obtained by Fourier inversion, according to eqn (3.7.6). This method, though correct, is tedious and formal.

Another method is to start from the Langevin equation for the mirror

$$I\ddot{\phi} + \zeta\dot{\phi} + b\phi = N(t) \tag{13.2.3}$$

where I is the moment of inertia of the mirror about the fiber axis, ζ is the friction constant of the mirror in the ambient gas, b is the torsional constant of the mirror, and N is the random torque. This is just the same as eqn (7.4.2),

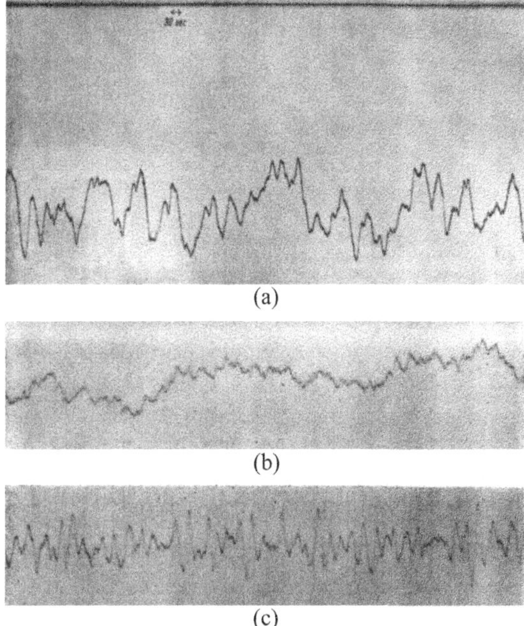

FIG. 13.2. Some of the results of Kappler's measurements of the Brownian motion of a mirror mounted on a torsion fiber. Traces (a) and (c) have similar pressures (4×10^{-3} torr and 1×10^{-3} torr respectively), but different mechanical properties. Traces (b) and (c) have the same mechanical properties, but (b) is at atmospheric pressure.

rewritten so that the names of the variable and the parameters are appropriate for the current problem. Fourier transform both sides of eqn (13.2.3); this yields

$$(-\omega^2 + i\beta\omega + \omega_0^2)\hat{\phi}(\omega) = \frac{1}{I}\hat{N}(\omega) \tag{13.2.4}$$

where $\beta = \zeta/I$, $\omega_0^2 = b/I$, and f and $\hat{N}$ are the Fourier transforms of ϕ and N respectively. Consequently

$$\langle | f(\omega) |^2 \rangle = \frac{1}{I^2}\langle | \hat{N}(\omega) |^2 \rangle \frac{1}{(\omega^2 - \omega_0^2)^2 + \beta^2\omega^2} \tag{13.2.5}$$

We shall assume that the random torque, $N(t)$, has the same statistical properties as the random force we have been studying heretofore. That is, it is a Gaussian Markov process with vanishingly small correlation time. Consequently it is a white noise process and $<| n(\omega) |^2>= 4\zeta kT$. Thus

$$G(\omega) = \frac{4kT\zeta/I^2}{(\omega^2 - \omega_0^2)^2 + \beta^2\omega^2} \tag{13.2.6}$$

Integration of eqn (13.2.6) with respect to ω verifies that $<\phi^2>$ is indeed kT/b, independent of pressure.

The next step is to determine the friction constant, ζ. When the pressure is moderate, so that continuum hydrodynamics holds for the motion of the mirror, the friction constant is given by $\zeta = 32\eta a^3/3$ for rotation about a diameter (Happel and Brenner 1973). Since the viscosity, η, is independent of the pressure for gases under these conditions, so is the friction constant.

At low pressures, on the other hand, the mean free path of the ambient gas is comparable to or larger than the dimensions of the mirror. In this circumstance $\zeta = 2m < c > p\rho I/kT$ (Uhlenbeck and Goudsmit 1929). Here m is the mass of a gas molecule, $< c >$ its mean velocity, p the pressure, and ρ the mass of the mirror per square centimeter. In the low pressure regime, the friction constant is directly proportional to the pressure. Incidentally, exactly the same crossover from a pressure proportional regime to a pressure independent regime occurs for the translational friction constant. It is not surprising that the pressure dependence of translational and rotational should be similar; both effects are due to the same mechanism of molecular collisions.

Equation (13.2.6) shows that, at low pressure, $G(\omega)$ has a maximum near $\omega = \omega_0$. This maximum becomes sharper as the pressure, and consequently ζ, decrease. This is qualitatively in accord with the trend seen in the traces from Kappler's measurements. It would be interesting to make a detailed comparison of the results of Kappler's experiments with eqn (13.2.6). With modern digitizing equipment and correlators, this should be a feasible experiment. Of course, with the tools available in the early 1930s, at the time of the original experiments, it would have been an extremely difficult task.

13.3 A Pneumatic Example

Another application of a similar character is to a pneumatic infra-red detector, also known as a Golay cell (Golay 1949). In this device, infra-red radiation falls on an absorber which is part of the wall of a cell containing a gas. Another wall of the cell is made of a very thin flexible mirror. The absorber heats up on absorbtion of infra-red radiation. There is good thermal contact between the absorber and the gas, so the temperature of the gas increases; consequently its pressure also increases. The pressure increase causes the flexible wall to deform. The deformation is detected by reflecting light from the mirror surface. In order to prevent a gradual overall heating of the cell, the incident infra-red radiation is chopped. A schematic view of such a pneumatic cell is shown in Fig. 13.3.

Clearly random impacts of the flexible surface by gas molecules, both internal and external, will engender random deformations of the surface, and hence random deflections of the probing light beam. Golay asserts that the mean displacement of the surface in working cells with no incident infra-red radiation is of the order of 10^{-10} in. He states that this is 'considerably smaller than the total r.m.s. displacement of the flexible mirror representing the surface tension

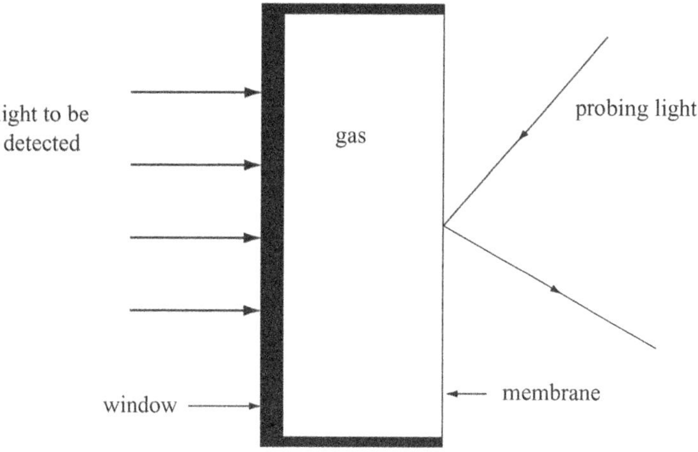

FIG. 13.3. Schematic depiction of the pneumatic infra-red detector, or Golay cell.

energy $\frac{1}{2}kT$ stored in it on the average." It is worth inquiring into the reason for this.

If we treat the flexible mirror as a membrane, the energy is

$$H = \frac{1}{2}\rho \int (\frac{\partial u}{\partial t})^2 dA + \frac{1}{2}\rho c^2 \int (\nabla u)^2 dA \tag{13.3.1}$$

where ρ is the mass density of the membrane, u its displacement, and c the velocity of waves on the membrane, $c^2 = \tau/\rho$,with τ the tension. Let us expand $u(\mathbf{r}, t)$ in normalized eigenfunctions of the operator ∇^2

$$u(\mathbf{r}, t) = \sum q_{m,n}(t)\Phi_{m,n}(\mathbf{r}) \tag{13.3.2}$$

(Since the membrane is two-dimensional, rhere will be two indices to the eigenfunctions.) Substituting this into eqn (13.3.1) yields

$$H = \frac{1}{2}\rho \sum_{m,n} (\dot{q}_{m,n}^2 + \omega_{m,n}^2 q_{m,n}^2) \tag{13.3.3}$$

where ω_{mn}^2 is c^2 times the appropriate eigenvalue.

It follows directly from eqn (13.3.3) and the equipartition theorem that $<| q_{mn} |^2> = kT/\rho\omega_{mn}^2$. As we might expect, low frequency, long wavelength disturbances contribute more heavily to $<| u^2 |>$ than do high frequency, short wavelength distubances. On the other hand, the density of eigenfrequencies increases with frequency as $\omega d\omega$. Thus $<| u^2 |>$ diverges! This is an ultraviolet catastrophe analogous in origin to the ultraviolet catastrophe of the Rayleigh–Jeans law in classical radiation theory. The high frequency, short wavelength behavior is not described correctly by the theory being applied.

In the case of radiation theory it was Planck's introduction of the quantization of radiative energy that saved the day. Nothing so drastic is necessary in the present case. A continuum model, such as the membrane model, might break down at distances of the order of atomic distances. The sum over frequencies should then be cut off at a frequency corresponding to such a wavelength, much as in the Debye theory of specific heat. A cutoff at such a frequency would indeed render the sum finite. However, there is another effect that gives a considerably lower cutoff for the high frequencies and is therefore more important.

The membrane model has no resistance to bending. In reality, however, the sheet of material being described has a finite thickness and cannot be bent with a radius of curvature much smaller than its thickness. Wavelengths shorter than the thickness of the sheet should be ignored. For the case of a vibrating string, the effect of this on Brownian motion has been worked out in detail (Balazs 1959). For a string $<| u^2 |>$ is finite, but $< (\partial u/\partial x)^2 >$ diverges. This is because the density of eigenfrequencies in one dimension varies only as $d\omega$. We can give a similar analysis for the case of a membrane.

For a vibrating membrane with stiffness, the energy can be written as

$$H = \frac{1}{2}\rho \int (\frac{\partial u}{\partial t})^2 dA + \frac{1}{2}\rho c^2 \int (\nabla u)^2 dA + \frac{1}{2} D \int (\nabla^2 u)^2 dA \qquad (13.3.4)$$

with the corresponding equation of motion (Landau and Lifshitz 1986, sec. 14)

$$\rho \frac{\partial^2 u}{\partial t^2} + \zeta \frac{\partial u}{\partial t} = \rho c^2 \nabla^2 u - D \nabla^4 u + F(\mathbf{r}, t) \qquad (13.3.5)$$

where we have added in the friction term, $\zeta \partial u/\partial t$, and the external force per unit area, $\mathbf{F}$, by hand. Note that D is not a diffusion coefficient but is defined by

$$D = \frac{Eh^3}{12(1-\sigma^2)}$$

with E the Young's modulus, σ the Poisson ratio, and h the thickness of the membrane.

The boundary conditions at the edge of the membrane must also be specified. Considering that the membrane closes off a gas filled space, one should reasonably require that it be clamped at the boundary. This means that $u = 0$ and $\partial u/\partial n = 0$ on the boundary; here n is the outward normal to the boundary in the plane of the membrane. These boundary conditions lead to an eigenvalue problem with some algebraic complications. Therefore we adopt the boundary conditions corresponding to a membrane supported at its boundary, $\partial^2 u/\partial n^2 = 0$, which are easier to apply. To be sure, they do not describe quite the same physical system that we originally started to discuss, but this will not affect our conclusions qualitatively.

The displacement, u, can be expanded in eigenfunctions of the operator $\nabla^2 + (D/\rho)\nabla^4$ as in eqn (13.3.2) and the result put in eqn (13.3.4). Equation (13.3.3) is obtained again, but here the eigenvaues are those appropriate to $\nabla^2 + (D/\rho)\nabla^4$. With the boundary conditions adopted, the eigenvalues are

$$\begin{aligned} \omega^2_{m,n} &= c^2 k^2_{m,n} + \frac{D}{\rho} k^4_{m,n} \\ \mathbf{k}_{m,n} &= \pi(\frac{m}{a}, \frac{n}{b}) \end{aligned} \tag{13.3.6}$$

Since $<| q_{mn} |^2>= kT/\rho\omega^2_{mn}$ and

$$\int \langle u^2 \rangle dA = \sum_{m,n} \langle | q_{m,n} |^2 \rangle \tag{13.3.7}$$

we have, for the mean square dispacement averaged over the area of the membrane,

$$\frac{1}{ab} \int \langle | u^2 | \rangle dA = \frac{kT}{ab\rho} \sum_{m.n} \frac{1}{\omega^2_{mn}} \tag{13.3.8}$$

We can estimate the value of the sum by turning it into an integral

$$\sum_{m,n} \frac{1}{\omega^2_{mn}} \rightarrow \frac{ab}{\pi^2} \int \frac{dk_x dk_y}{c^2 k^2 + (D/\rho) k^4} \tag{13.3.9}$$

where the integral is over the first quadrant of the $\mathbf{k}$ plane. The lower limit of the integral is not zero, but k_{min} because the sum begins with $m, n = 1$ not 0. Extending the integral to the origin would lead to an infra red catastrophe that has nothing to do with the physics of the problem, but is merely an artifact of approximating the sum by an integral. With this in mind, the evaluation of the integral is simple.

$$\frac{1}{ab} \int \langle | u^2 | \rangle dA = \frac{kT}{4\pi\rho c^2} \ln[\frac{1 + (D/\rho c^2) k^2_{\min}}{(D/\rho c^2) k^2_{\min}}] \tag{13.3.10}$$

It only remains to give a numerical estimate of this result.

We do not know the elastic constants of the membranes used in practice, so we shall just make reasonable guesses. Fortunately, the final result only depends weakly on these elastic constants since they occur as arguments of a logarithm. We take the Young's modulus as 3×10^{-10} dyne cm^{-2}, $\sigma = .4$. These are reasonable for a polymer. We take $h = .01$ cm, $\rho = .8$ g cm^{-1} and $c = 100$ m s^{-1}. The value of c is the most critical estimate because it occurs squared in the prelogarithmic factor. However, if one should want to use a different value of c, it is easy to do. In first approximation the change of the logarithmic factor may be neglected and the prefactor merely rescaled. We have taken $a = b = 1$ cm and $k^2_{\min} = 10$ cm^{-1}.

Putting all of these estimates into eqn (13.3.10), the mean square displacement averaged over the membrane is about 1×10^{-19} cm^2 so the root mean square displacement is about 3×10^{-10} cm. This number is very close to the experimental number of about 10^{-10} in reported by Golay. The precise value here computed should not be taken too seriously because of the many approximations of both a physical and mathematical nature used in arriving at it, in particular its sensitivity to c. Yet it does indicate that the pneumatic infra-red detector operates at close to its Brownian motion limit.

13.4 Electrical Systems

Fluctuations of current or voltage in electrical systems are the phenomena which gave rise to the name 'noise' for the subject of study in this chapter. The sources of electrical noise that we shall discuss here are due to the random thermal motion of the charge carriers in the electrical system. In the vast majority of cases, these charge carriers are electrons. Thus, in a strict sense, they do not really fall into the category of Brownian motion. The Brownian motion studied in this book is due to collisions of light ambient particles with heavy Brownian particles. Electrons are light particles and electron–electron collisions conserve momentum. Hence they will not affect the total current. Nevertheless, electrical thermal noise is due to fluctuations caused by collisions, collisions of the electrons with the heavy ions of the ambient lattice. This noise can be described by the same mathematical tools used in this book, Indeed, the similarities are so striking that many authors use the term 'Brownian motion' to apply to all mechanisms of random thermal motion. It is thus not out of place to discuss electrical systems in this book, especially in consideration of their practical importance.

Thermal noise due to the motion of electrons in metals is called *Johnson noise*. Its theory was first worked out by H. Nyquist (1928). Nyquist's theorem states that the random voltage generated across a resistor of resistance R at frequency ω is

$$\langle | V(\omega) |^2 \rangle d\omega = \frac{2kTR}{\pi} d\omega \tag{13.4.1}$$

Note that the spectrum of the random voltage is independent of frequency; the noise is 'white'.[27] To derive this result we consider two resistors, R_1 and R_2 connected in parallel. as in Fig. 13.4a (Lawson and Uhlenbeck 1950). The random voltage in resistor 1 delivers power V_1^2/R_2 to resistor 2, and the random voltage generated in resistor 2 delivers power V_2^2/R_1 to resistor 1. The system is in thermal equilibrium, so these powers must be equal. Consequently, since the resistances R_i are arbitrary, it follows that V_i^2/R_i is independent of the resistance of resistor i.

[27] Many authors give this result with a numerical factor of 4 instead of $2/\pi$. This is because they work with the spectrum per unit frequency interval $d\nu$ instead of per unit angular frequency interval $d\omega = 2\pi d\nu$.

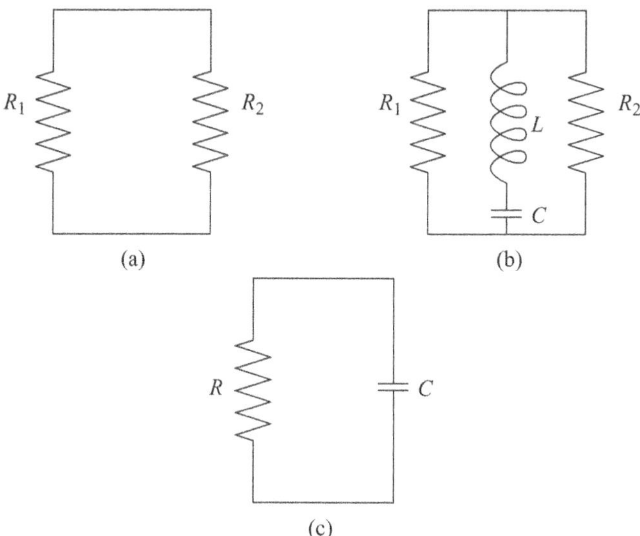

FIG. 13.4. The various electrical circuits used in the text for the discussion of the Nyquist theorem.

The resistors in the arrangement above must not only transfer equal amounts of power, on the average, but must also transfer equal amounts of power at each frequency. To see this we insert a loss free LC filter between the two resistors, as shown in Fig. 13.4b. This will reduce the power supplied to either of the resistances at the resonant frequency of the filter. This reduction must be the same in both directions; otherwise one would be able to violate the second law of thermodynamics. Therefore we must have

$$\langle | V(\omega) |^2 \rangle d\omega = Rf(T, \omega) d\omega \tag{13.4.2}$$

where f is some function, yet to be determined, that depends only on the temperature T and the frequency, ω.

To determine the function f we consider still a third circuit, that shown in Fig. 13.4c. Connect a lossless condenser across the resistor R. The energy of the condenser of capacity C is

$$E = \frac{1}{2} V_C^2 C \tag{13.4.3}$$

where V_C is the voltage across the condenser; it is, of course, also a fluctuating quantity and differs from V, the voltage generated by the resistor because of the voltage drop in the resistor itself. The principle of equipartition of energy ($\frac{1}{2}kT$ per quadratic degree of freedom) requires that

$$\langle E \rangle = \frac{1}{2} \langle V_C^2 \rangle C = \frac{1}{2} kT \tag{13.4.4}$$

Elementary circuit theory relates $V(\omega)$ to $V_C(\omega)$:

$$\langle V_C^2(\omega)\rangle = \langle V^2(\omega)\rangle \frac{1}{1+(\omega RC)^2} \tag{13.4.5}$$

Integrating over all freqencies, using eqn (13.4.2) for $< V^2(\omega) >$, we have

$$\frac{C}{2}\int_0^\infty \langle V_C^2(\omega)\rangle f(T,\omega)d\omega = \frac{RC}{2}\int_0^\infty \frac{f(T,\omega)d\omega}{1+(RC\omega)^2} = \frac{1}{2}kT \tag{13.4.6}$$

or

$$\int_0^\infty \frac{f(T,\omega)d\omega}{1+(RC\omega)^2} = \frac{kT}{RC} \tag{13.4.7}$$

Equation (13.4.7) can only be true if $f(T,\omega)$ is a constant, independent of ω; integration shows that this constant must be $2kT/\pi$.[28]

The discussion given here follows that of Wilson (1952; pp. 123 *et seq.*). The original derivation by Nyquist used a different method. The two resistors were coupled by a lossless transmission line, impedance matched to the two loads. The equipartition theorem was then applied to the energy stored in the transmission line at equilibrium in order to determine the function f. The method presented here has been adopted to avoid the necessity of knowledge of transmission line theory. The derivation of Nyquist has one advantage, however. The object generating the fluctuating voltage need not be merely a resistor. It can be, in general, an impedance. Then eqn (13.4.1) still holds if R is interpreted as the real part of the impedance, Z.

Of course, the noise voltage cannot have a constant spectrum up to arbitrarily high frequencies. On the one hand, at high frequencies, the electrical resistance of matter becomes frequency dependent. There are also quantum mechanical effects that modify the equipartition theorem at high frequencies ($\hbar\omega/kT \geq 1$). These limitations on the validity of eqn (13.4.1) are of no importance in most experimental situations.

Now refer again to Fig. 13.4c. If q denotes the charge on the condenser, then the equation describing this circuit, by Kirchoff's laws, is

$$R\frac{dq}{dt} + \frac{1}{C}q = V(t) \tag{13.4.8}$$

We have just seen that V is a random variable with white spectrum. Because it is the result of the uncoordinated motion of many electrons, we may reasonably assume that it is Gaussian. Therefore eqn (13.4.8) is a Langevin equation of precisely the form that we have studied extensively in previous chapters. In a similar manner, if we substitute an inductance, L, for the capacitance, C, in Fig. 13.2c, the circuit equation becomes

[28]It is clear that $f = 2\pi kT/R$ does indeed satisfy eqn (13.4.5). It is not so obvious that it is the only function (under very mild smoothness conditions) that satisfies the equation. However, it is true, and follows from the uniqueness theorem for Stieltjes transforms.

$$L\frac{di}{dt} + Ri = V(t) \tag{13.4.9}$$

Again, this is a Langevin equation of precisely the same form we have become accustomed to. Consequently , all of the conclusions that we have previously drawn for Brtownian motion can be applied to random charges and currents in thermal equilibrium. The only caveat is that our entire discussion has been restricted to linear circuits. Fluctuations of non linear systems, whether mechanical or electrical, bring in new considerations which are beyond the scope of this book (Van Kampen 1965).

These concepts can be illustrated by considering the current fluctuations in the coil of a galvanometer. In a galvanometer, there is damping due to the resistance, and consequently the system has a certain response time, τ. The mean current, averaged over this response time is

$$j = \frac{1}{\tau}\int_0^{\tau} idt \tag{13.4.10}$$

Since i is a random process, so is j. Now

$$\langle j^2 \rangle = \frac{1}{\tau^2}\int_0^{\tau}\int_0^{\tau} \langle i(t)i(t')\rangle dtdt' = \frac{2 < i^2 >}{\tau^2}\int_0^{\tau} dt \int_0^{t} dt' e^{-R/L(t-t')} \tag{13.4.11}$$

We have used eqn (5.1.2), with appropriate change of notation, to determine the correlation function of the current. Assuming that the observation time, τ, is long compared to L/R, the damping time, the integral is just $L\tau/R$. Thus $< j^2 >= 2kT/R\tau$.

To get an order of magnitude of the current fluctuations in the galvanometer, assume $R = 100\Omega$, $\tau = 2$ s, $T = 290$ K. Then $< j^2 >^{\frac{1}{2}} = 6.3 \times 10^{-12}$A. If we assume that the minimum current one can measure accurately is about four times the noise, then this minimum current is about 10^{-11}A. This has been verified experimentally.

13.5 Discussion

There are many other examples of mechanical and electrical instruments that are affected by thermal noise. We have attempted to give the main ideas behind the theory and a few illustrative examples. The main theoretical tool is quite simple; it is the theorem of equipartition of energy at equilibrium. Of course, to apply the theorem one must also know the theory of operation of the device in question. This is special to each device and no general recipe can be given.

Further examples of the limits on the precision of measurements caused by Brownian motion are treated in a review article by Barnes and Silverman (1938). The first person to have pointed out this limit and studied several examples was G. L. deHaas-Lorentz (1913), but she gave no numerical examples. The first study that showed in numerical detail that Brownian motion was the ultimate experimental limiting factor was by G. Ising (1926). He analyzed data on the fluctuations of the zero point of a galvanometer, ascribed by the original authors to microseismic causes, and showed that they fit very well the predictions of Brownian motion theory.

We mentioned early in this chapter that fluctuations due to Brownian motion, though always present, are not the only fluctuations present in physical systems. Indeed, unless special care is taken to eliminate other causes of random variation, Brownian motion may not be the most important source of noise. The general theory of extraction of signals from noise is a large one. There are many textbooks on the subject, mostly written from the perspective of communications engineering. Two articles that treat the subject from the viewpoint of experimental physics are (MacDonald 1948) and (McCombrie 1953), in addition to the Barnes and Silverman article referred to above.

14

DIFFUSION PHENOMENA

14.1 Brownian Motion in Configuration Space

In the previous chapters of this book we have studied mostly Brownian motion in the phase space of a single Brownian particle. The basic equations were the Fokker–Planck equation or the Langevin equation. Many of the experimentally measurable aspects of Brownian motion, however, only probe the position of the Brownian particle; they only depend on its behavior in configuration space. The full phase space description is not needed to describe these experiments. The Einstein theory described in Chapter 4 and the Smoluchowski theory described in Chapter 8 were, in fact, configuration space descriptions.

The most obvious behavior of this type is diffusion. Einstein's theory is essentially a theory of the diffusion of independently moving particles in space. Einstein's equation, eqn (4.4.2), is just the well known diffusion equation. Its *fundamental solution* or *source function* in n dimensions is

$$\frac{1}{(2\pi Dt)^{n/2}}e^{-(\mathbf{r}-\mathbf{r}_0)^2/2Dt} \tag{14.1.1}$$

The properties of this equation, variously called the heat equation or the diffusion equation, and methods for its solution under a wide variety of boundary conditions are commonly treated in textbooks on partial differential equations; we shall not go into them here.

Another example of a configuration space description is eqn (8.2.7), Smoluchowski's equation. This equation governs the diffusion of independent particle under the influence of an external field of force. For certain force fields, the equation is relatively easy to solve. Such fields include the constant field and the harmonically bound particle. For others, it can only be solved approximately. Again, we shall not be concerned in this book with specific solution techniques.

This chapter will discuss some of the applications that have been made of these configuration space considerations to cases of physical and chemical interest.

14.2 Diffusion Controlled Reactions

The rate of chemical reactions in condensed phases is often determined by the rate at which the reacting species get close enough together to react. Once they get within some critical distance of each other, the making and breaking of chemical bonds takes place rapidly. The process by which the reactive particles reach that critical distance is the Brownian motion of the particles themselves. Such reactions are called *diffusion controlled* reactions since the diffusive Brownian motion of the species is the rate determining step.

The theory of the rates of such diffusion controlled reactions was first developed by Smoluchowski (1916, 1917) though the problem he actually treated was that of colloid coagulation. We shall describe his methods first for ordinary chemical reactions, which are simpler, and then show how they can be extended to coagulation.

There are two reacting species, A and B, in solution. At first we consider species B to be stationary and species A to diffuse with diffusion coefficient D_A. When an A particle comes within a distance R of a B particle, the particles react and effectively disappear from the problem. For simplicity the particles are assumed to be spherical. For the time being, we shall assume that there are no forces between the A and B particles when they are farther apart than R. We ask for the rate at which A particles disappear.

Let the concentration of A and B particles be n_A and n_B respectively. The rate at which A particles disappear is then n_B times the flux of A particles across the sphere of radius R surrounding a B particle. If $\mathbf{j}$ be the current density of A particles this can be symbolized as

$$rate = -n_B \int \mathbf{j} \cdot d\mathbf{A} = 4\pi R^2 n_B j(R,t) \tag{14.2.1}$$

The integration is over the surface of a sphere of radius R about the B particle. Since there are no forces between the A and B particles if their mutual distance exceeds R, the current, $\mathbf{j}$, is given by

$$\mathbf{j} = -D_A \nabla n_A \tag{14.2.2}$$

Therefore we have to determine n_A near the surface of the sphere.

To do so we must solve the diffusion equation

$$\frac{\partial n_A}{\partial t} = D \frac{1}{r^2} \frac{\partial}{\partial r} (r^2 \frac{\partial n_A}{\partial r}) \tag{14.2.3}$$

with the initial condition $n_A(r,t) = n_A^0$, a constant, and the boundary conditions

$$\begin{aligned} &(a)\ n_A \to n_A^0 \qquad r \to \infty \\ &(b)\ n_A = 0 \qquad r = R \end{aligned} \tag{14.2.4}$$

Condition (a) requires that the concentration very far away from a sink (B particle) is not perturbed by the sink. Condition (b) insures that the sphere of radius R is indeed a sink. The A particle reacts, i.e. is no longer an A particle the instant it encounters the sphere. We have assumed that the distribution of A particles is spherically symmetrical about the B particle.

The solution to eqn (14.2.4) can be found as follows. First define the new independent variable

$$w(r,t) = rn_A(r,t) \qquad (14.2.5)$$

w satisfies the one-dimensional diffusion equation

$$\frac{\partial w}{\partial t} = D\frac{\partial^2 w}{\partial r^2} \qquad (14.2.6)$$

The boundary conditions are that w vanishes at R and that w is asymptotic to $n_A^0 r$ as r approaches infinity. The solution to (14.2.6) that satisfies these conditions can be constructed from the fundamental solution by the method of images.

At each point $r_0 > R$, we place a point source given by eqn (14.1.1) for one dimension. This should not be a unit point source however, but should be weighted by the initial condition at that point, $w^0 = n_A^0 r_0$. This expression must be integrated over all points $r_0 > R$. The quantity so constructed is a solution of eqn (14.2.6) that satisfies the initial condition. It does not, however, satisfy the boundary condition at R.

We can construct another solution of the diffusion equation by integrating over sources placed to the left of R, i.e. $s < R$, with weights chosen so that the weight at $s = 2R - r_0$ is just $n_A^0 r_0$. This guarantees that the weights at points at equal distances to the right and left of R are the same. This solution also satisfies the differential equation in the region $r > R$ and its initial value is zero there. This solution also does not satisfy the boundary condition at R.

The values of the two solutions just constructed are, however, just equal at R. Consequently their difference will vanish at R and will satisfy the differential equation and the initial condition in the region of interest, $r > R$. Reverting from w to n_A, we may thus write

$$\begin{aligned} n_A(rt) &= \frac{w(r,t)}{r} \\ &= \frac{n_A^0}{(4\pi Dt)^{\frac{1}{2}}}[\int_R^\infty e^{-(r-r_0)^2/2Dt} r_0 dr_0 \\ &- \int_R^\infty e^{-(r+r_0-2R)^2/2Dt} r_0 dr_0] \end{aligned} \qquad (14.2.7)$$

Elementary but tedious manipulations transform this into

$$n_A = n_A^0(1 - \frac{r}{R} + \frac{r}{R}\frac{2}{\sqrt{\pi}}\int_0^{(r-R)/(4Dt)^{\frac{1}{2}}} e^{-\xi^2} d\xi) \qquad (14.2.8)$$

Combining eqns (14.2.1), (14.2.2), and (14.2.8) gives, finally

$$\text{rate} = n_A n_B 4\pi DR(1 + \frac{R}{(\pi Dt)^{\frac{1}{2}}}) \qquad (14.2.9)$$

Thus the rate constant, the coefficient of $n_A n_B$, is not really a constant; it is time dependent.

Let us estimate the size of the time dependent term. If we assume $R = 2\times 10^{-7}$ cm, $D = 10^{-5}$ cm^2 s^{-1}, then the second term is less than one percent of the first one for times longer than 10^{-5} s. Consequently the time dependent correction is of little interest experimentally unless special fast reaction techniques are being used for the observations.

Up to now we have assumed that the B particles are fixed; in reality, they move just as do the A particles. Our calculation may be revised to take this into account in a very simple way. If two particles are diffusing independently of each other, then the joint probability is the product of the probability distributions for the individual particles. If $n_A(\mathbf{r}, t)$ is the distribution for particle A and $n_B(|\mathbf{r}+\mathbf{s}|, t)$ that for particle B, both of these are Gaussian functions (compare eqn (14.1.1)). Then the probability distribution of s is obtained by integrating their product over $\mathbf{r}$,

$$n_{AB}(s,t) = \int d\mathbf{r} \frac{e^{-r^2/4D_A t}}{(4\pi D_A t)^{3/2}} \frac{e^{-|\mathbf{r}+\mathbf{s}|^2/4D_B t}}{(4\pi D_B t)^{3/2}} \tag{14.2.10}$$

We have denoted the probability density for a pair of particles, an A and a B, to be a distance s apart by $n_{AB}(s,t)$. This integral is easily evaluated; it is

$$n_{AB} = \frac{1}{(4\pi(D_A + D_B)t)^{3/2}} e^{s^2/4(D_A+D_B)t} \tag{14.2.11}$$

Consequently eqn (14.2.10) is corrected to take account of the Brownian motion of B merely by replacing D_A by $D_A + D_B$.

The main result, eqn (14.2.9) without the time dependent term may be obtained more simply but less rigorously by assuming a steady state distribution of A particles about a B particle. This distribution, arising from solution of $\nabla^2 n = 0$, is $n = n^0(1 - R/r)$. Evaluating the current, $\mathbf{j}$, again yields the time independent part of eqn (14.2.9).

It should be emphasized that this model assumes that once two particles come within the critical distance of each other, they react immediately. If this is not the case, then the observed rate will be slower than the rate here computed. In other words, we have obtained an upper bound for the rate. The critical distance, R, is often assumed to be the sum of the radii of the reacting molecules, $R = R_A + R_B$, but this is not essential for the development of the theory. The Smoluchowski theory can be generalized to take account of a finite rate of reaction once the particles come together (Collins and Kimball 1949) by using partially absorbing rather than totally absorbing boundary conditions. Since the diffusion process and the reaction process proceed sequentially, their lifetimes should be additive, and to obtain for the observed rate constant, k, should be

$$\frac{1}{k} = \frac{1}{k_{int}} + \frac{1}{k_{Smo}} \tag{14.2.12}$$

where k_{int} is the intrinsic rate constant and k_{Smo} is the Smoluchowski rate constant. This is precisely what the detailed calculations yield.

Diffusion control is very common for reactions in condensed phases. The ideas presented in this section have therefore been very useful in the development of chemical kinetics. They have been elaborated upon and refined, and their chemical basis discussed in detail. To elaborate on these developments would take us too far from our primary subject of Brownian motion; yet we cannot ignore them altogether.

If we use the Stokes–Einstein relation $D_{\mathrm{j}} = kT/6\pi\eta R_j$ for the diffusion coefficients of the various species, then we can write the rate constant as

$$k = 4\pi(D_A + D_B)R = \frac{2kTR}{3\eta}(\frac{1}{R_A} + \frac{1}{R_B}) \tag{14.2.13}$$

If we further make the common assumption that $R = R_A + R_B$, we arrive at

$$k = \frac{2kT}{3\eta}\frac{(R_a + R_B)^2}{R_A R_B} \tag{14.2.14}$$

Assuming that the effective radii of the reacting species are independent of temperature, it follows that the activation energy of the rate constant, k, is the same as that of the fluidity, $1/\eta$. Since viscosities are more easily measurable than diffusion coefficients, this provides a convenient way to assess the agreement of the theory with experiment.

Reactions for which this simple theory may be expected to be adequate include free radical reactions, fluorescence quenching by heavy atoms, and reactions of molecular fragments generated by photolysis. The predictions of the theory are in general qualitative agreement with experiment, and the agreement becomes better when the correction for a finite intrinsic rate is taken into account. The interested reader is referred to a review by Noyes (1961) and a monograph by Rice (1985) for further discussion of elaborations of the theory and comparison with experiment.

14.3 The Effect of Forces

In the last section we treated the diffusion of free particles. There were no interactions between the particles until they came in contact, and then they disappeared. There are important cases, however, where the interactions between the particles are important and cannot be neglected. One class of such cases is the reaction between charged species: ions in solution. Here the Coulomb force between the ions plays an appreciable role. In principle, the way to proceed in these cases is straightforward. Instead of starting with the ordinary diffusion equation, eqn (14.2.3), start with the Smoluchowski equation

$$\frac{\partial n}{\partial t} = D\nabla \cdot (\nabla n - \frac{1}{kT}Kn) \tag{14.3.1}$$

This is eqn (8.2.7) written for three dimensions instead of the one-dimensional notation of Section 8.2. The boundary conditions are the same and the method of computing the rate from the solution of eqn (14.3.1) is the same as that already used in the last section.

The difficulty is that eqn (14.3.1) is very hard to solve when $\mathbf{K} \neq 0$ so that, in general, we cannot write down a specific formula analogous to eqn (14.2.8) and find the current from it. On the other hand, as we remarked in the last section, there is a simpler, though less rigorous way to get the steady state current. By using this method we get the asymptotic rate constant, but give up all information on the time dependent terms in the rate 'constant.' This was first done by Debye (1942)

In the steady state, eqn (14.3.1) can be integrated immediately once to yield

$$\frac{\partial n}{\partial r} + \frac{n}{kT}\frac{\partial U}{\partial r} = \frac{c_1}{r^2} \tag{14.3.2}$$

Here, U is the potential of the force $\mathbf{K}$ and c_1 is a constant to be determined. This equation states that the total flux of A through a sphere of radius r is independent of r in the steady state. Equation (14.3.2) is a linear first order inhomogeneous differential equation of standard type. Its solution is

$$n = c_1 e^{-U(r)/kT} \int_r^\infty e^{U(y)/kT} \frac{dy}{y^2} + c_2 e^{-U(r)/kT} \tag{14.3.3}$$

The requirement that $n \to n_A^0$ as $r \to \infty$ requires that $c_2 = n_A^0$; the condition $n(R) = 0$ requires that

$$c_1 = \frac{-n_A^0}{\int_R^\infty e^{U/kT} \frac{dy}{y^2}}$$

Consequently the total flux over the surface at distance R is

$$\text{flux} = n_A^0 4\pi (D_A + D_B) \frac{1}{\int_R^\infty e^{U(y)/kT} \frac{dy}{y^2}} \tag{14.3.4}$$

Here we have already replaced D by $D_A + D_B$ for the same reason described in the previous section.

Thus it is clear that the rate constant is the same as the rate constant given by eqn (14.2.13) except for the replacement of the factor R by $[\int \exp(U/kT)dy/y^2]^{-1}$. When $U = 0$ the integral is easily evaluated, and eqn (14.2.13) is recovered. One can, of course, further assume that $R = R_A + R_B$ as before. We may also use the Stokes–Einstein law to replace the diffusion coefficients by the radii of the particles and viscosity of the solvent, as in eqn (14.2.14).

Now suppose the reacting species, A and B are ionic species with charges z_A and z_B respectively. The concentration should be so low that the interaction U can be described adequately by the bare Coulomb force between the ions

$$U(r) = \frac{z_A z_B e^2}{\epsilon r}$$

where ϵ is the dielectric constant of the solvent. This will describe the interionic interactions in the system only in the case of very dilute solutions where Debye screening is negligible. The integral in eqn (14.3.4) can be evaluated in closed form

$$\int_R^\infty e^{z_A z_B e^2/\epsilon kTy} \frac{dy}{y^2} = \frac{z_A z_B e^2}{\epsilon kT}[e^{z_A z_B e^2/\epsilon kTR} - 1]$$

If we set $R = R_A + R_B$, then the formula (14.3.4) becomes

$$\text{flux} = n_A^0 f / \frac{(R_A + R_B)^2}{4R_A R_B} \frac{8kT}{3\eta}$$

$$f = \frac{z_A z_B e^2}{\epsilon kT} \frac{1}{R_A + R_B}[e^{z_A z_B e^2/\epsilon kT(R_A+R_B)} - 1]^{-1} \tag{14.3.5}$$

The quantity $z_A z_B e^2/\epsilon kT$ has the dimensions of length; it is called the Onsager length. It is the distance at which the Coulombic potential energy of a pair of ions equals their mean kinetic energy.

The factor f expresses completely the difference between the case of ionic reactants and neutral reactants in very dilute solution. The Onsager length for uni-univalent ion pairs in water at room temperature is about 7×10^{-8} cm. If we assume that $R_A + R_B$ is 5×10^{-8} cm, then f is approximately 1.9 for ions of opposite charge and 0.5 for ions of the same sign. Thus, for the conditions quoted, the charges enhance or diminish the rate constant of neutral species by roughly a factor of two. This is to be expected qualitatively. Particles of like sign attract each other, and will come together faster than if there were no interaction. Particles of unlike sign repel each other, and will come together more slowly.

These results only apply to exceedingly dilute solutions. In the more realistic case of solutions where screening begins to become important and the Debye–Hückel limiting law holds, we may use the potential of mean force due to screening for U

$$U(r) = \frac{1}{2} \frac{z_A z_B e^2}{\epsilon kT}[\frac{e^{\kappa R_A}}{1 + \kappa R_A} + \frac{e^{\kappa R_B}}{1 + \kappa R_B}] \frac{e^{-\kappa r}}{r} \tag{14.3.6}$$

where $\kappa = (4\pi e^2/\epsilon kT) \sum n_j z_j^2$ is the inverse of the Debye length, the mean thickness of the ionic atmosphere. This formula should hold at least for large r. But the main effect of the screening comes at large r, where the exponential factor decreases the effective strength of the Coulomb potential. Consequently we can use it to estimate the size of the effect.

Although the integral in eqn (14.3.4) cannot be evaluated exactly in terms of elementary functions, it may be evaluated numerically if desired, or evaluated by some other approximation method. The approximation given by Debye is equivalent to setting $\kappa = 0$ in $\exp(-\kappa r)$ in eqn (14.3.6), but retaining it elsewhere. The net result is to decrease the effect of ionic interaction on the rate constant. The screening due to the collective effect of all of the other ions in solution modifies the interaction between ions, screening and this decreasing it. Consequently the factor f will become closer to unity than it would be in the absence of screening.

Since the kinetics of reactions in solution is an important field, and since diffusion control is not unusual for such reactions, the Smoluchowski theory has been much used by workers in this field. It has worked very well and been very useful, but it is perhaps not surprising that extensions and improvements of the theory have been sought. Some of these have been, the nature of the proper boundary conditions, the effects of rotation (the reactive sites on a molecule are not normally spherically symmetrically distributed), the effects of initial conditions (reactive species need not have been produced homogeneously), and many body effects. It would take us too far afield to discuss these here, and the interested reader is referred to the monograph by Rice (1985) for a thorough discussion of these and other topics.

We should like, however, to emphasize that the Smoluchowski theory discussed so far does not take into account the fact that B molecules, say, are distributed throughout the system. They are depleting the A concentration in their neighborhoods by the same mechanism as discussed here for a specific B particle. Therefore the A particle concentration is changing because of chemical reaction far from the B particle under consideration. This will affect the diffusion rate and the flux of As at the position of the chosen B. *Mutatis mutandis*, the B concentration is affected similarly.

This has been studied by Grassberger and Procaccia (1982) and by Dunlap *et al.* (1994). These authors studied the survival probability of A particles in a random distribution of static traps. In a random distribution, there will be large, albeit rare, volumes devoid of traps. After long times, most of the A particles not in these regions will have been trapped, and the contribution of the rare large regions will dominate the survival probability and the mean-square displacement of particles. The authors quoted compute, in fact, averages of different quantities, but are in agreement on the underlying mechanism involved. The long time behavior is not diffusive.

Grassberger and Procaccia state, however, that if the traps are themselves mobile (as we have assumed in this and the preceding section), then diffusive asymptotic behavior will be established for long times. No demonstration of this is provided. It is presumably based on the idea that if the traps can diffuse, the trap-free volumes will eventually fill up with traps, and new trap-free volumes created. But these new trap-free spaces will have already had their A concentration reduced. This is a subtle effect that cannot be studied through the Smoluchowski equation, which treats the average concentration of the A particles. The effect depends on rare large fluctuations about this average.

14.4 The Coagulation of Colloids

The so-called Smoluchowski theory of diffusion controlled reactions was not, in fact, presented as a theory of a class of chemical kinetic phenomena by Smoluchowski. He was interested in providing an explanation of the observations of Zsigmondy (1917) on the coagulation of colloidal solutions of gold when electrolyte is added to the solution.

The colloid particles have ions adsorbed on their surfaces. They therefore repel each other and the colloid is stable. On addition of electrolyte, a double layer which screens the repulsion of the charges is formed. The particles begin to coagulate at a rate depending on the concentration of electrolyte. If the electrolyte concentration is changed, a concentration is reached at which the coagulation occurs rapidly. Addition of more electrolyte does not further increase the coagulation rate. What was measured was the total concentration of colloid particles (of all sizes) as a function of time after the electrolyte is added.

The coagulation is supposed to occur by the following mechanism. Let n_j be the concentration of colloid particles containing j atoms. Then the rates of change of the ns are given by

$$\frac{dn_k}{dt} = \frac{1}{2} \sum_{j+l=k} k_{jl} n_j n_l - n_k \sum_l k_{kl} n_l \tag{14.4.1}$$

The first sum represents the growth of n_k due to coalescence of smaller colloid particles; the factor of 1/2 is to correct for double counting of encounters. The second term represents the loss of n_k due to coalescence of particles of size k with particles of other sizes. The k_{ij} factors are the rate constants for these steps. These rate constants are taken to be the Smoluchowski diffusion controlled rates given by eqn (14.2.13). It was to derive an explicit expression for these ks that Smoluchowski developed the diffusion control theory.

Equations (14.4.1) are an infinite set of coupled nonlinear ordinary differential equations; it is not possible to find a solution in closed form. To simplify the situation we assume that the interaction radii between particles of size j and k are all the same. A consequence is that all of the diffusion coefficients are the same, since $D \propto 1/R$. This is doubtless a drastic oversimplification but it simplifies the equations; all the k_{ij}s are equal under this hypothesis. Thus we have the equations

$$\frac{dn_k}{dt} = \kappa[\frac{1}{2} \sum_{l+j=k} n_j n_l - n_k \sum_l n_l] \tag{14.4.2}$$

where κ denotes the common value of the rate constants. Experimentally, the total number of particles of all sizes, $N = \sum n_k$, was measured. This can be done by stopping the coagulation by adding an agent such as gelatin, or by measuring in a flow system with an ultramicroscope. Summing both sides of eqns (14.4.2) over k

$$\frac{dN}{dt} = -\frac{1}{2}\kappa N^2 \tag{14.4.3}$$

The solution to this equation is

$$N(t) = \frac{N(0)}{1 + N(0)\kappa t/2} \tag{14.4.4}$$

where $\kappa = 8\pi DR$, D and R being the diffusion coefficient and radius of reaction of a single particle. The time dependence of n_j for small j can now be calculated, if desired, from eqn (14.4.2).

The general validity of these predictions of the Smoluchowski theory has been verified experimentally. Of course, we cannot expect the details to be correct in all particulars because of the simplifications made to get the equations into tractable form.

14.5 Taylor Diffusion

Suppose liquid is flowing in a laminar flow down a tube or channel. We would intuitively think that impurities placed in the fluid would be carried downstream with the flow velocity appropriate to their position in the flow. In fact, although the mean position of the impurities does move with the mean flow velocity, there is an appreciable dispersion about this mean. The dispersion increases linearly with the time, as though the impurity particles were diffusing. However, the numerical coefficient of this term proportional to the time is not the molecular diffusion coefficient of the impurity.

This phenomenon is known as *Taylor diffusion* (Taylor 1953, 1954). Taylor first considered it in connection with the problem of measuring the velocity of fluid flow in buried pipes. One method of doing this was to inject an electrolyte into the flow and measure the conductivity of the fluid at the position of electrodes attached to the pipe downstream. It is necessary to be able to relate the conductivity-time curve thus obtained to the fluid flow profile in order to determine the velocity.

Taylor's theory predicted that, after some induction period, the distribution of impurity or solute was Gaussian about its mean. The mean impurity position itself traveled with the mean fluid velocity averaged over the profile. The width of the Gaussian was proportional to the time. If the x direction denotes the direction of flow, Taylor's result for the dispersion of the profile for flow through a tube of circular cross-section was

$$< \delta x^2 > = 2D_{\text{eff}}t$$

$$D_{\text{eff}} = \frac{a^2 u^2}{48 D_m} \tag{14.5.1}$$

a is the radius of the tube, U is the mean velocity of the parabolic velocity profile (Poiseuille flow), and D_m is the molecular diffusion coefficient. The numerical factor 48 depends on the cross-section of the tube, and hence on the velocity profile. Westhaver (1947) had previously obtained the same result in connection with a different problem.

Thus is a surprising result. Intuition initially leads to the conjecture that the impurity concentration profile should reflect the velocity profile, since the impurity particles are carried along with the flow, aside from molecular diffusion, which is very small. Furthermore, the effective diffusion coefficient varies *inversely* as the molecular diffusion coefficient. Taylor states 'This theoretical conclusion seemed so remarkable that I decided to set up apparatus to find out whether the predictions of the analysis could be verified experimentally.' He did set up a clever, simple experiment that did verify the predictions of the theory.

The physical explanation of Taylor diffusion lies just in the Brownian motion of the impurity particle, the molecular diffusion alluded to in the previous paragraph. Contrary to the implication given there, molecular diffusion cannot be neglected. As impurity particles are carried downstream by the flow, their diffusion transverse to the flow causes them to sample regions of different flow velocities. Diffusion in the direction of the flow does indeed give a negligible contribution to the dispersion, but diffusion normal to the flow enables the particles to sample regions of sensible velocity differences. The particles make random walks in the direction normal to the flow. After sufficient time has elapsed for the particles to sample the entire velocity distribution, their velocity distribution will mirror the fluctuations induced by the Brownian motion. The coupling of the macroscopic flow and the Brownian motion enhances the diffusion caused by the latter.

In thinking about this physical picture, it is essential to keep in mind that the system is *bounded* in directions normal to the flow. If the system were unbounded, then the impurity particles would never be able to sample the entire velocity profile in finite time, and the dispersion of the concentration profile in the flow direction would depend on time in a different way (Ben-Naim *et al.* 1992).

An excellent review of the Taylor and related phenomena from the point of view of random processes has been given by Van den Broeck (1988). A monograph of Brenner and Edwards (1993) gives an admirable account of the theory from the point of view of transport processes.

We shall derive Taylor's results using the method of elimination of fast variables described in Section 8.3. In a sense this is just a systematization of Taylor's original method; in the original it was not easy to follow the motivation for many of the steps. First we need to find the analog of the Smoluchowski equation for a Brownian particle in a flowing fluid. This is very easily done with the techniques already developed in Chapter 8, so we shall merely give an outline here.

The frictional force between a particle moving with velocity $\mathbf{v}$ in a fluid moving with velocity $\mathbf{u}$ (which may be a function of position) is $-\zeta(\mathbf{v}-\mathbf{u})$ instead of just $-\zeta\mathbf{v}$. When this is used to compute the A coefficient in the Fokker–Planck equation (Section 3.4) the effect is to make the A coefficient $\mathbf{u}-\mathbf{v}$ instead of $\mathbf{v}$. Taking this change into account in the elimination of the slow variables (Section 8.4), the Smoluchowski equation becomes

$$\frac{dn}{dt} + \mathbf{u} \cdot \nabla n = D_m \nabla^2 n - \frac{1}{\zeta} \nabla \cdot \mathbf{K} n \tag{14.5.2}$$

This differs from the Smoluchowski equation for a still liquid only in the convective term. Such a convective term is precisely what one would add on intuitive grounds. The main point of a detailed derivation is to show that there are no additional terms generated by the flow. Equation (14.5.2) is a well known equation, especially in the case $\mathbf{K} = 0$; it is known as the *convective diffusion equation.*

Instead of considering flow in a pipe of circular cross-section, as Taylor did, we shall treat the case of simple shear flow between two parallel planes; the calculation is a little simpler. There are two bounding plates, at $z = 0$ and $z = a$, parallel to the xy plane. Fluid is flowing in the x direction with a velocity $\mathbf{v}(z) = V\mathbf{e}_x z/a$, a linear velocity profile. Into this flowing fluid we introduce an impurity with a concentration profile $n(x, z)$. In the model treated here, the y coordinate is irrelevant, and we shall ignore it henceforth. We now apply the approximation method of elimination of the fast variables (Section 8.3) to eqn (14.5.2).

The fast variable in this case is the coordinate normal to the flow direction, z. The slow variable is x, for we want to describe the behavior of the concentration profile as a function of x on time scales long compared to the time it takes the impurities to diffuse a distance a. Practically speaking, this means we work with long tubes of small radius, or in the model treated here, large plates with small interplanar spacing. The small parameter in the problem is D_m/Va. This can be looked on as the ratio of the transit time in the z direction, D_m/a^2 to the time taken to cover the same distance in the x direction, a/V. The boundary conditions are that $\partial n/\partial z = 0$ at $z = 0$ and $z = a$. This means that impurity particles get reflected at the boundaries.

We can apply the method of Section 8.3 directly. The slow variable is x; the fast variable is z. The operator L is $-V(y/a)\partial/\partial x$; the operator Λ is $D\partial^2/\partial y^2$. Remember that we are neglecting the second x derivative term. $n(x, z, t)$ is the quantity called ψ in Section 8.3 and we shall denote by $c(x, t)$ the quantity denoted by n there.

$$c(x, t) = \int_0^a n(x, z, t) dz \tag{14.5.3}$$

From eqn (8.3.11) we find

$$n^{(0)}(x, z, t) = \frac{1}{a} c(x, t) \tag{14.5.4}$$

Equation (8.3.15) becomes

$$D_m \frac{\partial^2 n^{(1)}}{\partial z^2} = -\frac{Va}{D_m}\left(\frac{1}{2} - \frac{z}{a}\right) \tag{14.5.5}$$

This equation is easily solvable by two quadratures. The solution satisfying the boundary conditions and also the condition $\int n^{(1)} dz = 0$ (compare eqn (8.3.9)) is

$$n^{(1)} = \frac{Va}{D_m}\left(\frac{1}{24} + \frac{1}{6}\left(\frac{z}{a}\right)^2 - \frac{1}{4}\left(\frac{z}{a}\right)^3\right)\frac{dc}{dx} \tag{14.5.6}$$

Finally, when eqns (14.5.4) and (14.5.6) are substituted into eqn (8.3.16)

$$\frac{\partial c}{\partial t} + \bar{V}\frac{\partial c}{\partial x} = D_{\text{eff}}\frac{\partial^2 c}{\partial x^2}$$
$$D_{\text{eff}} = \frac{\bar{V}^2 a^2}{30 D_m} \tag{14.5.7}$$

Here $\bar{V}$ is the mean fluid velocity averaged over a cross-section of the flow; $\bar{V} = V/2$ for the flow considered here.

Although we have carried out the computation for a linear velocity profile, it is clear that the steps are the same for any velocity profile. The only thing that will change in the final answer is the numerical factor in D_{eff}. For example, if we had considered a parabolic rather than a linear profile (but still between parallel plates) we would have obtained the factor 210 instead of the factor 30 in eqn (14.5.7).

Thus the concentration profile averaged over a cross section of the flow behaves as though it were an ordinary diffusing concentration profile centered at the moving point $\bar{V}$. The effective diffusion coefficient varies *inversely* as the molecular diffusion coefficient. At first sight this looks surprising, but the direction of the change can be understood physically. The higher the molecular diffusion coefficient, the more rapidly the impurity particle samples the different regions of the velocity profile. Therefore the velocity fluctuations that the particle encounters in its Brownian meandering through the z direction get averaged out more rapidly. This manifests itself in a lower apparent diffusion coefficient. This picture is similar to the picture of the phenomenon of motional narrowing in electron resonance spectra.

Our derivation has neglected the $\partial^2 n/\partial x^2$ term in the original equation of motion for the sake of simplicity. If this term is retained, eqn (14.5.7) is again obtained with the effective diffusion coefficient now given by

$$D_{\text{eff}} = \frac{\bar{V}^2 a^2}{30 D_m} + D_m \tag{14.5.8}$$

In practical situations, the second term is negligible compared to the first. The derivation of eqn (14.5.8) is left as an exercise for the reader.

Taylor diffusion arises in engineering applications, and is also currently used for the measurement of molecular diffusion coefficients. Molecular diffusion coefficients in liquids are quite small. A conventional diffusion experiment takes a long time, and experimental conditions must be maintained constant over this long time. The Taylor diffusion coefficient is much larger, and one can do a Taylor experiment in several hours that might takes several days if done by older techniques. Of course, inelastic light scattering is also a fast and convenient method for measuring diffusion coefficients, but is not so useful for molecules of size comparable to the solvent size. For then the scattering powers of solvent and solute are comparable, and the scattering contrast suffers.

While we have treated here the experimentally applicable case when the configuration space of the fast variable is bounded, it is interesting to ask, what happens when the boundary is removed, and the system becomes infinite in all directions. Our method does not apply to this case; here there is no clear separation of time scales for motion parallel and normal to the flow. This model has been treated by Ben-Naim *et al.* (1992), who considered velocity profiles of the form

$$V_x(z) = V(0) \mid z \mid^{\beta} \text{sgn} z$$

Here we shall give a qualitative physical argument for the time dependence of the root-mean- square displacement.

The displacement transverse to the velocity direction is purely diffusive; its rms displacement is of order $z \propto (Dt)^{\frac{1}{2}}$. Therefore the velocity of a particle at time t is, roughly $v_x(z) \propto v_0(Dt)^{\beta/2}$, so that its x displacement at time t is $x \propto v_0 t(Dt)^{\beta/2}$. The exponent describing the dispersion is thus $1 + \beta/2$. Recall that for Taylor diffusion the exponent is always unity; the only thing that changes as one changes the velocity profile is the numerical coefficient.

There are practical situations where the effects of boundaries may not be important. One such is the diffusion of solutes in groundwater flowing through stratified material (Matheron and de Marsily 1980). If the relevant observation times are short compared to the time necessary to sample the entire flow profile, then our asymptotic limit does not apply. Incidentally, Matheron and de Marsily also studied the effect of randomness in the transverse strata (as one might expect in a hydrological context) on the results. Another example is the depolarization of spins by random walks on a lattice, a phenomenon which is mathematically analogous to the problems discussed here (Czech and Kehr 1986; Mazo and Van den Broeck 1986).

15

ROTATIONAL DIFFUSION

15.1 Rotational Diffusion

So far we have been discussing only translational diffusion, the motion of the center of mass of a particle through Brownian motion. There is also rotational diffusion, the change of orientation of a particle due to random torques exerted on it by the surroundings. Rotational diffusion is important for the study of dielectric relaxation, fluorescence depolarization, NMR line widths, and, in short, almost any phenomena that has to do with the relaxation of some polarized quantity.

We shall treat in this section only the rotation of a sphere in which is embedded a vector, $\mathbf{n}$. This vector may represent a physical quantity such as an electric moment, a magnetic moment or a transition moment. The objects of physical interest are usually correlation functions of functions of θ, the angle between $\mathbf{n}(0)$ and $\mathbf{n}(t)$. Since θ lies between 0 and π, there is no interest in computing $< \delta\theta^2 >$ (at first sight the analogous quantity to $< \delta x^2 >$ for translation). Rather, we will want averages of periodic functions of θ; in particular the Legendre functions, $P_n(\cos\theta)$, turn out to be useful quantities.

Although Einstein (1906b) briefly treated the motion of a sphere about a fixed diameter, it was Debye (1929) who first gave a theory of rotational diffusion in a constant external field based on the Smoluchowski equation. Let θ and ϕ be the polar angles of the vector $\mathbf{n}$ with respect to some coordinate system fixed in space. We require the distribution function $f(\theta, \phi, t)$, the distribution function per unit solid angle for $\mathbf{n}$ to lie in the direction (θ, ϕ) within the solid angle $d\Omega$. The equation governing this distribution function can be derived in the same way the Fokker–Planck equation was derived in Section 3.4. The random variables are the polar angles; we are working in the limit where inertia can be neglected so the angular velocities do not enter as dynamical variables. This was the method used by Debye in his original treatment; the method involves some tedious spherical trigonometry to elucidate the relations between the three sets of angles involved (the polar angles of $\mathbf{n}(t)$, of $\mathbf{n}(t+dt)$, and the angle between $\mathbf{n}(t)$ and $\mathbf{n}(t+dt)$). We shall avoid this by taking the more intuitive approach outlined in Section 8.2, following the approach of Coffey *et al.* (1996, sect. 1.15).

The Smoluchowski equation has the form of an equation of continuity

$$\frac{\partial f}{\partial t} + \nabla \cdot (\mathbf{j}_d + \mathbf{j}_f) = 0 \tag{15.1.1}$$

where $\mathbf{j}_d$ and $\mathbf{j}_f$ are the probability current densities due to diffusion and due to the field, respectively. We shall take the current density due to diffusion as given by

$$\mathbf{j}_d = -D_R \nabla f \tag{15.1.2}$$

The current density due to the field can be identified from the Langevin equation.

The instantaneous direction of the vector $\mathbf{n}$ satisfies the equation

$$\frac{d\mathbf{n}}{dt} = \omega \times \mathbf{n} \tag{15.1.3}$$

This is a purely kinematical relation which holds for any vector fixed in the body, and has nothing to do with whether or not the particle is undergoing Brownian motion. The Langevin equation is

$$I\frac{d\omega}{dt} = -\zeta_R \omega + \mathbf{N}(t) + \mu \mathbf{n} \times \mathbf{E} \tag{15.1.4}$$

The first term on the right is the frictional force on the rotating sphere; ζ_R is the rotational friction constant. The second term is the random torque, which we assume to have the same stochastic properties as we have previously assumed for the random force in the translational case. For the third term we have assumed that there is an external electric field, $\mathbf{E}(t)$ which act on a dipole of moment μ in the direction of $\mathbf{n}$.

In the non-inertial limit, i.e. when the inertial torque, the angular acceleration, is small compared to the other torques, we may set the left-hand side of eqn (15.1.4) equal to zero. Hence

$$\omega = \frac{1}{\zeta_R}(\mathbf{N}(t) + \mu \mathbf{n} \times \mathbf{E}) \tag{15.1.5}$$

Placing this in eqn (15.1.3), and using the properties of the triple vector product $(\mathbf{n} \times \mathbf{E}) \times \mathbf{n}$ yields

$$\frac{d\mathbf{n}}{dt} = \frac{\mathbf{N}(t) \times \mathbf{n}}{\zeta_R} + \mu \mathbf{E} \cdot (\mathbf{1} - \mathbf{nn}) \tag{15.1.6}$$

So, just as in the qualitative method of Section 8.2, we have a Langevin equation in $\mathbf{n}$ alone, not involving ω.

The current density has two parts, as indicated in eqn (15.1.1). The diffusive part, $\mathbf{j}_d$, is as given by eqn (15.1.2) and arises from the random torque contribution to $d\mathbf{n}/dt$. The part due to the field is

$$\mathbf{j}_f = f(\frac{d\mathbf{n}}{dt})_{field} = \frac{\mu}{\zeta_R}\mathbf{E} \cdot (\mathbf{1} - \mathbf{nn}) f \tag{15.1.7}$$

To get the Debye equation in explicit form, it is only necessary to write the Laplacian operator in the space of angles (for the diffusion term) and the divergence operator in the space of angles (for the field dependent term) (Margenau and Murphy 1943, sec. 5.2 –5.4). The final result is

$$\frac{\partial f}{\partial t} = \frac{kT}{\zeta_R}[\frac{1}{\sin\theta}\frac{\partial}{\partial\theta}(\sin\theta\frac{\partial f}{\partial\theta}) + \frac{1}{\sin^2\theta}\frac{\partial^2 f}{\partial\phi^2}] + \frac{\mu E}{\zeta_R}\frac{1}{\sin\theta}\frac{\partial}{\partial\theta}(\sin^2\theta f) \qquad (15.1.8)$$

Note that, as would be expected, the component of $\mathbf{E}$ along $\mathbf{n}$ plays no role. Only the components E_θ and E_ϕ are relevant to the present problem. In fact, we shall only be interested in solutions that are independent of the azimuthal angle ϕ because ϕ does not enter into the potential energy of the dipole in the field so that it would be difficult to do experiments described by ϕ dependent solutions. Consequently we set $\partial^2 f/\partial\phi^2 = 0$ so the equation we have to work with is

$$\frac{\zeta_R}{kT}\frac{\partial f}{\partial t} = \frac{1}{\sin\theta}\frac{\partial}{\partial\theta}(\sin\theta\frac{\partial f}{\partial\theta}) + \frac{\mu E}{kT}\frac{1}{\sin\theta}\frac{\partial}{\partial\theta}(\sin^2\theta f) \qquad (15.1.9)$$

Now the dimensionless parameter $\mu E/kT$ is much less than unity; it is of order 10^{-4} for a dipole moment of 4×10^{-18} esu, a field of 1 esu (300 v cm^{-1}), and 300 K. We may therefore restrict ourselves to approximate solutions to eqn (15.1.9) valid for small values of this parameter.

There are two interesting cases to consider. In the first case, we assume the system has been exposed to the field for a long time and is at equilibrium in the field at $t = 0$. At this instant, the field is suddenly turned off, $E = 0$ for $t > 0$. What is the mean dipole moment of the system thereafter? At $t = 0$, the distribution is

$$f(t=0) = \frac{1}{2}(1 + \frac{\mu E\cos\theta}{kT})$$

This is the Boltzmann distribution expanded to first order in the small parameter $\mu E/kT$.

This suggests that we look for solutions of eqn (15.1.9) of the form

$$f = \frac{1}{2}(1 + \frac{\mu E\cos\theta}{kT}g(t))$$

Since $E = 0$ for $t > 0$, this hypothesis yields the differential equation

$$\frac{dg}{dt} = -\frac{2kT}{\zeta_R}g \qquad (15.1.10)$$

with solution

$$g = e^{-t/\tau} \qquad (15.1.11)$$

where $\tau = 2kT/\zeta_R$ is called the Debye relaxation time. The mean polarization of the system is then $\int \mu\cos\theta f \sin\theta d\theta$. The integral is easily evaluated with the result

$$< \mu \cos\theta >= \frac{\mu^2 E_0}{3kT} e^{-t/\tau} \tag{15.1.12}$$

The point to be made here is that the polarization decays exponentially in time with a relaxation time $2kT/\zeta_R = 2D_R$.

The second important case we consider here is a system exposed to an oscillating field $E = E_0 \cos \omega t$ for a long time, so that any transients have had time to decay. Again we ask for the polarization as a function of time, and we assume that $f = 1 + \mu E/kT \cos\theta g(t)$, similar to the relaxation case. But now, because of the time dependence of E, the equation to be solved is

$$\frac{dg}{dt} = -\frac{2kT}{\zeta_R} g + \cos \omega t \tag{15.1.13}$$

with solution

$$g(t) = \frac{1}{1+\omega^2\tau^2}(\omega\tau \sin \omega t + \cos \omega t) \tag{15.1.14}$$

Again, the detailed form of $< \mu \cos\theta >$ is easy to work out; it is

$$< \mu \cos\theta >= \frac{\mu^2 E_0}{3kT}\frac{1}{1+\omega^2\tau^2}(\cos \omega t + \omega\tau \sin \omega t) \tag{15.1.15}$$

The primary observation is that the mean polarization has a component in phase with the field and a component out of phase with the field. The phase shift, δ, is given by $\tan\delta = -\omega\tau$. It is customary to write

$$< \mu \cos\theta >= Re(\alpha E_0 e^{i\omega t})$$

and regard $\alpha = \alpha' - i\alpha''$ as a complex polarizability.[29] Since the dielectric constant, ϵ, is $\epsilon = 1 + 4\pi\alpha$, the dielectric constant can also be considered to be a complex quantity. The real part measures the energy stored in the dielectric during one cycle of the field. The imaginary part measures the energy dissipated over a cycle; the field does net work on the out of phase component of the polarization over a cycle. For the particular model studied here , ϵ is given by

$$\epsilon(\omega) - 1 = (\epsilon'(0) - 1)\frac{1}{1+\omega^2\tau^2} - i\omega\tau(\epsilon'(0) - 1)\frac{1}{1+\omega^2\tau^2} \tag{15.1.16}$$

We have assumed in this calculation that the only source of polarization is the orientation of the permanent dipole of the molecule. Real molecules have electronic polarization also. This can be taken into account by adding the term $\alpha_\infty E_0$ to the right-hand side of eqn (15.1.2). α_∞ is the electronic polarizability,

[29] The minus sign in the definition is conventional.

assumed to be the high frequency polarizability,; hence the subscript. This addition changes eqn (15.1.16) by substituting $\epsilon - \epsilon_\infty$ for $\epsilon - 1$ in the three places it appears in (15.1.16).

If $\omega\tau$ is eliminated between the equations for ϵ' and ϵ'', a plot of ϵ'' vs. $\epsilon' - 1$ (or $\epsilon' - \epsilon_\infty$ in the more realistic case) is a semicircle with center on the ϵ' axis. Such a plot is known as a Cole–Cole plot. A useful way to check theory against experiment is to construct a Cole–Cole plot of the data and see if it conforms to the predictions of the theory. Very often the data does not fall on such a simple plot. Sometimes the experimental points do fall on a semicircle, but the center does not lie on the ϵ' axis. Sometimes the locus of experimental points do not even fall on a semicircle.

This is interpreted as implying that the system is not described by a single relaxation time. Rather a distribution of relaxation times is required. This may come about for several reasons. Molecules are not spherical in general, and the rotatory diffusion tensor may be anisotropic. For an ellipsoidal particle, the theory has been given by Perrin (1934).

Furthermore, there is no necessity that the dipole moment of the molecule should have the direction of one of the principal axes of the diffusion tensor. And there may be a variety of local environments of the rotating Brownian particles, which would give rise to a variety of relaxation times. The long range dipole–dipole interactions between the rotating polar Brownian particles are neglected. High frequencies are not treated correctly because of the neglect of inertial terms. Clearly the model here adopted, that of Debye, does not capture all of the complexities of real materials. We refer the reader to specialized treatises on dielectrics for a discussion of these effects (Böttcher 1952).

15.2 Fluorescence Depolarization

Another application of the theory of rotational Brownian motion is to the depolarization of fluorescence of molecules in solution. This was first worked out by Perrin (1926a,b). Suppose a molecule is excited by light polarized along the transition moment between the ground and excited states. When the molecule decays radiatively to the ground state, the emitted light is also polarized along the direction of the transition moment. Consequently, if the molecule is stationary so that the direction of the transition dipole does not change, the emitted light will be completely polarized along the direction of the polarization of the incident light. But if the molecule rotates between the absorption and emission, then the emitted light will only be partially polarized because the transition dipole, which is fixed in the molecule, has changed direction. This is the phenomenon of fluorescence depolarization.

This scenario has been simplified to bring out the essence of the phenomenon. In reality, in a fluid the chromophore molecules do not have their transition moments aligned in any particular spatial direction since the molecules are randomly oriented. Thus, even if the chromophores are stationary, the fluorescence will be partially polarized. There may also be other intramolecular depolarizing mechanisms. We shall be concerned here solely with the depolarization due to rotational Brownian motion for spherical particles.

The geometry is shown by Fig. 15.1.

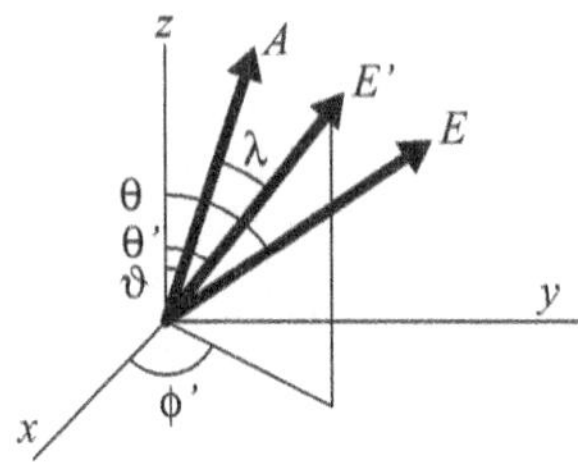

FIG. 15.1. The relative orientation of dipoles in a fluorescence depolarization experiment. A: absorbing dipole at zero time. E': emitting dipole at zero time. E: emitting dipole at some later time, t. In order to keep the diagram uncluttered, the azimuthal angles of A and E are not shown.

The incident light is polarized along the z direction and incident along the x direction. The emitted light is observed along the y direction. (θ, ϕ) are the polar angles of the emission transition moment at time t and (θ', ϕ') are the polar angles at time zero. We admit that the transition moment of the emission need not be the same as that of the absorption; we may be observing a fluorescent decay to an excited state. Hence we introduce (λ, μ), the polar angles of the emitting dipole with respect to the absorbing one. λ is fixed by the geometry and quantum mechanics of the transitions of the chromophore; it does not change with time. The depolarization ratio is defined by

$$p = \frac{I_{\parallel} - I_{\perp}}{I_{\parallel} + I_{\perp}} \tag{15.2.1}$$

$I_{\parallel}$ and $I_{\perp}$ are the intensities of fluorescent radiation polarized parallel and perpendicular, respectively, to the polarization of the incident beam.

Now $I_{\parallel} \propto \cos^2\theta$ and $I_{\perp} \propto \sin^2\theta \sin^2\phi$ for a given molecule. Hence for the sample as a whole $< I_{\parallel} > \propto < \cos^2\theta >$ and $< I_{\perp} > \propto \frac{1}{2}(1 - < \cos^2\theta > \ )$ since $< \sin^2\phi > = \frac{1}{2}$. Thus

$$p = \frac{3 < \cos^2\theta > -1}{< \cos^2\theta > +1}$$

$$\frac{1}{p} - \frac{1}{3} = \frac{2}{3}(\frac{3}{2} < \cos^2\theta > -1)^{-1} = \frac{2}{3}\frac{1}{< P_2(\cos\theta) >} \tag{15.2.2}$$

The Legendre polynomial $P_2(\cos\theta)$ is $(4\pi/5)^{\frac{1}{2}} Y_{2,0}(\theta, \phi)$. The problem of determining p thus reduces to that of computing the average of a spherical harmonic. To find this average, we need the solution of the Smoluchowski equation that reduces to a delta function, $\delta(\theta-\theta')\delta(\phi-\phi')/\sin\theta$ for $t=0$; that is, the Green's function of the equation.

The eigenfunctions of the operator on the right-hand side of the Smoluchowski equation, (15.1.8) with $E=0$ are well known to be the spherical harmonics, so that the Green's function is

$$G(\theta,\phi\mid\theta',\phi')=\sum_{l=0}^{\infty}\sum_{m=-l}^{l}Y_{l,m}(\theta,\phi)Y^*_{l,-m}(\theta',\phi')e^{-l(l+1)Dt} \tag{15.2.3}$$

This is the probability density that must be used to compute $< P_2 >$. One easily finds that

$$< P_2(\cos\theta) >=(\frac{4\pi}{5})^{\frac{1}{2}}Y^*_{2,0}(\theta',\phi')e^{-6Dt} \tag{15.2.4}$$

since $l=2$. The probability of fluorescent decay in time interval dt around t is

$$\frac{1}{\tau}e^{-t/\tau}dt \tag{15.2.5}$$

where τ is the mean lifetime of the excited state of the chromophore. It should not be confused with the τ of the previous section, the Debye dielectric relaxation time. Averaging eqn (15.2.4) over time then yields

$$< P_2(\cos\theta) >=(\frac{4\pi}{5})^{\frac{1}{2}}Y^*_{2,0}(\theta,\phi)\frac{1}{1+6D\tau} \tag{15.2.6}$$

This must be further averaged over all orientations of the *absorption* moment with the weight factor $3\cos^2\vartheta\sin\vartheta d\vartheta d\epsilon/4\pi$ where (ϑ,ϵ) are the polar angles of the absorption oscillator. The factor of $\cos^2\vartheta$ arises because the probability of absorbtion is proportional to $\cos^2\vartheta$.

From the addition theorem for spherical harmonics,

$$P_2(\cos\theta)=\frac{4\pi}{5}\sum_{m=-2}^{2}Y_{2,m}(\vartheta,\epsilon)Y^*_{2,-m}(\lambda,\mu) \tag{15.2.7}$$

Averaging this expression, as just described,

$$< P_2(\cos\theta) >=\frac{2}{5}P_2(\cos\lambda)=\frac{1}{5}(3\cos^2\lambda-1) \tag{15.2.8}$$

So, finally,

$$\begin{aligned}\frac{1}{p}-\frac{1}{3}&=(\frac{1}{p_0}-\frac{1}{3})(1+6Dt)\\ \frac{1}{p_0}-\frac{1}{3}&=\frac{5}{3}\frac{2}{3\cos^2\lambda-1}\end{aligned} \tag{15.2.9}$$

The depolarization ratio, p, can be measured directly. The fluorescence lifetime, τ, can be measured in a separate experiment in which it is not necessary to use polarized light. In many cases it is safe to assume that the angle λ is zero. When it is not safe to assume this, λ can often be measured by measuring the fluorescence depolarization in oriented samples, say in a crystal or a film, where the molecules are not free to rotate. Thus the rotatory diffusion coefficient can be measured by fluorescence depolarization. For a discussion of experimental methods and further theoretical considerations, see Weber (1966).

Another optical method for measuring rotatory diffusion coefficients, especially for large molecules, is flow birefringence. For this, it is essential that the molecules not be spherical. We shall not give details of the theory, but only outline the basic physical idea. A solution containing molecules that are optically and hydrodynamically anisotropic is caused to flow. Most often the fluid is placed between rotating concentric cylinders. Because the molecules are hydrodynamically anisotropic, for example ellipsoidal, they will be oriented by the flow. However, their rotational Brownian motion will partially destroy the orientational order imposed by the flow. The degree of orientation can be measured by observing the system through crossed polarizers placed at the bottom and top of the cylinder normal to the axis of rotation; this is why the optical anisotropy is needed. The degree of orientation is dependent on the rotatory diffusion coefficients of the molecules. If the rotatory diffusion is slow, the flow is effective in orienting the molecules. If the rotatory diffusion is rapid, then the Brownian motion dominates the orienting tendency of the flow and the flow is ineffective.

This is a steady state experiment, so that only the solution of the time independent, steady state, Smoluchowski equation is needed for its theoretical description. However, the equation must be supplemented by terms describing the torques on the molecules due to the velocity gradients in the flow. For a more thorough treatment, the reader is referred to the review of Cerf and Sheraga (1952).

15.3 Non-Spherical Brownian Particles

Brownian particles are not all spheres. In the previous two sections we have treated the case of a spherical rotor for the purpose of exposing the physical ideas with the minimum of mathematical complication. In the case of a general body, the very description of rotation takes some care. This is not peculiar to Brownian motion; even in analytical dynamics, the application of Newton's laws of motion in the absence of fluctuations and friction, the kinematics of rotation is complicated. The complication is caused by two factors. First, there is no vector specifying the orientation of a rigid body whose time derivative is the angular velocity. Second, rotations about different axes do not commute.

The orientation of a rigid body in space with respect to a system of reference fixed in space requires three degrees of freedom. A system of reference can be fixed in the body and the orientation in space characterized by the nine direction cosines of the body-fixed axes with respect to space-fixed axes. There are six othonormality relations between the nine direction cosines, so only three are independent, just the number required. But nine objects with six constraints make for an awkward set of independent variables.

A rotation can be characterized by the axis of rotation and the angle of rotation. The axis is specified by two variable, the polar angles of the axis with respect to the space fixed system. Consequently there are three variables available. The space of these three variables can be visualized as a ball of radius π. A vector from the origin to a point in this ball points in the direction of the axis of the rotation being described; the length of the radius vector is the angle of rotation about this rotation. This angle is restricted to be less than π because a rotation of $\pi + \alpha$ about an axis in the direction of the unit vector $\mathbf{n}$ is the same as a rotation of $\pi - \alpha$ about $-\mathbf{n}$. For this reason, points at the opposite ends of a diameter on the ball represent the same physical rotation, and so must be identified with each other. This means that the configuration space of rotations is not just a ball, but is topologically a more complex object.

Another set of three angles which can be used as independent coordinates to characterize a rotation are the Euler angles. These are defined by aligning the body so that the space-fixed and body-fixed axes coincide, and then rotating the body to the position one wishes to describe by three successive rotations about specific axes (Edmonds 1957). These coordinates are defined in Appendix D.

Still another set of coordinates useful for the description of rotation are the Cayley–Klein parameters (Goldstein 1951). Although purely kinematical in origin, they are closely related to the description of spin in quantum mechanics. Favro (1960, 1965) has used this set of variables to study the rotational diffusion of Brownian particles.

In analytical dynamics, a rotor is characterized by its moment of inertia tensor. The moment of inertia is a symmetric tensor and thus has six independent components. A coordinate system fixed in the body can always be chosen so that this tensor is diagonal. The three diagonal components are called the principle moments of inertia. If all three principal moments are equal, the rotor is called a *spherical top*. If two are equal, the third different, it is called a *symmetrical top*, and if no two are equal it is called an *asymmetric top*.

Since we are neglecting inertial effects, the moment of inertia is irrelevant to our treatment of rotation, but the friction constant tensor takes its place. The same names are used to describe the symetries of the friction constant as are used for the moment of inertia. The rotational diffusion coefficient tensor is essentially the matrix inverse of the friction constant. Hence these same names are used to describe the classes of rotational diffusion tensors. The Fokker–Planck equation for rotation including angular velocities, that is, including inertial terms, has been studied by Hubbard (1972). However, we shall here restrict ourselves to the Smoluchowski limit, as we have done earlier in this chapter. In essence, we have already written down the case of the spherical top. The case of the asymmetric top is quite complicated; the rotational diffusion equation has 23 terms when expressed in terms of Euler angles. No exact solutions are known, but some approximate solutions for the Green's functions have been developed

(Favro 1965). We shall discuss here the remaining case, that of the symmetric top.

If we choose the body-fixed system of axes to be the principle axes of the rotatory diffusion tensor, then the Smoluchowski equation for the freely rotating body is

$$\frac{\partial f}{\partial t} = \sum_i \frac{\partial}{\partial \xi_i} D_i \frac{\partial}{\partial \xi_i} \tag{15.3.1}$$

where the ξ_i are the angles of rotation about the body fixed axes. The equation has is simplest form in this coordinate system because D is diagonal, but the ξ_i are not convenient variables. We need variables that describe the orientation relative to space-fixed axes. These can be taken to be the Euler angles.

Unfortunately there are several competing conventions for the definition of the Euler angles. We use here the definition given by Edmonds (1957); it is a common convention in works dealing with molecular or nuclear rotation but differs from the one often used in works on the analytical dynamics of rigid bodies. α is the angle between the space fixed z axes (called the vertical axis) and the line of nodes, the intersection of the space fixed x, y plane and the body fixed x', y' plane. β is the angle of rotation about the line of nodes, and γ is the angle of rotation about the body-fixed z' axis. This is hard to follow without a diagram, so that a diagram is given in Appendix D.

The derivatives with respect to the variables ξ_i must be expressed in terms of the derivatives with respect to the Euler angles α, β, γ. If we set $D'_x = D'_y = D_1$, $D'_z = D_2$, then eqn (15.3.1) is transformed into

$$\begin{aligned}\frac{\partial f}{\partial t} = D_1[&\frac{\partial^2}{\partial \beta^2} + \cot\beta \frac{\partial}{\partial \beta} + \frac{1}{\sin^2\beta}\frac{\partial^2}{\partial \alpha^2} + \\ &(\cot^2\beta + \frac{D_3}{D_1})\frac{\partial^2}{\partial \gamma^2} - 2\frac{\cot\beta}{\sin\beta}\frac{\partial^2}{\partial\alpha\partial\gamma}]f\end{aligned} \tag{15.3.2}$$

An outline of this transformation is given in Appendix D. Note that the operator on the right- hand side of this equation is, aside from an overall numerical factor, just the Schrödinger equation for the rotation of a symmetric top. The eigenfunctions and eigenvalues of this operator have been well studied by workers in the field of the rotational states of molecules or nuclei. Favro (1965) discusses how these results can be used to obtain the Green's function for the rotor.

The relaxation times are the reciprocals of the eigenvalues of the operator on the right- hand side of eqn (15.3.2). The eigenvalues and eigenfunctions are labeled by three integers (separation constants, which would be called quantum numbers in the Schrödinger eqnuation context). Favro shows that the relaxation times are given by

$$\frac{1}{\tau_{lm}} = D_1 l(l+1) + (D_3 - D_1)m^2$$

$$l = 0, 1, \cdots; \quad -l \leq m \leq l \tag{15.3.3}$$

and is independent of the third separation constant. In other words, there will be two relaxation times for the relaxation of a vector quantity corresponding to $l = 1$. The spherical top, $D_1 = D_3$, has only one relaxation time associated with $l = 1$ as we saw in Section 15.1. This extra relaxation time in the symmetric top case will manifest itself experimentally in the dielectric loss, say, if the direction of the dipole moment does not coincide with one of the principal axes of the diffusion tensor.

Equation (15.3.2) holds for the free rotor. If an external torque is present the eqnuation must be supplemented with another term

$$-\frac{\partial}{\partial \xi_i}(\mathbf{N} \cdot \zeta^{-1})_i f$$

where $\mathbf{N}$ is the external torque. This is analogous to the $\nabla \cdot \mathbf{j}_f$ of eqn (15.1.7) which was, of course, for a particular kind of torque. This additional term must also be expressed in terms of the Euler angles by the method outlined in Appendix D.

15.4 Concluding Remarks

We have discussed several special cases of rotational diffusion in the non inertial limit. There are numerous other examples which also could have been given. For example a particle could be rotating in a periodic potential; this is a model for hindered internal rotation in molecules such as ethane, or for rotation of molecules in crystals. At low temperatures ($kT \approx$ barrier height) the quantum mechanical nature of the motion in the potential is important. At high temperatures, however, a classical description may become adeqnuate; Then, considerations similar to those given here, suitably modified to take into account the different nature of the torque, will be a useful description.

As we have seen, the description of rotational motion is more complicated than the description of translational motion. Exact solutions are very hard to come by and exact solutions are important because they form the foundation of perturbation methods. Thereforeit is not surprising that considerable effort has been put into the development of algorithms for effective numerical computation. In particular, methods based on continued fraction representations of the solutions are effective. The books by Risken (1989) and Coffey *et al.* (1996) contain treatments of this topic.

16

POLYMER SOLUTIONS

16.1 A Model for Dilute Solutions of Polymers

Polymer molecules are long chain molecules made of repeating units. In this chapter we consider only linear homopolymers where all of the units are the same. This class of molecules will amply illustrate the problems and techniques of the theory. Consideration of branched and cross-linked polymers, as well as polymers consisting of two or more types of units, can be left for more specialized monographs. Each unit of the polymer chain will contain several atoms, but we shall assume for the purpose of a hydrodynamic mesoscopic treatment that all of the atoms in a unit can be considered as one object, forgetting about its composite nature.

We shall be interested in the transport properties of dilute solutions of such polymeric molecules, properties such as diffusion and viscous flow. For this purpose we assume that the solvent is a structureless hydrodynamic continuum. This assumption is consonant with the assumption that the polymeric repeat unit may be treated as a hydrodynamic unit.

Suppose the polymer is composed of N units. The number of degrees of freedom of the chain is, however, not $3N$ but something smaller because the units are held together by chemical bonds of fixed bond length and fixed bond angle. In real macromolecules the bond lengths and angles are not rigorously fixed. Bonds do stretch and contract, and angles do bend, but it is considered a good approximation to treat them as fixed. Consequently the point representing the configuration of the polymer molecule in the $3N$ dimensional configuration space of the molecule actually lies on some hypersurface of lower dimension. For example, if the model of treating the polymer as a freely jointed chain with no excluded volume is adopted, the representative point ofthe molecular model must lie on a $2N+1$ dimensional subspace satisfying the constraints $\mid \mathbf{r}_i - \mathbf{r}_{i+1} \mid = const.$ Such a subspace is called a *submanifold.* An m-dimensional submanifold of an n-dimensional Euclidean space is the locus of points in the n-dimensional space satisfied by the $n-m$ equations $f_j(x_1, x_2, ..., x_n) = 0$. In the cases we consider, these are the equations of constraint of fixed bond length and bond angle.

Our object in this chapter is to illustrate how the theory of Brownian motion enters into and affects the theory of the transport properties of dilute macromolecular solutions. Consequently we shall not attempt to develop here a realistic model which takes all of the many degrees of freedom into account. Instead, we shall illustrate the main points of the theory by means of a simple model which is a caricature of a more realistic model due to Rouse (1953) and Zimm (1956).

The theory of Rouse did not take hydrodynamic interaction into account; the theory of Zimm did. For the meaning of the term *hydrodynamic interaction*, see Section 16.2.

The basic model of Rouse treats the chain as a string of n beads joined by freely jointed Hookean (linear) springs. The springs do not represent the chemical bonds joining repeat units of the polymer. As stated earlier, the chemical bonds are stiff enough so that they can be treated as of constant length. The springs represent a set of contiguous repeat units, sufficiently numerous that two consecutive sets of such repeat units may be considered to be statistically independent. Models indicate that 6 to 8 repeat units will fulfill this requirement sufficiently accurately in many cases, but this depends on the stiffness of the chain, that is, the potential hindering relative rotation of neighboring bonds. Thus n, the number of beads, is less than N, the number of monomer units. These springs have zero equilibrium length, and a root-mean-square displacement b. Each of these springlike segments is called a *Kuhn statistical segment.*

It is, in fact, Brownian motion that insures the existence of the springs. Even though the position vectors of close by monomer units are correlated by bond length and bond angle constraints, this correlation has rather short range (compare the discussion leading to eqn (4.2.4)). Thus the end to end distance of such a segment ought to have a Gaussian distribution. A Gaussian probability distribution can be looked on as the Boltzmann factor for a quadratic potential energy, a Hooke's law spring. This potential (actually a potential of mean force) is primarily entropic, not energetic, in origin. Nevertheless, it is the potential of a real force, similar to the situation in rubber, where the restoring force in a stretched sample is of primarily entropic origin. It is the Brownian motion of the monomer units of the statistical segment that causes these units to explore the various parts of their configuration space. In this sense, it is the Brownian motion that is the origin of the elastic force.

In addition to the elastic restoring force that acts on a given bead, there are other forces. There is the friction of the solvent on the bead; this will be discussed in the next section. There is also the Brownian motion force of the solvent on the beads (to be distinguished from the Brownian force giving rise to the springs, discussed in the preceding paragraph). The Rouse– Zimm theory takes all of these forces into account in describing the motion of the beads.

Our caricature of the Rouse–Zimm theory contains only two beads and one spring joining them. It is a dumbbell model. This truncated model still contains all of the essential ingredients that occur also in the more complex problem with n beads and $n-1$ springs. It bypasses the need for a complex normal mode analysis of the motion of the model. This is our justification for discussing it. One could embellish the model by assuming different masses and different friction constants for the two beads. We do not believe that this is called for here because since such an embellishment would add little to the illustration of the effect of Brownian motion on macromolecular solution dynamics. Of course, the model is much too simplified to do justice to the description and analysis of experiments, and no attempt should be made to use it for such a purpose. The complete Rouse–Zimm theory is very popular for the analysis of real data.

16.2 Hydrodynamic Interaction

The forces on the chain units due to the surrounding fluid are the random force due to the fluctuations in the bombardment by solvent molecules, and the frictional force, which we assume is proportional to the relative velocity between the chain unit and the surrounding fluid. These are just the assumptions we have made earlier in this book and do not add anything new to the story. Previously, however, when we considered a single particle immersed in the fluid, the relative velocity was the difference between the particle velocity and the fluid velocity in the absence of the particle. In the case of a polymer chain the velocity in the absence of the chain element being considered is perturbed by the presence of the other chain elements. These elements present in the fluid change the velocity of the fluid at distant points, and it is this perturbed or changed velocity that is the fluid velocity with respect to which the relative velocity must be reckoned. If we denote the velocity of the fluid at the position of element l by $\mathbf{v}_l$ and the velocity of the element itself by $\mathbf{u}_l$ (both with respect to the external reference system), then the frictional force is given by

$$\mathbf{F}_l = \zeta(\mathbf{v}_l - \mathbf{u}_l) \tag{16.2.1}$$

But $\mathbf{v}_l$ itself has yet to be determined. We must consider the perturbations of the local fluid velocity caused by the several chain elements. According to linearized hydrodynamics, a force, $\mathbf{F}_k$, exerted on the fluid at $\mathbf{R}_k$ will induce a velocity $\delta\mathbf{v}_l$ at position $\mathbf{R}_l$ given by

$$\begin{aligned} \delta\mathbf{v}_l &= -\mathbf{T}_{lk}\cdot\mathbf{F}_k \\ \mathbf{T}_{lk} &= \frac{1}{8\pi\eta_0 R_{lk}}[\mathbf{1} + \frac{\mathbf{R}_{lk}\mathbf{R}_{lk}}{R_{lk}^2}] \end{aligned} \tag{16.2.2}$$

$\mathbf{T}_{lk}$ is called the *Oseen tensor* (Burgers 1938; Happel and Brenner 1973). Mathematically speaking, it is the Green's function of the linearized Navier–Stokes equations for an incompressible fluid. A derivation of the expression for $\mathbf{T}$ is given in Appendix E. Note that the subscripts on $\mathbf{T}$ are not tensor indices but refer to particles.

We digress to illustrate how the Oseen tensor can be used to give a derivation of Stokes' law. Assume we have a sphere being held stationary in a fluid of uniform velocity at infinity, $\mathbf{v}$. What should the force, $\mathbf{F}$, be to maintain the sphere stationary? Of course we know from solution of the Navier–Stokes equations that the answer is $\mathbf{F} = -6\pi\eta a\mathbf{v}$ where a is the radius of the sphere and η is the viscosity of the fluid. Stick boundary conditions are assumed; this means, the fluid velocity relative to the sphere velocity vanishes on the surface of the sphere.

> We try to determine $\mathbf{F}$ by replacing the sphere by a point force situated at its center. The induced velocity at the position $\mathbf{r}$ on the surface of the sphere (or rather where the surface had been before removal) is then given by eqn (16.2.2). And this should be equal to $\mathbf{v}$ in order that the relative velocity of the sphere and the fluid should vanish. Clearly it cannot be equal to $\mathbf{v}$ everywhere at the surface of the sphere. But perhaps we can make the relative velocity vanish on the average; that is, when averaged over the surface. The average of the Oseen tensor over the surface of a sphere is $(8\pi\eta_0 a)^{-1}(4/3)\mathbf{1} = \mathbf{1}/6\pi\eta_0 a$. Consequently the required force is $\mathbf{F} = -6\pi\eta_0 a\mathbf{v}$, Stokes' law.

Note that this is only an approximate derivation of Stokes' law, for the fluid dynamical boundary conditions were not satisfied exactly, but only on the average. Thus a point force is not strictly hydrodynamically equivalent to a sphere, but is only approximately so. The approximation becomes better the larger r/a. Here we applied it for $r/a = 1$.

Zwanzig *et al.* (1968) have given an example of the errors that can arise when using the Oseen tensor to simulate the effect of a sphere. They computed the diffusion coefficient of a rigid pearl necklace model of a polymer, a model in which the monomer units are situated along a straight line and, in contrast to the Rouse model, the bonds are of fixed length. When the Oseen interaction is used to describe the hydrodynamic interaction in this model, singularities develop for the diffusion coefficient for a range of friction constants which overlaps the physically interesting range. Thus calculations of quantities in which finite size bodies are approximated by points should not be trusted when the calculation depends on short distances in an important way.

In spite of this flaw, the approximation of using the Oseen tensor to simulate a sphere is quite standard in polymer solution theory. Improvements taking the finite size of the element perturbing the fluid have been suggested, and do, in fact, do a better job. But they are more complicated than the Oseen tensor and difficult to use in the multiparticle computations involved in the treatment of polymer dynamics. Consequently we shall use the Oseen tensor to describe hydrodynamic interactions in this chapter, in spite of its deficiency. In Chapter 17 we shall see a context in which it is necessary to go beyond the Oseen approximation.

Now let us return to our discussion of chain dynamics. If we denote the fluid velocity at the position of particle l unperturbed by any chain elements by $\mathbf{v}_l^0$, then the set of velocities $\mathbf{v}_l$ are given by the set of equations

$$\mathbf{v}_l = \mathbf{v}_l^0 - \sum_{k \neq l} \mathbf{T}_{lk} \cdot \mathbf{F}_k \qquad (16.2.3)$$

To use these equations, the $\mathbf{F}$s must be expressed in terms of the $\mathbf{v}$s using eqn (16.2.1). In our case there are only two beads, and hence only two equations of type (16.2.3).

In most problems of polymer solutions, the Oseen tensor in the form (16.2.2) is too difficult to use. The reason is that the components of the tensor depend on the configuration of the macromolecule which is ,of course, variable. The exceptions are models of rigid macromolecules (for which the flexible Rouse–Zimm model is inappropriate); we shall not consider such examples here. For flexible chains it is customary to replace the full Oseen tensor with its equilibrium average. The interbead distances, $R_{i,j}$, are distributed with an isotropic Gaussian distribution, since the springs are Hookean. Consequently $< \mathbf{T}_{ij} >$ is

$$< \mathbf{T}_{ij} >= \frac{1}{6\pi\eta} < \frac{1}{R_{ij}} > \mathbf{1} \tag{16.2.4}$$

The value of the Gaussian average of $< 1/R_{ij} >$ can easily be computed, yielding

$$< \mathbf{T}_{ij} >= \frac{1}{3\pi\eta} \frac{1}{(2\pi N b^2)^{\frac{1}{2}}} \mathbf{1} \equiv \tau \tag{16.2.5}$$

This is the approximate preaveraged expression for the Oseen tensor that is used in almost all applications and that we shall also use.

16.3 The Equation of Motion

We adopt as the level of description for this chapter the level of the Smoluchowski equation; that is, we neglect inertial effects. This is similar to what we did in studying diffusion in the last chapter. Here it is even more justified. The time scale of collisions by molecules of the surrounding medium is the same as for the previous case. Thus velocities will be thermalized and relax on a rapid time scale. But because of the large mass and interconnections of the polymer molecule, the configuration will change appreciably more slowly. The neglect of inertial effects means that the force on the system due to the Brownian motion is balanced by the other forces. This can be formulated by the equations

$$\begin{aligned} -\zeta[\dot{\mathbf{r}}_1 - (\mathbf{v}_1^0 + \delta\mathbf{v}_1^0)] - kT\nabla_1 \ln f - \kappa(\mathbf{r}_1 - \mathbf{r}_2) &= 0 \\ -\zeta[\dot{\mathbf{r}}_2 - (\mathbf{v}_2^0 + \delta\mathbf{v}_2^0)] - kT\nabla_2 \ln f + \kappa(\mathbf{r}_1 - \mathbf{r}_2) &= 0 \end{aligned} \tag{16.3.1}$$

Here $\delta\mathbf{v}_1$ and $\delta\mathbf{v}_2$ are the perturbations in the fluid velocity at the position of beads 1 and 2 respectively due to the other bead. The $\nabla \ln f$ terms are the diffusion force, entropic in nature, due to Brownian motion, and the last terms represent the forces on the beads caused by the springs. f is the distribution function of the coordinates $\mathbf{r}_1$ and $\mathbf{r}_2$. If there are any external forces on the macromolecule, they must be also be included in the left-hand side of eqn (16.3.1). We omit them here since we shall not be treating any cases, such as dielectric dispersion or electrophoretic mobility, where they might be important.

It is convenient to change variables to center of mass and relative coordinates.

$$\mathbf{R} = \frac{\mathbf{r}_1 + \mathbf{r}_2}{2}$$
$$\mathbf{q} = \mathbf{r}_1 - \mathbf{r}_2 \tag{16.3.2}$$

Then eqns (16.3.1) become

$$-\zeta(\dot{\mathbf{R}} - \frac{\mathbf{v}_1^0 + \mathbf{v}_2^0}{2} - \frac{\delta\mathbf{v}_1 + \delta\mathbf{v}_2^0}{2} - \frac{kT}{2}\nabla - R\ln f = 0$$
$$-\zeta(\dot{\mathbf{q}} - \mathbf{v}_1^0 + \mathbf{v}_2^0 - \delta\mathbf{v}_1^0 + \delta\mathbf{v}_2^0 - 2kT\nabla_q \ln f - 2\kappa\mathbf{q} = 0 \tag{16.3.3}$$

We want to eliminate $\dot{\mathbf{q}}$ and $\dot{\mathbf{R}}$ from eqn (16.3.3). This can be done by noting that the flow field, $\mathbf{v}$, does not change much over a region of molecular dimensions. Consequently we may approximate the fluid velocity at the position of bead i by $\mathbf{v}_i = \mathbf{v}_0 + \mathbf{r}_j \cdot \nabla\mathbf{v}$. These are the leading terms in a Taylor expansion about the origin, assumed to be in the molecule. Thus, from eqn (16.3.3)

$$\dot{\mathbf{R}} = \mathbf{v}_0 + \mathbf{R} \cdot \nabla\mathbf{v} - \frac{(\mathbf{1} + \zeta\tau)}{2}\frac{kT}{2}\nabla_R \ln f$$
$$\dot{\mathbf{q}} = \mathbf{q} \cdot \nabla\mathbf{v} + (1 - \zeta\tau)(\frac{-2kT}{\zeta}\nabla_q \ln f - \frac{2}{\zeta}\kappa\mathbf{q}) \tag{16.3.4}$$

τ is the approximate form of the Oseen tensor given by eqn (16.2.5).

We are finally at the point where we can write the equation of motion for f. f is a probability distribution function. Therefore it should obey an equation of continuity which we can write in the two equivalent forms

$$\frac{\partial f}{\partial t} = -\nabla_1 \cdot \dot{\mathbf{r}}_1 f - \nabla_2 \cdot \dot{\mathbf{r}}_2 f$$
$$\frac{\partial f}{\partial t} = -\nabla_R \cdot \dot{\mathbf{R}} f - \nabla_q \cdot \dot{\mathbf{q}} f \tag{16.3.5}$$

Substituting $\dot{\mathbf{q}}$ and $\dot{\mathbf{R}}$ from eqn (16.3.4), we finally arrive at

$$\frac{\partial f}{\partial t} = -\nabla_R \cdot [\mathbf{v}_0 + \mathbf{R} \cdot \nabla\mathbf{v} + \frac{kT}{2\zeta}(1 + \zeta\tau)\nabla_R \ln f]f$$
$$- \nabla_q \cdot [(\mathbf{q} \cdot \nabla\mathbf{v})f - (1 - \zeta\tau)(\frac{2kT}{\zeta}\nabla_q f + \frac{2\kappa}{\zeta}\mathbf{q}f)] \tag{16.3.6}$$

Equation (16.3.6) is the Smoluchowski equation for the model in question.

As far as we know, the full Kramers–Klein equation, which takes inertial effects into account, has never been written down explicitly for this or any other polymer model. No doubt the reason for this is that any inertial effects on transport processes are not expected to be important until the frequencies are much greater than (time scales much shorter than) are available for modern experiments. For example, Chang and Mazo (1976) have studied the effect of solvent friction on the line widths in the vibrational spectrum of polymers in dilute solution. They used a linear chain model based on actual molecular bonds and not on the entropic chains of the Rouse–Zimm model. Vibrational spectroscopy is indeed a high frequency phenomenon, and the high frequency limit of the Oseen tensor is needed to describe the hydrodynamic interaction (eqn (16.2.2) gives the low frequency limit). The conclusion was that the line widths were not molecular weight dependent and consequently threw little light on the non local properties of the chain.

In the next section we shall study the application of the submanifold model to a typical and important polymer solution property, the intrinsic viscosity.

16.4 Diffusion and Intrinsic Viscosity

Since the formalism developed in the preceding three sections is so general, it is useful to see how it can be applied to specific examples. The examples of the translational diffusion and intrinsic viscosity will be considered here. For the translational diffusion coefficient, we note that the diffusion coefficient for one bead alone is $D_{\text{single}} = kT/\zeta$. For two beads, if there were no interactions between them but they were constrained to rigidly move together, we would expect $D_{\text{double}} = kT/2\zeta$. However, there are the spring force and the hydrodynamic interaction; these will mediate the forces on the beads, and give rise to an altered diffusion coefficient.

To calculate this altered diffusion coefficient we consider eqn (16.3.6) for the case of a quiescent fluid, $\mathbf{v} = 0$, and integrate over the relative coordinate, $\mathbf{q}$. The result is an equation of motion for the distribution function of the center of mass of the system, which we call φ.

$$\varphi(\mathbf{R}, t) = \int f(\mathbf{q}, \mathbf{R}, t) d\mathbf{q} \tag{16.4.1}$$

This equation is

$$\frac{\partial \varphi}{\partial t} = \frac{kT}{2\zeta}(1 - \zeta\tau)\nabla^2\varphi \tag{16.4.2}$$

from which we read off immediately that

$$D = \frac{kT}{2\zeta}(1 - \zeta\tau) = \frac{kT}{2\zeta}(1 - \frac{\zeta}{3\pi\eta(4\pi b)^{\frac{1}{2}}}) \tag{16.4.3}$$

where we have use the preaveraged Oseen tensor, eqn (16.2.6) with $N = 2$.

To get an estimate of the magnitude of the correction term, we use Stokes' law for ζ, $\zeta = 6\pi\eta a$, where a is the radius of the bead. The correction term is then $a/(b\sqrt{\pi})$. In the more elaborate form of the model, with many beads and many springs, the correction from the simple case where the motions of the beads are not supposed to interact (called in polymer jargon the *free draining* case) is larger and more important.

Notice that the diffusion coefficient for the Hookean dumbbell is smaller than that for the free draining dumbbell. Unfortunately it is difficult to give a physical argument that this should be the case. In some orientations of the interbead vector, $\mathbf{q}$, relative to the direction of instantaneous motion of the dumbbell one can see intuitively that the motion of one bead will inhibit the motion of the other, decreasing the diffusion rate, In other orientations, the motion of one bead will enhance the motion of the other, increasing the rate. We must average over all orientations, as we have done by integrating over the direction of $\mathbf{q}$. It is hard to see intuitively which configurations dominate in determining the sign of the average, but the theoretical development gives an unambiguous answer, as we have just seen.

Of course, since we deliberately restricted ourselves to two beads, our computation gives no hint of the dependence of the diffusion coefficient on the molecular weight, or degree of polymerization. The more complete theory of Zimm does have this information. for example, in the limit of very strong hydrodynamic interaction, $D \propto N^{-1/2}$ instead of N^{-1} which is predicted for the free draining case.

We now move on to the subject of intrinsic viscosity. This is a concept mostly used in the polymer field, and so we must first define it. Let η_0 be the viscosity of the solvent, and c be the concentration of solute in a solution. The viscosity of the solution will in general be concentration dependent, and we may write

$$\eta = \eta_0(1 + [\eta]c + \cdots) \tag{16.4.4}$$

The quantity $[\eta]$ defined by this equation is called the *intrinsic viscosity*. Note that it does not have the dimensions of viscosity; its dimensions are those of inverse concentration, length^3/mass.[30] An equivalent definition of the intrinsic viscosity is

$$[\eta] = \lim_{c \to 0} \frac{\eta - \eta_0}{\eta_0 c} \tag{16.4.5}$$

In other words, $[\eta]$ is the limiting slope of the curve of η/η_0 as $c \to 0$.

As a point of history, it seems that the first solution model for which the intrinsic viscosity was calculated theoretically was a system of a suspension of spheres in a continuous fluid (Einstein 1906a, 1911).[31] The concentration unit used in this work was the volume fraction of spheres, a dimensionless quantity. In terms of this concentration unit, Einstein's result was $[\eta] = 2.5$.

[30]The concentration unit conventionally used in experimental work on polyner viscosities is deciliters per gram.

[31]This work comprised Einstein's doctoral dissertation.

Einstein's calculation for spheres had nothing to do with the theory of Brownian motion. However, Brownian motion does enter into the calculation of the intrinsic viscosity of solutions of dumbbells and more complex Rouse–Zimm models because of the internal degrees of freedom of the molecular models. To see how this occurs, we need an expression for the change of the solution viscosity due to the solute molecules in terms of the parameters of the molecular model. Since the viscosity is the proportionality coefficient between the stress tensor in the flowing fluid and the velocity gradient tensor, we must express the stress tensor in terms of the model. We do this following Kramers (1946); see also Bird *et al.* (1977). An alternative derivation based on calculating the excess rate of dissipation caused by the solute particles can be found in the book by Yamakawa (1971).

The momentum transferred across a plane moving with the local fluid velocity in a fluid by a dumbbell suspended in the flow arises from two mechanisms: if the dumbbell intersects the plane, there is a force in the spring that transfers momentum across the plane. Secondly, the dumbbell may move across the plane, carrying its momentum with it. The second of these contributions is familiar from the kinetic theory of dilute gases, where it contributes the entire effect. It arises from the perturbation of the velocity distribution function from its equilibrium form due to the flow. Since we are computing in the Smoluchowski limit, where the velocity distribution has already relaxed to its equilibrium form, this mechanism will give no contribution to the viscosity. It does give an isotropic contribution to the normal stress, but nothing to the tangential stresses, and hence nothing to the viscosity.

So we concentrate on the contribution due to the forces exerted across the plane. Place a plane moving with the local fluid velocity at an arbitrary point in the fluid. Place in the solution an imaginary cube of side $1/n^{1/3}$, where n is the number density of particles (dumbbells). n is supposed to be very small, so that the dumbbells do not interact with each other, but $1/n^{1/3}$ should still be large with respect to b, the root mean square length of a spring. There will be, on the average, one particle in the box. The box is oriented so that the plane is parallel to a face; call the direction perpendicular to the plane the x axis. The plane is halfway between the two faces perpendicular to x. The conditional probability that the molecule intersects the plane, given that there is one molecule in the box, is $\mid \mathbf{q} \cdot \mathbf{e}_x \mid n^{1/3}$, where $\mathbf{e}_x$ is a unit vector in the x direction. This is the ratio of the projection of the spring onto the x axis to the length of the box side.

There is a tension in the spring, $\mathbf{K} = -\kappa \mathbf{q}$ exerted across the plane by the particle in the left-hand side of the box (the negative x region) on that in the right (the positive x region). Hence the mean x component of the force exerted across the plane is

$$-n^{1/3} \int \mathbf{q} \cdot \hat{\mathbf{e}}_x \mathbf{K}(\mathbf{q}) f(\mathbf{q}) d\mathbf{q}$$

Dividing by the area of the plane face, $n^{-2/3}$, gives the force per unit area across the plane

$$-n\hat{\mathbf{e}}_x \cdot \int \mathbf{qK}fd\mathbf{q} \tag{16.4.6}$$

But this must be identified with $\mathbf{e}_x \cdot \pi$, where π is the stress tensor in the fluid. This is, in fact, the definition of the stress tensor: π_{ij} is the component of the force in the j direction exerted across a surface normal to the i direction. Consequently we may write

$$\pi_p = -n \int \mathbf{qK}fd\mathbf{q} = -n < \mathbf{qK} >= n\kappa < \mathbf{qq} > \tag{16.4.7}$$

π_p is the contribution of the dissolved polymer to the stress tensor of the solution. Of course, there is the contribution of the solvent also.

The viscosity is defined by

$$\pi = \eta(\nabla\mathbf{v} + (\nabla\mathbf{v})^T - \frac{2}{3}\mathbf{1}\nabla \cdot \mathbf{v}) \tag{16.4.8}$$

We consider our fluid to be incompressible in the fluid mechanics sense, which means that $\nabla \cdot \mathbf{v} = 0$. Furthermore we consider simple shear flow, defined by $\mathbf{v} = \mathbf{e}_x y\dot{\varepsilon}$. $\dot{\varepsilon}$, a constant, is the magnitude of the velocity gradient and is known as the rate of shear. Thus only the xy component of the stress is important for computing the viscosity.

We attempt to solve eqn (16.3.6) for f correct to order $\dot{\varepsilon}$. That is, we treat the flow term as a perturbation and solve to first order in that perturbation. Then we use eqn (16.4.7) to get $\pi_{p,xy}$ to first order in $\dot{\varepsilon}$. Since $\pi_{p,xy} = \delta\eta\dot{\varepsilon}$ we then have the desired expression for $\delta\eta = \eta - \eta_0$, where η_0 is the viscosity of the pure solvent.

The next step is thus to solve eqn (16.3.6). First we integrate eqn (16.3.6) over the center of mass coordinate, $\mathbf{R}$, for we need only the $\mathbf{q}$ dependence. Furthermore, we are interested in steady flow, so that $\partial f/\partial t = 0$. The equation to be solved is therefore

$$\nabla_q[-q_y\hat{\mathbf{e}}_x\dot{\varepsilon}\psi - (1-\zeta\tau)\frac{2kT}{\zeta}(\nabla_q\psi + \frac{\kappa}{kT}\mathbf{q}\psi)] = 0 \tag{16.4.9}$$

where $\psi(\mathbf{q})$ is the result of integrating $f(\mathbf{R},\mathbf{q})$ over $\mathbf{R}$. Let ψ^0 be the Boltzmann distribution for $\mathbf{q}$, i.e.

$$\psi^0 = (\frac{\kappa}{2\pi kT})^{3/2}e^{-\kappa q^2/2kT} \tag{16.4.10}$$

and write $\psi = \psi^0(1 + \dot{\varepsilon}h)$. Then eqn (16.3.6) becomes, to first order in the perturbation parameter,

$$\nabla_q \cdot (q_y \hat{\mathbf{e}}_x \psi^0 + (1-\zeta\tau)\frac{2kT}{\zeta}\nabla_q h\psi^0) = 0 \qquad (16.4.11)$$

Using the explicit form of ψ^0, this can be simplified to

$$-q_x q_y \frac{\kappa}{kT} + (1-\zeta\tau)\frac{2kT}{\zeta}(\nabla^2 h + \mathbf{q}\cdot\nabla_q h\frac{\kappa}{kT}) = 0 \qquad (16.4.12)$$

It is easy to verify that a particular solution to this equation is

$$h = \frac{q_x q_y}{(1-\zeta\tau)}\frac{\zeta}{4kT} \qquad (16.4.13)$$

We now use this solution to carry out the second part of the program to calculate $\delta\eta$.

Inserting the solution (16.4.13) in eqn (16.4.7) yields

$$\frac{\pi_{p,xy}}{\dot{\varepsilon}} = \delta\eta = \eta - \eta_0 = \frac{n\kappa\zeta}{4kT(1-\zeta\tau)}\int q_x^2 q_y^2 \psi^0 d\mathbf{q} \qquad (16.4.14)$$

Carrying out the integration indicated in the preceding equation gives

$$[\eta] = \frac{\zeta}{1-\zeta\tau}\frac{kT}{\kappa}\frac{N_{\mathrm{Av}}}{100M} \qquad (16.4.15)$$

Avogadro's number, N_{Av}, the molecular weight, M, and the factor of 100 arise from converting from the number density, n, to the concentration in grams per deciliter. Of course, this result can be expressed in terms of the mean square length of the spring instead of the spring constant using the relation $b^2 = kT/\kappa$, as follows immediately from the equipartition theorem.

The more realistic Rouse–Zimm theory follows essentially the same path as that which we have outlines here. The displacements of the individual springs can be described as a set of coupled, damped, harmonic oscillators. The Smoluchowski equation for the distribution of the configuration of the beads can be simplified by a principal axis transformation and then various transport phenomena studied. By treating only the simple model of the dumbbell we have tried to illustrate how the Brownian motion interacts with the fluid flow and the intramolecular degrees of freedom without getting lost in the details of the more complicated, albeit more realistic, models.

For polymer science, an important property of the intrinsic viscosity is its dependence on the polymer's molecular weight or degree of polymerization. In the limit of very strong hydrodynamic interaction the Zimm theory predicts $[\eta] \propto N^{1/2}$. Intermediate values of the hydrodynamic interaction are also covered by the theory. The reader who wishes to pursue the subject further should be well prepared to read further.

16.5 Historical Remarks and Additional Reading

What we have done in the previous sections of this chapter is to slight the polymer science aspects of the topics we have treated in order not to obscure the Brownian motion aspects. In this concluding section I hope to compensate for this in part, by pointing the reader to where he or she can find more information.

The development already given relies heavily on the treatment given by Bird *et al.* (1977). Apparently the first person to suggest that a useful simple model for a polymer chain in solution might be a dumbbell was Kuhn (1934).

These models, treat the case of rather flexible molecules. In the opposite limit, the molecule may be assumed to be a body which completely impedes the flow of fluid through it. Einstein's calculation of the intrinsic viscosity of rigid sphere solutes is the first example of this genre of model. Many authors generalized Einstein's calculations to particles of non spherical shape, in particular to ellipsoids; a review of this material has been given by Frisch and Simha (1956). Kuhn and Kuhn (1945) were apparently the first to point out that, for particles with degrees of freedom other than translation, Brownian motion had to be taken into account in formulating a theory of their transport properties. These degrees of freedom could be rotational, affecting the orientation of a rigid particle with respect to a fluid flow field, or configurational, as in flexible macromolecules.

In the 1940s the problem of transport in dilute polymer solutions became popular, due, no doubt, to the growing industrial importance of polymers, and the realization that the scientific problems they posed were both interesting and tractable (though difficult). Debye and Bueche (1948) and Kirkwood and Riseman (1948) presented models based on chain models for the polymer molecules, as was indeed the earlier calculation of Kramers (1945). The calculations of Debye and Bueche and Kirkwood took hydrodynamic interaction into account, while that of Kramers did not.

Shortly thereafter, Kirkwood (1949, 1953) published a rather general theory of irreversible processes in dilute polymer solutions. This theory was based on the idea explained in Section 16.1; the motion of the macromolecule takes place on a submanifold of its $3N$ dimensional configuration space. This submanifold inherits a metric from the ambient $3N$ dimensional space and is thus a Riemannian space. All of the powerful apparatus of Riemannian geometry can be brought to bear on studying the motion of the polymer segments on the submanifold. Unfortunately this very elegant theory was marred by an error which was pointed out and corrected by Ikeda (1956) and by Erpenbeck and Kirkwood[32] (1963). The corrected theory has not, to the best of my knowledge, been used for the calculation of any specific transport coefficient. A particularly clear exposition of this theory has been presented by Yamakawa (1971).

The theories of Rouse and Zimm, followed next. This is the theory which we have presented here applied to the dumbbell model, a much simpler system than those considered by the originators of the theory. The Rouse–Zimm theory

[32] Posthumously

contains almost all of the essential physics of dilute polymer solutions' transport properties and is much easier to use than the previously mentioned theories. It has therefore been taken over as the theory of choice when applications to the interpretation of experiments are at issue. An excellent account of the theory has also been given in the book of Yamakawa (1971). It is interesting to note that Zimm, in his original paper, stated that 'Our study may be considered to be an application of Kirkwood's general formalism to a particular example.' We do not believe that this characterization does justice to the ideas introduced by Rouse and Zimm.

With respect to the subject of constraints, an interesting paper concerning Brownian motion when the Brownian particle is constrained to move on a submanifold of its full phase space should be mentioned (Van Kampen 1986). Suppose a system with N degrees of freedom must move in a space of $M < N$ dimensions; generally this is an M dimensional submanifold of the N dimensional space. This could happen because of boundary conditions, because of conservation laws, or because of constraints. Examples of constrains could be constant bond length and constant bond angle in molecules, or constant distance between all particles which is what one means by a rigid body. Further suppose that the motion of the system in the original N dimensional space is described by a Smoluchowski or a Kramers–Klein equation. What is the equation of motion of the system when described by coordinates restricted to the M dimensional submanifold?

There is no general answer one can give to this question. The answer depends on the dimensionality of the original space, N. It also depends on the type of constraint. When faced with a situation where a reduction in dimensionality to a submanifold is required, the only safe procedure is start from the assumed Smoluchowski or Kramers–Klein equation of motion in the ambient N space and carry out the reduction to the M space explicitly. The result may or may not be a Smoluchowski or Kramers–Klein equation in the reduced dimensionality space. In fact, it was just this circumstance that led to the error in Kirkwood's original submanifold theory (mentioned three paragraphs above). Kirkwood postulated an equation of Smoluchowski type on the submanifold of independent degrees of freedom of the polymer. In fact, starting from a Smoluchowski equation for the $3N$ degrees of freedom before applying constraints, and then projecting this onto the submanifold of reduced dimensionality generates a term involving components of the fluid velocity vector lying outside of the tangent space to the submanifold, and also terms in the friction tensor involving components both inside and outside of that tangent space.

The constraints of constant bond length and constant bond angle or rigid body constraints are in fact idealizations. In physical systems constraints of this type are ways of approximating the effects of interparticle forces that keep bonds of almost fixed length, angles at almost fixed values, or bodies almost rigid. Ideally, the problem should be set up in the full $3N$ dimensional space with the potentials of these forces included, as well as the noise. Then, in order to introduce the idealized constraints, the limit where the potentials are infinitely confining should be taken. In practice one does not proceed this way. The idealized limit of replacing stiff potentials by constraints is taken before adding the noise. It is not known, in general, when these two orders of operations result in the same equation of motion. It is known (Helfand 1979) that the order of turning stiff potentials into constraints and evaluating the partition function in equilibrium statistical mechanics do not commute. Caution is called for.

In this chapter we have concentrated on the Brownian motion aspect of the kinetic theory of macromolecules and have given scant justice to the polymer science aspects. Furthermore, we have restricted ourselves to transport in dilute solutions, so dilute that the polymer molecules do not interact with one another. In recent years, the emphasis in the theory of polymers has shifted to concentrated solutions and to polymer melts. In addition, the topic of *excluded volume* has drawn much attention. Excluded volume means that two segments of a polymer molecule cannot occupy the same region of space at the same time; the chain cannot intersect itself All of the theories we have mentioned so far ignore this effect. Thus their applicability are restricted to so called θ-solvents. In poor solvents the repulsive forces between solvent molecules and polymer segments give rise to an effective mean attraction between polymer segments. On the other hand, the excluded volume effect arises from a repulsive interaction between segments. θ-solvents are poor solvents for which these attractive and repulsive forces just cancel each other out as far as their effect on chain dimensions is concerned. The reader wishing to follow up further the topic of excluded volume is referred to any textbook on polymer science. The book by Yamakawa (1971) discusses the effect of excluded volume on polymer dynamics, and that by Bird *et al.* (1977) has a very brief discussion and some references. The book by de Gennes (1979) also has a good qualitative discussion. The monograph of Doi and Edwards (1987) is also an excellent source of information on theories of polymer dynamics.

17

INTERACTING BROWNIAN PARTICLES

17.1 Effects of Concentration

So far in this book we have been considering the motion of a single Brownian particle in a medium. We have studied this problem from several points of view, stochastic and molecular, but nevertheless we have never considered more than one Brownian particle at a time. Of course, almost any experimental realization of Brownian motion will involve many Brownian particles. Generally the Brownian particles are so dilute that they do not influence each of other's motion to any appreciable extent. This assumption underlies everything studied so far in this book.

It is now time to generalize our considerations to consider more concentrated solutions of Brownian particles, solutions where the particles come close enough to interact from time to time. These interactions can be of two kinds. There can be direct intermolecular interactions including excluded volume effects (since excluded volume is caused by repulsive intermolecular forces). It also includes hydrodynamic interactions, of the type discussed in Section 16.2. As we shall see, however, it is not sufficient to approximate the Brownian particles as point centers of force. It is necessary to use the hydrodynamic interaction appropriate for the boundary conditions at the actual particle-medium interface, for example at the surface of a sphere for spherical Brownian particles. To find this hydrodynamic interaction in a multi particle system is a difficult hydrodynamical problem.

Thus there are two sets of problems associated with the description of Brownian motion in concentrated solutions. The first is to find the appropriate equations of motion describing the evolution of the system. These equations, whatever their form, will include hydrodynamic interaction in some guise. The second set of problems is to derive the appropriate interactions.

In our study of single Brownian particles, we noted the separation of time scales between the random motion of the medium particles and the motion of the Brownian particle. This separation is due to the great disparity in mass between the two types of particles. To be sure, we found that there were slow motions in the medium also, the collective hydrodynamic modes. These modes interacted with the Brownian particle and this gave rise to considerable complication in deriving the evolution equations, the Fokker–Planck equation or the Langevin equation.

These complications are surely not erased by the introduction of additional Brownian particles, but a new complication of similar nature arises. If we focus our attention on a small number of Brownian particles, and consider the others as part of the medium, then part of the medium will move on the fast (solvent particle) time scale, while part will move on the slow (Brownian particle) time scale. This must be taken into account in finding the evolution equations for the small set of selected particles. To the best of my knowledge, this program of simultaneously accounting for both the slow motion of the neighboring Brownian particles and the slow hydrodynamic modes of the medium has never been carried out. In this chapter we present the present state of knowledge in this area.

17.2 The Fokker–Planck Equation

What is the equation of evolution for the distribution function of a single Brownian particle in a solution containing many Brownian particles? What is the equation of evolution for the distribution function of n Brownian particles? These are the questions that we shall discuss in the current section. With respect to the first question, intuitively we expect that the equation of evolution will be the familiar Fokker–Planck equation, but that the value of the parameters, the coefficients $\mathbf{A}_{ij}$ and $\mathbf{B}_{ij}$ will be modified. For the second question, intuition is of little help. We need a formal development to learn what the necessary modifications are.

We begin as usual by writing down the Hamiltonian of the system. It is useful to make a distinction in the notation between the Brownian particles and the medium particles; we shall denote the positions and momenta of the Brownian particles by capital letters, $\mathbf{R}, \mathbf{P}$, and those of the medium particles by lower case letters, $\mathbf{r}, \mathbf{p}$. The mass of the Brownian particles is M and that of the medium particles is m. It is not difficult to generalize the development here to include the possibility of several kinds of medium particles being present, but we shall stick to the one component solvent case. The Hamiltonian is

$$H = \frac{P_0^2}{2M} + \sum_{j=1}^{N} \frac{P_j^2}{2M} + \sum \frac{p_k^2}{2m} + U \tag{17.2.1}$$

Here we have $N+1$ Brownian particles, and U is the interaction potential between all of he particles present. It contains Brownian–Brownian, solvent–solvent, and Brownian–solvent interactions.

The Liouville operator for the system can be written in the form

$$\begin{aligned} L &= L_F + L_B \\ L_F &= -i[\sum_{j\neq 0}(\frac{\mathbf{P}_j}{M}\cdot\nabla_{R_j} + \mathbf{F}_j\cdot\nabla_{P_j}) + \sum_k(\mathbf{p}_k\cdot\nabla_{r_j} + \mathbf{f}_k\cdot\nabla_{p_k})] \\ L_B &= -i(\frac{\mathbf{P}_0}{M}\cdot\nabla_{R_0} + \mathbf{F}_0\cdot\nabla P_0) \end{aligned} \tag{17.2.2}$$

L_B is the Liouville operator of Brownian particle 0 and L_F is the Liouville operator of the rest of the system. Clearly this treats particle 0 unsymmetrically from the rest of the Brownian particles present. The reason for doing this is that we are ultimately interested in the one-body distribution function, so it is useful to concentrate on one particular particle. The Liouville equation for the entire system is

$$i\frac{\partial \rho_{N+n+1}}{\partial t} = L\rho_{N+n+1} \tag{17.2.3}$$

We introduce the one-body distribution function in the usual manner

$$\rho_1(\mathbf{R}_0, \mathbf{P}_0) = \int \rho_{N+n+1} d^n\mathbf{r} d^n\mathbf{p} d^N\mathbf{R} d^N\mathbf{P} \tag{17.2.4}$$

From this point on, the development is similar to that of Section 10.3. We introduce a projection operator, $\wp$,

$$\wp f = \eta \int f d^n\mathbf{r} d^n\mathbf{p} d^N\mathbf{R} d^N\mathbf{P} \tag{17.2.5}$$

where the function η is independent of $\mathbf{P}_0$ and is normalized to unity. The complementary projection operator is $1-\wp$. Introducing the 'relevant' and 'irrelevant' functions

$$\begin{aligned} \wp\rho_{N+n+1} &= \eta\rho_1 = f \\ (1-\wp)\rho_{N+n+1} &= g \end{aligned} \tag{17.2.6}$$

The methods of Section 10.3 lead to the equation of motion for f

$$i\frac{\partial f}{\partial t} = \wp L f + \wp L e^{-i(1-\wp)Lt)} g(0) - i\wp L \int_0^t e^{i(1-\wp)L(t-s)}(1-\wp)Lf(s)ds \tag{17.2.7}$$

As in Section 11.2, we take $g(0) = 0$.

The terms in L_B are, on the average, $(m/M)^{\frac{1}{2}}$ times smaller than those in L_F. There are some terms in L_F, those due to Brownian particle motion and Brownian particle–Brownian particle interaction, that are also small, However, we are going to expand in powers of L_F and there is no requirement that one expand everything that depends on a small parameter in powers of that parameter. There are known situations where that is, in fact, a bad thing to do. It gives rise to secular terms; compare the discussion in Section 8.3. Consequently, if we expand the exponential operator in eqn (17.2.7) and keep only the lowest order non-vanishing term, we find

$$i\frac{\partial f}{\partial t} = \wp L f - i\wp L \int_0^t e^{-i(1-\wp)L_F(t-s)}(1-\wp)Lf(s)ds \tag{17.2.8}$$

There is a free choice of the function η, subject to the requirements stated after eqn (17.2.5). We choose η to be the equilibrium canonical distribution function for the system of $N+n$ particles. The momentum of particle zero does not appear in η; particle zero is treated as a fixed center of force as far as η is concerned.

The specific forms of L_F , L_B, and η can be inserted into eqn (17.2.8) and manipulations similar to those of Section 11.2 carried out yielding

$$\frac{\partial \rho_1}{\partial t} + \frac{\mathbf{P}_0}{M} \cdot \nabla_{R_0} \rho_1 = \int_0^t < \mathbf{F}_0(t-s)\mathbf{F}_0(0) > : \nabla_{P_0} \cdot (\nabla_{P_0} + \frac{\mathbf{P}_0}{MkT}) \rho_1(s) ds \tag{17.2.9}$$

In this equation the time evolution of the average $< \mathbf{F}(t-s)\mathbf{F}(0) >$ is to be computed with Brownian particle zero held fixed; all of the other Brownian particles are free to move under the influence of their interactions. Assuming that the force correlation function has a short correlation time, the integral can be replaced by the ordinary Fokker–Planck collision term as discussed in Section 11.2. It is usually assumed that this can be done, but there is no good proof of it.

If we had expanded the exponent in eqn (17.2.7) bringing down all of the Brownian particle terms, i.e. if we had redefined L_B, then we would have obtained an equation superficially identical to eqn (17.2.9) but with an altered prescription for the computation of the force correlation. In the altered prescription, all of the Brownian particles would be held fixed in the dynamical computation of $\mathbf{F}$, and the result then averaged over all configurations of the Brownian particles. This would lead to a constant term in the force correlation, and consequently a secular term in the integral equation (17.2.9). This problem is avoided by the method we have used.

The force–force correlation function appearing in eqn (17.2.9) has contributions from several sources. First of all there are the forces due to the solvent molecules exerted on the Brownian particle. It is these forces which generate the force–force correlation function in the case of an infinitely dilute solution. These forces will, however, be modified by the additional Brownian particles in the solution which will affect the motions and distribution of the solvent particles. It is this effect, the motion of Brownian particles influencing the motion of medium particles which in turn exert forces on other Brownian particles, which we have called hydrodynamic interaction in Chapter 16. Secondly, there are direct intermolecular interactions between the Brownian particles, for example excluded volume. Both of these types of effects will depend on the concentration of Brownian particles. Hence the force–force correlation function will be concentration dependent.

The derivation given here is due to Mazo (1969a). It is subject to the same criticism given in Section 11.3 for the infinite dilution case, compounded by the fact that there are some additional slow components in the random force on the Brownian particles, those caused by the slow motions of the other Brownian particles. Attempts to remove these problems have been made by Deutch and

Oppenheim (1971), but to my knowledge no study of the multiparticle case similar to the mode coupling approach outlined in Section 11.4 has yet been carried out. Such a study should show how the hydrodynamic interaction arises from molecular considerations.

An analogous development can be carried out for the distribution functions of several Brownian particles, (Mazo, 1969c; Deutch and Oppenheim, 1971). We shall not go into the details of the derivations here, contenting ourselves with stating that the resulting Fokker–Planck equation is just what one might expect intuitively; for the sake of simplicity we write it down for the case $n = 2$.

$$\frac{\partial \rho_2}{\partial t} + (\frac{\mathbf{P}_1}{M} \cdot \nabla_{P_1} + \frac{\mathbf{P}_2}{M} \cdot \nabla_{P_2})\rho_2 = \sum_{i,j=1}^{N} \nabla_{P_i} \cdot \zeta_{ij} \cdot (\nabla_{P_j} + MkT\nabla_{P_j})\rho_2 \quad (17.2.10)$$

One can see that the formal treatments of multi-Brownian particle systems carried out so far yield very sensible results, in accord with what would have been guessed. Yet there is still work to be done to put these developments on as firm a footing as for the case of a single Brownian particle.

17.3 The Multiparticle Smoluchowski Equation

In previous chapters we have seen that the full Fokker–Planck equation contains more detail than is necessary for many applications. If we are willing to wait until the momentum distribution has practically equilibrated then a closed description in the configuration space of the Brownian particles rather than in their phase space is possible. This is the description afforded by the Smoluchowski equation.

In Chapter 8 we showed how the Smoluchowski equation can be derived from the Fokker– Planck equation by the technique of elimination of fast variables. The same technique can be applied to the multiparticle case. Murphy and Aguirre (1972) and Aguirre and Murphy (1971, 1973) have carried this out in detail, starting from the many body Fokker–Planck equation (17.2.10). If we adjoin, for the sake of generality, external force terms, $\mathbf{F}_{j,\text{ext}}$ which we have omitted from eqn (17.2.10) and denote the N particle distribution function in configuration space by f_N, the N-body Smoluchowski equation takes the form

$$\frac{\partial f_N}{\partial t} = \sum_{i,j=1}^{N} \nabla_i \cdot \mathbf{D}_{ij} \cdot (\nabla_j - \frac{\mathbf{F}_{j,\text{ext}}}{kT}) f_N \quad (17.3.1)$$

The diffusion tensors $\mathbf{D}_{ij}$ are defined as follows: consider the $3N \times 3N$ matrix

$$\mathbf{Z} = \begin{bmatrix} \zeta_{11} & \cdots & \zeta_{1N} \\ \ldots & \ddots & \ldots \\ \zeta_{N1} & \cdots & \zeta_{NN} \end{bmatrix}$$

where each ζ_{ij} is itself a three by three matrix, the friction constant matrix for particles i and j, which appear in eqn (17.2.10). Define the matrix $\mathbf{D}$ by $\mathbf{D} = kT\mathbf{Z}^{-1}$. $\mathbf{D}$ can also be partitioned into three by three matrices

$$\mathbf{D} = \begin{bmatrix} \mathbf{D}_{11} & \cdots & \mathbf{D}_{1N} \\ \ldots & \ddots & \ldots \\ \mathbf{D}_{N1} & \cdots & \mathbf{D}_{NN} \end{bmatrix} \tag{17.3.2}$$

These are the $\mathbf{D}_{ij}$ that appear in eqn (17.3.1).

For $N = 2$, a short calculation yields the explicit equations

$$\begin{aligned} \mathbf{D}_{11} &= kT(\zeta_{11} - \zeta_{12}\zeta_{22}^{-1}\zeta_{21})^{-1} \\ \mathbf{D}_{12} &= kT(\zeta_{12} - \zeta_{11}\zeta_{21}^{-1}\zeta_{22})^{-1} \\ \mathbf{D}_{21} &= kT(\zeta_{21} - \zeta_{22}\zeta_{12}^{-1}\zeta_{11})^{-1} \\ \mathbf{D}_{22} &= kT(\zeta_{22} - \zeta_{21}\zeta_{11}^{-1}\zeta_{12})^{-1} \end{aligned} \tag{17.3.3}$$

It is of interest to get explicit expressions for the **D**s for the idealized case when the only force on the Brownian particles is the hydrodynamic interaction, and this hydrodynamic interaction is given by the Oseen approximation. For the sake of simplicity we assume that all the Brownian particles are identical and have the same friction constant ζ. For the following derivation we shall adopt the notational convention that bold face characters without subscripts will refer to vectors or tensors in the $3N \times 3N$ dimensional configuration space of the Brownian particles while those with subscripts will refer to three-dimensional vectors or tensors. The vector $\mathbf{v}^0 = (\mathbf{v^0}_1, \mathbf{v}_2^0, ..., \mathbf{v}_N^0)$ where each $\mathbf{v}_j^0$ is the fluid velocity that would be present at the position of particle j if that particle were not present. $\mathbf{v} = (\mathbf{v}_1, \mathbf{v}_2, ..., \mathbf{v}_N)$ are the actual velocities at those positions, and $\mathbf{T}$ is the matrix of Oseen tensors $\mathbf{T}_{ij}$. The Oseen tensor $\mathbf{T}_{ii}$ is not defined, of course; for the purpose of our notation we define it to be zero.

The velocity vector is then given by

$$\mathbf{v} = \mathbf{v}^0 - \zeta\mathbf{T} \cdot \mathbf{v} \tag{17.3.4}$$

The forces on the various Brownian particles are

$$\mathbf{F} = \zeta\mathbf{v} = \mathbf{Z} \cdot \mathbf{v}^0 \tag{17.3.5}$$

Solving these two equations for **Z**,

$$\mathbf{Z}^{-1} = \frac{1}{\zeta}(\mathbf{1} + \zeta\mathbf{T}) \tag{17.3.6}$$

Reverting now to the 3×3 matrices, from the definition of the diffusion coefficient matrices, $\mathbf{D}_{ij}$, we find that

$$\mathbf{D}_{ii} = \frac{kT}{\zeta}\mathbf{1}; \quad \mathbf{D}_{ij} = kT\mathbf{T}_{ij} \;\; (i \neq j) \tag{17.3.7}$$

It should be clear that the first T in the expression for $\mathbf{D}_{\mathrm{ij}}$ stands for temperature while the second, with subscripts, stands for the Oseen tensor.

Note that the diffusion tensors are now position dependent; $\mathbf{D}_{ij}$ depends on $\mathbf{R}_i$ and $\mathbf{R}_j$. This remarkably simple result holds only in the Oseen approximation; when the hydrodynamic interaction between the Brownian particles is described by a more accurate description it is no longer necessarily even true that $\mathbf{D}_{ij}$ depends only on the coordinates of particles i and j. The fact that the diffusion tensors are not constants means that the Smoluchowski equation, (17.3.1), will be more difficult to solve.

17.4 The Diffusion Coefficient

The title of this section refers to *the* diffusion coefficient. In fact, even in the case of a single solute species in a solvent there is more than one diffusion coefficient. First, there is the coefficient that occurs in Fick's law, relating the diffusion flux to the concentration gradient. This is called the *mutual diffusion* coefficient. Second, there is the coefficient that describes motion of a single, tagged, solute particle in a homogeneous solution without concentration gradient. This is called the *self-diffusion* coefficient. In previous chapters we did not make a distinction between these quantities because we only considered solutes at infinite dilution; at infinite dilution the self-diffusion and mutual diffusion coefficients are numerically equal. Since we are now considering interactions between solute particles, we are forced to distinguish between the two concepts. They differ at non-vanishing concentration. A discussion of the diffusion coefficients on the basis of irreversible thermodynamics is given in Appendix F.

When eqn (17.2.9) is converted to a conventional Fokker–Planck equation by assuming that the force correlation function on the right-hand side has a short correlation time (compare the transition from eqn (11.2.5) to eqn (11.2.7)) the friction constant thus defined corresponds to motion of particle 0 in a homogeneous solution. This is because the initial condition chosen for eqn (17.2.7) was that of the remaining particles in equilibrium with particle 0. This is clearly not appropriate for describing a gradient in concentration of the Brownian particles. Consequently, the diffusion coefficient appearing in the Smoluchowski equation is the self- diffusion coefficient, albeit at non-zero concentration.

Had we wanted to describe a system with a concentration gradient, we should choose as initial condition the surroundings in local equilibrium, that is with a built in concentration dependence of the local chemical potentials and hence a built in concentration gradient (Mazo 1969b).

The generalized wave number and frequency dependent diffusion coefficient has been defined in eqn (11.4.13). This definition also holds at finite concentrations. The diffusion coefficient that is measured in a conventional concentration gradient diffusion experiment is given by

$$D = \lim_{z\to 0}\lim_{q\to 0} D(\mathbf{q}, z) \tag{17.4.1}$$

$D(\mathbf{q}, z)$ can be measured in a light or neutron scattering experiment. Both D and $D(\mathbf{q}, z)$ will be concentration dependent when interactions between Brownian particles are important.

When interactions are described in the simplest approximation, that discussed at the end of the last paragraph, purely hydrodynamic interactions given by the Oseen approximation, there is no concentration dependence (Altenberger and Deutch 1973). To see this we consider the Smoluchowski equation (17.3.1) without the external force terms. Integrate over the positions of all the particles except those of particle zero. Now note that

$$\nabla_{R_i} \cdot \mathbf{D}_{ij} = \nabla_{R_j} \cdot D_{ij} = 0 \tag{17.4.2}$$

This can easily be verified by direct computation, or it can be seen physically by noting that in the model being used, the fluid is incompressible, i.e. $\nabla \cdot \mathbf{v} = 0$. Since $\mathbf{v} = \mathbf{T} \cdot F$, eqn (17.4.2) follows. Consequently, after integration by parts, we arrive at the ordinary diffusion equation with diffusion coefficient kT/ζ, exactly the same as for a single particle.

The conclusion to be drawn from this short computation is that the description of the hydrodynamic interaction is very important for describing the concentration dependence of the diffusion coefficient. The Oseen approximation describes the interaction well at large distances between the particles (assumed spherical), but is not a good approximation at short distances. Clearly the short distance behavior is vital to a correct description. And of course, the effect of direct intermolecular forces between the Brownian particles must also be considered. Many workers have contributed to this problem. It is a very difficult one, and numerically different answers can be found in the literature. It is now fairly widely agreed that the problem has reached a definitive solution, at least for hard sphere Brownian particles, in the work of Batchelor (1976) and Felderhof (1978). We shall present the method of Felderhof here because it can be extended to the case of the generalized diffusion coefficient, $D(\mathbf{k}, z)$; the method of Batchelor does not appear to lend itself to such generalization. We shall concentrate here on the mutual diffusion coefficient. The self diffusion coefficient can be treated in a similar manner.

First note that the mutual diffusion coefficient can be written in the form

$$D_c = \frac{1}{N_{Av}\zeta} c_0 (\frac{\partial \mu_0}{\partial c_0})_{T,P} \tag{17.4.3}$$

where ζ is the friction constant and μ_0 and c_0 are the chemical potential and concentration of the Brownian solute, respectively. The derivation is the same as that of the Einstein relation, eqn (4.5.4); here, however, the solution is not considered necessarily dilute-ideal, and the friction coefficient may be concentration dependent. Use

$$\begin{aligned} \mu_0 &= \mu_0^* + RT \ln \gamma_0 c_0 \\ \zeta &= \zeta_0 (1 + \kappa c_0) \end{aligned} \tag{17.4.4}$$

and write in eqn (17.4.3)

$$\lim_{c_0 \to 0} \left(\frac{\partial \ln \gamma_0}{\partial c_0}\right)_{T,P} = \frac{B_2}{\bar{V}_0} \tag{17.4.5}$$

where B_2 is the second osmotic virial coefficient of the solute, γ_0 is its activity coefficient and $\bar{V}_0$ is its partial molar volume. Thus, to first order in the concentration, the diffusion coefficient is given by

$$D = \frac{kT^0}{\zeta} \left[1 + c_0(B_2/\bar{V}_0 - \kappa)\right] \tag{17.4.6}$$

We start with the Smoluchowski equation (17.3.1) for N Brownian particles in a solvent. The forces $\mathbf{F}_j$ that occur in it are the intermolecular forces existing between the various Brownian particles; there are no external forces present in the current application. The diffusion tensors, $\mathbf{D}_{ij}$, are functions of the positions of all of the Brownian particles; this is the source of the difficulties of the problem.

At low particle density we shall assume that the diffusion tensors can be written in the form

$$\mathbf{D}_{ij} = D_0 \mathbf{1} \delta_{ij} + \mathbf{A}_{ij} + \mathbf{B}_{ij} \tag{17.4.7}$$

where $\mathbf{D}_0$ is the diffusion coefficient of a Brownian particle at infinite dilution, and $\mathbf{A}_{ij}$ and $\mathbf{B}_{ij}$ are hydrodynamic interaction tensors defined by

$$\begin{aligned} \mathbf{u}_1 &= \frac{1}{kT}(D_0 + \mathbf{A}_{12}) \cdot \mathbf{F}_1 + \frac{1}{kT}\mathbf{B}_{12} \cdot \mathbf{F}_2 \\ \mathbf{u}_2 &= \frac{1}{kT}\mathbf{B}_{21} \cdot \mathbf{F}_1 + \frac{1}{kT}(D_0 \mathbf{1} + \mathbf{A}_{21}) \cdot \mathbf{F}_2 \end{aligned} \tag{17.4.8}$$

The $\mathbf{A}_{ij}$ and $\mathbf{B}_{ij}$ tensors depend on the coordinates of particles i and j alone; they are generalizations of the Oseen tensor which take proper account of the boundary conditions at the surface of the solute particle. Physically what has been assumed here is that at low densities the hydrodynamic interaction between particles can be well represented by that between pairs of particles. Three body and higher order hydrodynamic interactions certainly exist. However, at low concentrations, when the solute particles are far apart on the average, configurations where three bodies are close enough to exert sensible interactions are rare compared to those in which two bodies alone are in proximity. We are going to calculate the variation of the diffusion coefficient linear in concentration; compare eqn (17.4.4). If we were to go to the next order term, quadratic in concentration, then three body hydrodynamic interactions would have to be taken into account also.

From the Smoluchowski equation, an equation for the distribution function of N particles, we need to obtain the equation of evolution for the one particle distribution function $n_1(\mathbf{r})$. This can be done by integrating eqn (17.3.1) over the coordinates of all but one particle, particle 1, say. Because the $\mathbf{D}_{ij}$'s are position dependent, the equation so obtained will not be a closed equation for n_1, but will also contain $n_2(\mathbf{r}_1, \mathbf{r}_2)$. Well, then we can derive an equation for n_2 by integrating eqn (17.3.1) over the coordinates of all but two particles. Unfortunately this equation contains n_3, the three body distribution function. And so on. The procedure of integrating, holding fixed larger and larger groups of particles generates an entire hierarchy of equations, not a small closed set of equations. This is a common occurrence in many body problems in theoretical chemistry and physics. Examples are the BBGYK hierarchy for molecular distribution functions in the theory of fluids, the hierarchy of equations for Green's functions in quantum field theory, and the hierarchy of equations for reduced density matrices in quantum mechanics.

The only known way to reduce the hierarchy to a closed equation or small set of equations is to introduce some approximation. In the present case there is a small parameter, the concentration of solute, although this parameter does not occur specifically in eqn (17.3.1). We shall try to perform a development in this parameter. To obtain the diffusion coefficient correct to order c_0 we must have n_2 correct to its lowest non vanishing order, c_0^2. This means we must have n_1 also correct to this order.

The equations for n_1 and n_2 are

$$\begin{aligned}\frac{\partial n_1}{\partial t} &= D_0 \nabla_1 \cdot (\nabla_1 n_1 + \frac{1}{kT} \int \nabla_1 U(r_{12}) n_2(\mathbf{r}_1, \mathbf{r}_2) d\mathbf{r}_2) \\ &+ \nabla_1 \cdot \int \mathbf{A}_{12} \cdot [\nabla_1 n_2 + \frac{1}{kT} \nabla_1 U(r_{12}) n_2] d\mathbf{r}_2 \\ &+ \nabla_1 \cdot \int \mathbf{B}_{12} \cdot [\nabla_2 n_2 + \frac{1}{kT} \nabla_2 U(r_{12}) n_2] d\mathbf{r}_2 \end{aligned} \tag{17.4.9}$$

and

$$\begin{aligned}\frac{\partial n_2}{\partial t} = & D_0 \nabla_1 \cdot [\nabla_1 n_2 + \frac{1}{kT} \nabla_1 U(r_{12}) n_2] + \\ & D_0 \nabla_2 \cdot [\nabla_2 n_2 + \frac{1}{kT} \nabla_2 U(r_{12}) n_2] + \\ & \nabla_1 \cdot \mathbf{A}_{12} \cdot [\nabla_1 n_2 + \frac{1}{kT} \nabla_1 U(r_{12}) n_2] + \\ & \nabla_2 \cdot \mathbf{A}_{21} \cdot [\nabla_2 n_2 + \frac{1}{kT} \nabla_2 U(r_{12}) n_2] + \\ & \nabla_1 \cdot \mathbf{B}_{12} [\nabla_2 n_2 + \frac{1}{kT} \nabla_2 U(r_{12}) n_2] + \\ & \nabla_2 \cdot \mathbf{B}_{21} \cdot [\nabla_1 n_2 + \frac{1}{kT} \nabla_1 U(r_{12}) n_2] + \\ & \text{terms involving } n_3(\mathbf{r}_1, \mathbf{r}_2, \mathbf{r}_3) \end{aligned} \tag{17.4.10}$$

The $\mathbf{A}_{ij}$ and $\mathbf{B}_{ij}$ tensors are those occurring in the hydrodynamic interaction, eqn (17.4.8). U is the potential of intermolecular force. Consistent with what we have just said about the order of magnitude of terms, we neglect all terms involving n_3 in eqn (17.4.10). This is our required low density expansion of the equation of motion. To this order, n_1 and n_2 are decoupled from all of the higher equations of the hierarchy.

Note that if we work only to zeroth order in the concentration, we can ignore eqn (17.4.10) entirely and neglect the terms involving n_2 in eqn (17.4.9). Then eqn (17.4.9) becomes just the ordinary diffusion equation with diffusion coefficient D_0. This is precisely what should happen in this limit.

Equations (17.4.9) and (17.4.10) have time independent equilibrium solutions

$$\begin{aligned} n_1(\mathbf{r}) &= n_0 \\ n_2(\mathbf{r}_1, \mathbf{r}_2) &= n_0^2 g_0(r_{12}) \\ g_0(r_{12}) &= e^{-U(r_{12})/kT} \end{aligned} \tag{17.4.11}$$

Let us consider deviations from the equilibrium solution, $n_1(\mathbf{r}, t) = n_0 + \delta n$, $n_2 = n_0^2 g_0 + \delta n_2$. Placing these definitions in eqn (17.4.9), and keeping terms only up to linear order in n_0, we obtain

$$\begin{aligned} \frac{\partial \delta n}{\partial t} &= D_0 \nabla_1 \cdot \nabla_1 \delta n + \frac{n_0}{kT} \int \nabla_1 U g_0 \delta n(\mathbf{r}_2, t) d\mathbf{r}_2) \\ &+ n_0 \nabla_1 \cdot \int \mathbf{A}_{12} g_0 d\mathbf{r}_2 \cdot \nabla_1 \delta n(\mathbf{r}_1, t) \\ &+ n_0 \nabla_1 \mathbf{B}_{12} g_0 \cdot \nabla_2 \delta n(\mathbf{r}_2, t) d\mathbf{r}_2 \end{aligned} \tag{17.4.12}$$

If the intermolecular potential, U, is short range, then we may write, without sensible error, $\delta n(\mathbf{r}_2, t) = \delta n(\mathbf{r}_1, t)$ and $\nabla \delta n(\mathbf{r}_2, t) = \nabla \delta n(\mathbf{r}_1, t)$ when δn occurs under an integral sign with U. Furthermore, we may approximate

$$\delta n(\mathbf{r}_2, t) = \delta n(\mathbf{r}_1, t) + (\mathbf{r}_2 - \mathbf{r}_1) \cdot \nabla_1 \delta n(\mathbf{r}_1, t) \tag{17.4.13}$$

Hence eqn (17.4.12) becomes

$$\frac{\partial \delta n}{\partial t} = D \nabla^2 \delta n \tag{17.4.14}$$

with D given by

$$D = D_0 + n_0 \Big(\int (e^{-U/kT} - 1) d\mathbf{r}_2 + \int \mathbf{A}_{12} g_0 d\mathbf{r}_2 + \int \mathbf{B}_{12} g_0 d\mathbf{r}_2 \Big) \tag{17.4.15}$$

The first term in $D - D_0$ is the thermodynamic correction, B_2, of eqn (17.4.4); the remaining two terms are due to the hydrodynamic interaction.

To evaluate these terms, one must of course know $\mathbf{A}_{ij}$ and $\mathbf{B}_{ij}$. For the model of particles imbedded in a continuum fluid, this is a problem of hydrodynamics. It would take us too far afield to develop the necessary tools for attacking this problem here; we shall only outline the results that have been found. Batchelor, whose method was quite different from that which we have just sketched, obtained the $\mathbf{A}_{ij}$ and $\mathbf{B}_{ij}$ functions by extracting results from the literature on several different placements of the two solute particles relative to the ambient flow, $\mathbf{u}$. This, together with analytic asymptotic results, enabled him to get accurate numerical values for the $\mathbf{A}_{ij}$ and $\mathbf{B}_{ij}$ functions. Felderhof and coworkers (Cichocki *et al.* 1988) have derived series expansions for these functions for spheres in powers of a/R, the ratio of the sphere radius to the interparticle distance. Batchelor's result is[33]

$$\begin{aligned} D &= D_0(1 + 1.45\varphi) \text{ mutual diffusion} \\ D_c &= D_0(1 - 1.83\varphi) \text{ self-diffusion} \end{aligned} \tag{17.4.16}$$

Felderhof originally obtained 1.56 instead of 1.45 (Felderhof, 1978). This difference was caused by an insufficiently accurate calculation of the $\mathbf{A}_{ij}$ and $\mathbf{B}_{ij}$ functions. In subsequent papers, appearing after the work of Cichocki *et al.* a more accurate series expansion was used and the quoted result is now 1.454 for the mutual diffusion coefficient (Felderhof and Vogel 1992). For the self-diffusion coefficient, Batchelor's result was confirmed with a slight improvement in accuracy (Cichocki and Felderhof 1988, 1989). These results are all for hard sphere solutes with stick boundary conditions.

Note that the value 1.454 is the difference between the thermodynamic correction, B_2, and the hydrodynamic correction, κ . $B_2/\bar{V}_0 = 8$ for hard spheres, and K_2 is roughly 6.5. Hence relatively small inaccuracies in κ can lead to much larger percentage inaccuracies in D.

The relation of these results to experiment is not so clear. Experiments are done on real molecules, which are not hard spheres. The Brownian particle should have a large mass, and hence should be a polymer. In order to make the polymers simulate hard spheres as well as possible, they should be compact, which means they should be in a poor solvent. Thus there is not a great range of concentrations available to experiment. Globular proteins in solutions of high ionic strength are often compact, even though soluble. Gallagher and Woodward (1989) have studied the diffusion coefficient of a small protein, bovine pancreatic trypsin inhibitor, by light scattering. They use the hard sphere theory, not as a model for their system, but as a benchmark to try to estimate what other effects might be occurring. In their case, they were particularly interested in whether or not dimers were forming.

[33] Although we have not discussed the analysis of the self-diffusion coefficient, we have included the result here for reference.

But even if there is no dimerization, there are additional effects to be taken into account in the theory having to do with the effect of the Debye cloud surrounding the Brownian particle on the hydrodynamic interaction. It would be relatively easy to insert the screened Coulomb potential instead of the hard sphere potential in eqn (17.4.15). However, there ought to be additional effects; the electric field of a charged Brownian particle will tend to hold counterions close to it, thus altering the flow field of the solvent medium. Such effects are well studied for small electrolytes. To the best of my knowledge, this mechanism has not been studied for the case of Brownian particles.

The diffusion coefficients of suspensions of silica spheres in hexane have been studied by Kops-Werkhoven and Finaut (1981). The silica spheres were about 16 nm in radius and were covered by a layer of C_{18} hydrocarbon chains bonded to the surface in order to stabilize the suspension. Thus the silica spheres were not precisely 'hard', but are expected to reasonably well described by the hard sphere model. The conclusion from the experiments is that the linear departure from the infinite dilution value of the mutual diffusion coefficient was described by the Batchelor - Felderhof theory within experimental error.

Marcus *et al.* (1996) have studied the self diffusion of poly(methyl methacrylate) spheres of approximately 1 μm diameter in a thin cell of about 3 μm thickness; the distance from the particle boundary to the walls was thus roughly one particle diameter. This experimental system is free from the electrolyte effects just mentioned. However, the system is too thick to be treated as two-dimensional but too thin to be modeled as three-dimensional without explicit consideration of wall effects. Their data indicates that the coefficient of φ in eqn (17.4.16) is approximately -48 instead of -1.83. The walls will have an appreciable effect on the hydrodynamic interaction, but whether this can explain a factor of 26 in the coefficient is not clear. A theoretical discussion of such quasi-two-dimensional systems has been given by Schofield *et al.* (1996), but more remains to be done here.

We now go on to discuss the wave number-frequency dependent diffusion coefficient of a Brownian particle. This is the quantity that is measured in a dynamic light scattering experiment. Such an experiment measures the correlation function of the electric field scattered by a sample. For a solution of Brownian particles, of much greater mass than the solvent particles, the polarizability of the Brownian particles exceeds that of the solvent many fold. Consequently the scattering power of the Brownian particles dominates that of the solvent, and practically all of the scattering from the solution may be considered to come from the solute. The correlation function of the scattered electric field is proportional to the correlation function of the particle density

$$S(\mathbf{q}, z) = < n_q(z) n^*_{-q}(0) > = \frac{< n_q n^*_{-q} >}{z + q^2 D(\mathbf{q}, z)} \tag{17.4.17}$$

This equation is the definition of $D(z, \mathbf{q})$ as explained in Section 11.4. The Fourier component of the particle density is defined by

$$n_q(t) = \frac{1}{\sqrt{N}}\sum_{j=1}^{N} e^{i\mathbf{q}\cdot\mathbf{r}_j(t)}$$

$$< n_q(t)n^*_{-q}(0) > = \frac{1}{N}\sum_{j=1}^{N}\sum_{l=1}^{N} e^{i\mathbf{q}\cdot(\mathbf{r}_j(t)-\mathbf{r}_l(0))} \quad (17.4.18)$$

and the Laplace transform of the correlation function must be taken..

The calculations of the wave number dependent diffusion coefficient are much more involved than that of the zero wave number limit. The reason is that, as we can see from eqn (17.4.17), the two body correlation function, n_2, is necessary to calculate $S(\mathbf{q}, z)$. n_2 must be known to order n_0^3 in order to calculate $D(\mathbf{q}, z)$ to order n_0 (or φ). The reason is that a factor of n_0^2 cancels out between $< n_{\mathbf{q}}(z)n^*_{-\mathbf{q}} >$ and $< n_{\mathbf{q}}n^*_{-\mathbf{q}} >$ on the left and right sides respectively of eqn (17.4.15). We shall here only give references to the detailed analysis. The basic theory and specialization to the hard sphere system are given by Felderhof and Vogel (1992a, 1992b). A version of this calculation with a somewhat less accurate treatment of the hydrodynamic interaction is given by Dhont (1996). It is worthwhile pointing out that Dhont (and some other authors) do not work with $D(\mathbf{q}, z)$, but with a quantity we shall here call D' defined by

$$S(\mathbf{q}, t) = S(\mathbf{q})e^{-q^2 D'(\mathbf{q},t)t} \quad (17.4.19)$$

The relation between $D'(\mathbf{q}, t)$ and $D(\mathbf{q}, z)$ is well defined but not simple. They are not a Laplace transform pair.

17.5 The Viscosity

The only other transport coefficient of Brownian particle suspensions that is of much importance is the viscosity. In this section we revert to using the volume fraction, φ, as the measure of Brownian particle concentration. As mentioned in Chapters 1 and 16, Einstein (1906a, 1911) found that the viscosity of a dilute suspension of spheres is given by $\eta = \eta_0(1 + 2.5\varphi)$. This was a deduction from a purely hydrodynamic calculation; Brownian motion theory played no role in it. However, if we want to push the calculation to higher order, $\eta = \eta_0(1 + 2.5\varphi + k\varphi^2 + \cdots)$, Brownian motion does enter into the determination of k.

It is not surprising that Brownian motion plays no role in the intrinsic viscosity of structureless particles, but does affect terms of higher order in the concentration. We have seen in Chapter 16, in our discussion of the intrinsic viscosity of polymers, that it is the relative motion of frictional elements in the polymer chain that gives rise to the dependence of intrinsic viscosity on degree of polymerization. Similarly, it is the analogous relative motion of pairs, triplets, etc., of structureless particles, that yield the higher order effects in their solutions.

The treatment of higher order corrections to Einstein's intrinsic result is complicated. Many authors over many years tried to go beyond leading order in the volume fraction; the answers obtained varied considerably. We shall discuss here what we believe to be the best extant modern treatment for the model of impenetrable rigid bodies immersed in a continuum fluid (Wajnryb and Dahler 1997). The formal theory is not restricted to spherical objects, but implementation of the general theory for non spherical bodies is very difficult. Several different boundary conditions have been considered. Since a major part of the theoretical development is fluid dynamical, and not dependent on Brownian motion theory, we intend to merely sketch the ideas involved instead of presenting a detailed treatment of the theory.

Suppose we have N rigid Brownian particles immersed in a continuum fluid of viscosity η_0. The fluid obeys the Navier–Stokes equations for an incompressible fluid in the region exterior to the particles; we shall call this the exterior region.

$$\begin{aligned} \eta_0 \nabla^2 \mathbf{v}^0 - \nabla p + \mathbf{F}^0(\mathbf{r}) &= 0 \\ \nabla \cdot \mathbf{v}^0 &= 0 \end{aligned} \tag{17.5.1}$$

where $\mathbf{v}^0$ and $\mathbf{F}^0$ are the velocity of the fluid and the external force density on the fluid in the exterior region. Suitable boundary conditions, depending on the assumed nature of the fluid–particle interface are imposed at the boundary of the exterior region and the particles. Since we are not going to do any detailed calculations, we need not be specific. The reader may wish to think in terms of the usual 'stick' boundary conditions: the relative velocity of fluid and particle vanishes at the boundary.

Suppose that we could, in fact, solve these equations for a specific configuration of the Brownian particles. Such a solution would be useless since the Brownian particles are not fixed in space, but undergo Brownian motion. We should have to solve the equations for a great many configurations of the suspended particles and average over the solutions. Once we start averaging over spatial configurations, the distinction between the external region and the region inside the Brownian particles (where the Navier–Stokes equations do not hold) becomes meaningless. We can write the averaged version of eqn (17.5.1) as

$$\begin{aligned} \eta_0 \nabla^2 < \mathbf{v} > - \nabla < p > + \mathbf{F}^0(\mathbf{r}) + < \mathbf{F}(\mathbf{r}) > &= 0 \\ \nabla \cdot < \mathbf{v} > &= 0 \end{aligned} \tag{17.5.2}$$

Here the carets denote the average velocity, pressure, etc., and $< \mathbf{F}(\mathbf{r}) >$ is the average force density exerted by the suspended particles on the matter present at the point $\mathbf{r}$, while $\mathbf{F}^0$ is the density of external force, if any. $< \mathbf{F} >$ contains both the intermolecular forces of repulsion and the hydrodynamic interaction. It thus depends on $< \mathbf{v} >$, so that eqn (17.5.2) is not a closed equation.

The next step is to find the relation between $< \mathbf{F}(\mathbf{r}) >$ and $< \mathbf{v}(\mathbf{r}) >$. This is a long and complicated exercise carried out in detail by Wajnryb and Dahler, who construct a linear operator $\mathbf{B}$, connecting them

$$< \mathbf{F}(\mathbf{r}) >= -\int \mathbf{B}(\mathbf{r},\mathbf{r}') < \mathbf{v}(\mathbf{r}') > d\mathbf{r}' \tag{17.5.3}$$

It is in this construction that Brownian motion theory appears. We shall return to the genesis of **B**, but for the moment let us go on to show how the modified viscosity arises. $\mathbf{B}(\mathbf{r},\mathbf{r}')$ turns out to depend only on the difference of its arguments in a translationally invariant system such as a fluid. Its Fourier transform therefore depends on only one wave vector, **q**. In Fourier space eqn (17.5.3) can be expressed as

$$< \mathbf{F}(\mathbf{q}) >= -\mathbf{B}(\mathbf{q})\cdot < \mathbf{v}(\mathbf{q}) > \tag{17.5.4}$$

where we have committed a slight abuse of notation by using the same symbol for a quantity and its Fourier transform. The argument of the function will make it clear which is meant in a given situation. The quantity $\mathbf{B}(\mathbf{q})$ has not been constructed as one unitary expression, but rather as a cluster expansion in the number of particles involved in the interaction. Wajnryb and Dahler show that, in each order, s, in the number of particles

$$\mathbf{B}_s(\mathbf{q}) = \eta_0 b_s \varphi^s q^2 < \mathbf{v}(\mathbf{q}) > +O(q^4) \tag{17.5.5}$$

where b_s is a constant, an algorithm for whose calculation is given by the theory.

Now eqn (17.5.2) can also be written in Fourier space as

$$\begin{aligned} -\eta_0 q^2 < \mathbf{v}(\mathbf{q}) > -i\mathbf{q}p(\mathbf{q}) + < \mathbf{F}^{(0)}(\mathbf{q}) > - \\ \eta_0 q^2 \textstyle\sum_s b_s \varphi^s < \mathbf{v}(\mathbf{q}) > +O(q^4) = 0 \end{aligned} \tag{17.5.6}$$

Neglecting the $O(q^4)$ terms, we see that eqn (17.5.6) is equivalent to a Navier–Stokes equation in which the bare fluid viscosity has been replaced by an effective viscosity

$$\eta = \eta_0(1 + \sum_s b_s \varphi^s) \tag{17.5.7}$$

This is exactly what we wanted to show.

Wajnryb and Dahler give results for b_1 and b_2 for several boundary conditions; we quote here only their result for stick conditions. For b_1, Einstein's value of 2.5 is obtained; this was only to be expected. For b_2, the authors obtained the value 5.9148; this is to be compared to the value of 6.2 obtained by Batchelor (1977). Wajnryb and Dahler claim that the difference is not due to error in the formalism of either set of authors, but that their own formulation expedites accurate numerical implementation.

Now let us return to eqn (17.5.4). The deduction of this equation, together with an explicit form for **B**, is the core of the entire development. It requires many pages of calculation and the introduction of several new intermediate definitions. The present author despairs of rephrasing it in a form which is both comprehensible and short enough to be included in this chapter. We shall therefore restrict ourselves to some qualitative remarks.

On the one hand, it should not be surprising that $< \mathbf{F} >$ can be linearly related to $< \mathbf{v} >$. The force $\mathbf{F}$ (before averaging) will depend linearly on the fluid velocity through the friction coefficient or better, the N body friction matrix. This linearity will be retained under the averaging. Brownian motion has a tendency to render uniform the distribution of the relative positions of the particles, to homogenize them, while the deforming motion of the fluid has a tendency to make certain configurations more probable than others. This is the physical reason for the effect of Brownian motion on viscosity of a suspension. This effect does not show up in the $O(\varphi)$ term, since this term only depends on a single particle; that is why Einstein's calculation had no reference to Brownian motion. It will, however, make a contribution to the $O(\varphi^2)$ term, and higher terms.

In eqn (17.5.5) $\mathbf{B}(\mathbf{q})$ was asserted to be proportional to q^2, plus higher order terms in q that we neglected. This means we replaced a non local integral operator (compare eqn (17.5.3)) by a local operator, for q^2 in wave number space is equivalent to the Laplacian operator in ordinary 3-space. In other words, neglecting higher order terms in q is the same as saying that the frictional forces which occur at particle–fluid interfaces for a given particle configuration, stay well localized after the averaging over configurations. This is, of course, an approximation, but it is only in this sense that the suspension has an ordinary Newtonian viscosity. That is, if it were not approximately correct, the Navier–Stokes equations for the suspension would have a non-local stress tensor. It is not out of the question that this might occur at higher concentrations.

All the work discussed here refers to the zero frequency viscosity, that pertinent to the steady state flow of a fluid. If the fluid flow oscillates with a frequency ω, then the viscosity, even of a pure fluid, has a frequency dependence, $\eta_0(\omega)$. The effective viscosity of a suspension then contains a frequency dependence in addition to that of the neat solvent. An interesting discussion of this in the high frequency limit, where the inverse frequency is much shorter than the Brownian relaxation time has been given by Cichocki and Felderhof (1988).

This is as far as we shall take the question of the viscosity of suspensions. The interested reader is encouraged to consult the original papers of Batchelor and of Wajnryb and Dahler for the technical details of the theory.

17.6 Concluding Remarks

In this chapter we have studied aspects of how the interaction between Brownian particles in solution at finite concentration affect the properties of the solution or suspension. The reader will have noticed that the developments we discussed were based on extensions of the Fokker–Planck equation or the Smoluchowski equation to the many particle situation. These extensions were made on an intuitive basis; the general structure of the dynamical equations known from the one body case were kept and terms were added expressing the interactions with the same general structure. This involved generalizing the friction constant and the diffusion constants, which may have been tensors in ordinary space, to make them also tensors in the space of particle indices. To illustrate these remarks we recall how the operator $D\nabla^2$ in the one body Smoluchowski equation is replaced by a sum of operators of the form $\nabla_i D_{ij}(R_{ij})\nabla_j$ in the many body case.

In Section 17.2, we showed how such a generalization could be formally derived from the Liouville equation. We must emphasize, however, that this formal generalization has not been carried out with the same depth as for the single particle case studied in Chapter 11. In particular, the coupling of hydrodynamic modes to single particle modes, pair modes, or in general few body modes, has not been investigated. This is a fruitful subject for research.

One of the principle applications of the study of interacting Brownian particles has been to dynamic light scattering. We have mentioned this topic only briefly, and it is proper to give some references for the interested reader. The standard text on this subject is Berne and Pecora, (1976). An interesting article of Hess and Klein (1983) discusses the Brownian motion of interacting particles from the point of view of generalized hydrodynamics, that is hydrodynamics with space and time dependent transport coefficients. A more recent review by Klein (1997) brings much of this material more up to date.

Although the theories discussed have been fairly general with respect to the intermolecular potential between the Brownian particles, and hence with respect to the particles' shape, such numerical calculations as have been reported have been for the case of spheres. Furthermore, the theories are designed to give the transport coefficients as a series in the Brownian particles' density or volume fraction. Such series may not converge rapidly, so stopping with the lowest order correction to the infinite dilution limit may only be comparable with experiment at relatively low concentrations. The extension of Brownian motion theory to moderate instead of just low, concentrations is another field in which very little has been done, but whose pursuit would be quite worthwhile.

18

DYNAMICS, FRACTALS, AND CHAOS

18.1 Brownian Dynamics

The previous chapters in this book have treated the statistical properties of Brownian motion. This chapter is devoted to several topics related to the properties of individual Brownian trajectories and only indirectly to ensembles of such trajectories. Of course, statistics will enter in some guise or other during the course of the development.

Statistical mechanics used to be characterized as a science based on desperation. The standard methods of analytical mechanics could not be used to study macroscopic molecular systems because these systems contained too many particles for known methods to be tractable. A statistical treatment of such many body systems replaced the study of individual cases differing by their initial conditions.

With the advent of fast computing machines, it was realized that this is no longer the case. The equations of motion of systems of several hundred particles can be integrated numerically. The motion of a single such system can be followed through phase space for reasonable periods of time before excessive numerical error accumulated, and useful information about the system under study can be calculated from the phase space trajectory. All this can be done efficiently and relatively inexpensively at the present time. This method of numerical investigation is called the *molecular dynamics* method.

There is a problem, though, when the molecular dynamics method is applied to study Brownian motion or in general, mixtures of large and small molecules. The problem arises from the fact that the motion of the large and small molecules occur on different time scales, as we have emphasized many times already in this book. When a finite difference numerical integration scheme is used to solve the equations of motion numerically, the time step in the numerical routine must be short enough to capture the fastest events that contribute importantly to the time evolution of the system. In the kind of system we are discussing here, this time is of the order of the collision time of the small particles. On the other hand, we are primarily interested in phenomena occurring on the time scale of the large particles. Nevertheless, we must calculate these quantities using the short time step dictated by the light particle time scale. To get trajectories of useful length on the heavy particle time scale we must compute exceedingly long trajectories on the light particle's time scale.[34] This makes standard molecular dynamics time-consuming and expensive for systems with widely disparate time scales.

A variant of molecular dynamics, called *Brownian dynamics*, has been developed to circumvent this difficulty (Turq, Lantelme, and Friedman 1977; Ermak and McCammon, 1978; Fixman, 1978). In this method the object is not to integrate Newton's or Hamilton's equations for the system. Instead, the Langevin equation is adopted as the basic equation of motion and an algorithm is developed for integrating it directly. The positions and momenta of the small (solvent) molecules thus never appear in the problem and only the large particle time scale is relevant.

The price paid for the elimination of the fast time scale is the necessity for simulating the random force in the Langevin equation. More information about the random force than just its mean and variance is required. Of course, in such a method it is assumed from the start that the Langevin equation is correct on the time scale considered. It is not a method for assessing the validity of the Langevin equation.

As we have seen in Chapter 5, a stochastic differential equation such as the Langevin equation has different properties than differential equations for sure (non random) functions. Standard numerical methods for sure differential equations must be modified to apply to stochastic differential equations. In particular, convergence and stability questions must be reexamined. Convergence refers to how rapidly a numerical solution approached the exact solution as the size of the time step is decreased; stability refers to the propagation of errors due to the inevitable inaccuracies in the numerical solution. We shall not enter into stability questions here.

The method we shall expound is the one proposed by Lax (1966, sec. 3); for an introduction to the mathematical literature, see Gard (1988; sec. 7.2). A useful introductory review has been given by Wolf (1996).

Let us begin by writing the general Langevin equation (5.2.1), which we express in the form of eqn (5.22)

$$x(t) - x(0) = \int_0^t f(x(s), s)ds + \int_0^t g(x(s), s)dW(s) \tag{18.1.1}$$

where W is the Wiener process and f and g are sure functions of their arguments. We want to solve this equation numerically over some finite time period, T, $0 \leq t \leq T$. If this were an ordinary instead of a stochastic equation, we would turn it into a finite difference equation by breaking up the interval of interest in t

[34]A conceptually similar situation sometimes occurs in chemical kinetics. The overall rate of a reaction with a complex mechanism is dominated by the rate of the slowest step. But the rate of this step is affected by the concentrations of species created or destroyed in fast steps. Rate equations having this property are called *stiff* equations. They require special algorithms for efficient integration.

into N intervals of length $\Delta t = h$ with $Nh = T$. Writing $\bar{x}_n$ for the approximation to $x(nh)$ given by the finite differences, we write, in the simplest method

$$\bar{x}_{n+1} = \bar{x}_n + hf(\bar{x}_n, nh) + g(\bar{x}_n, nh)\Delta W \tag{18.1.2}$$

where $\Delta W = W_{n+1} - W_n$. Since W has independent increments, it is not necessary to subscript ΔW. We would expect, and it is a standard result of numerical analysis for sure equations, that the error of this approximation is $O(h^2)$. However, in the stochastic case when W is the Wiener process we have seen that ΔW is of order $h^{1/2}$, not h. If we want an accuracy of order h in our finite difference approximation to eqn (18.1.1) we must use a better approximation scheme than (18.1.2). In the stochastic case, we are, of course speaking of average errors, and the common average used in this stochastic numerical analysis error estimation is root-mean-square error.

The scheme (18.1.2) is derived from eqn (18.1.1) by expanding x in a Taylor's series in h superficially keeping only terms of order h. We just saw, however, that this expansion really contained a term of order $h^{1/2}$, which is larger than h for small h. We may therefore expect that a more accurate treatment of the Wiener process term will generate addition contributions of order h. A scheme suggested by Lax (1966) gets at these contributions by iterating eqn (18.1.1) once. The f term is already of order h, so nothing needs to be changed in it. The g term is written

$$g(x(t+s)) = g(x(t)) + g'(x(t))\Delta x(s) \tag{18.1.3}$$

where the prime refers to differentiation with respect to x. But we know that the dominant term in Δx is the one of order $h^{1/2}$, so we insert this in eqn (18.1.3) to obtain

$$\bar{x}_{n+1} = \bar{x}_n + f(\bar{x}_n, nh) + g(\bar{x}, nh)\Delta W + gg'(\Delta W)^2 \tag{18.1.4}$$

It is, of course, possible to derive more accurate schemes, analogous to Runge–Kutta schemes for sure differential equations, but these have not been used in the Brownian dynamics context. Equation (18.1.4) is the simplest finite difference approximation to the Langevin equation that can be profitably used.

The Langevin equation we have been considering here is one with multiplicative noise. Thus the question of interpretation discussed in Section 5.3 arises again. In writing eqn (18.1.2) as first approximation over one time increment, we used the Ito rule, but in going from (18.1.2) to (18.1.4) by iteration we introduced an extra integral. It turns out that the scheme (18.1.4) converges, not to the Ito solution of eqn (18.1.1), but to the Stratonovich solution of eqn (18.1.1). According to the discussion of Section 5.3, this is usually the solution desired. If a numerical approximation to the Ito solution of (18.1.1) is desired, it is necessary to modify the drift term, $f \to F = f + \frac{1}{2}gg'$ and to look for the Stratonovich solution of the so modified equation.

Let us now take the abstract functions, f and g, and replace them by physically meaningful specific forms. There are two useful and interesting situations. In the first, we are interested in the relaxation of the momentum of the Brownian particle, or system of Brownian particles. Then the many body Langevin equation is of the form (Deutch and Oppenheim 1971)

$$\mathbf{v}_i = -\sum_j \zeta_{ij} \cdot \mathbf{v}_j dt + \mathbf{F}_{\text{mol},i} dt + \mathbf{F}_{\text{hyd},i} dt + d\mathbf{F}_{\text{rand},i} \tag{18.1.5}$$

ζ_{ij} is the friction tensor between particles i nd j, $\mathbf{F}_{\text{mol},i}$ is the intermolecular force on particle i, $\mathbf{F}_{\text{hyd},i}$ the hydrodynamic force on i, and $\mathbf{F}_{\text{rand},i}$ the random force. The random force can be written

$$d\mathbf{F}_{\text{rand},i} = \sum_j \alpha_{ij} \cdot dW_j \tag{18.1.6}$$

where $\mathbf{W}_i(t)$ is a Wiener process with zero drift and unit diffusion coefficient, and

$$\sum_l \alpha_{il} \cdot \alpha_{jl} = kT\zeta_{ij} \tag{18.1.7}$$

The writing of the random force in this form merely amounts to a convenient normalization, following Ermak and McCammon. The Wiener processes are correlated in the usual way,

$$< d\mathbf{W}_{\mathrm{j}}(t) \cdot d\mathbf{W}_k(s) >= 2kT\mathbf{1}\delta_{jk}\delta(t-s)$$

As we have seen in the previous chapter, the friction constants are not, in fact constants when many Brownian particles are considered. They depend on the relative positions of the particles. Equation (18.1.6) implies that will be the case also for the random forces (and thus the αs). Hence eqn (18.1.6) is of the form of eqn (18.1.1) with g not a constant; the velocity random process is a multiplicative stochastic process.

The second situation is one in which we are interested only in the diffusive motion of the system after the momentum relaxation has occurred; we might call this the Smoluchowski regime. Here we neglect the inertial term in the Langevin equation and write

$$-\zeta \cdot d\mathbf{x} + \mathbf{F}_{\text{mol}} dt + \mathbf{F}_{\text{hyd}} dt + d\mathbf{F}_{\text{rand}} = 0 \tag{18.1.8}$$

This is a simpler set of equations than eqn (18.1.5), but still refers to a multiplicative stochastic process.

To solve for a particular sample path of (18.1.8), the scheme (18.1.4) is adopted. The changes $\Delta\mathbf{W}$ are, of course, only known statistically, so, to determine a particular realization $\Delta\mathbf{W}_i(t)$ must be chosen from a set of random variables having a Gaussian distribution with zero mean and variance $2kTt\mathbf{1}$.

There are standard algorithms for choosing random numbers with such a distribution. Then eqn (18.1.4) provides a straightforward recipe for computing $\mathbf{r}_{n+1}$ from $\mathbf{r}_n$. This gives a broken line approximation to a particular realization of the random process. To compute something that can be compared to experiment or to the result of an analytical theory, this procedure must be repeated for a large number of sample paths, the desired quantity calculated for each of them, and the results averaged.

We shall describe a particular application that is related to the theory of diffusion controlled reactions, the analytical theory of which was discussed in Section 14.2 (Northrup, Allison, and McCammon 1984). Recall that, in that section we discussed the spherically symmetric free diffusion of reactants to a spherical reactive center, a sink at which reaction occurred immediately upon contact. We now want to generalize the problem by allowing the reactive center to be non spherical, to have a non-uniform distribution of reactivity over its surface, and to have a nonzero probability of leaving a contact without reaction. It is too difficult to make these generalizations analytically, but possible to do so numerically.

Take the reactive center and surround it by a sphere or radius b. The region $r > b$ will be called the outer region; the region $r < b$ will be called the inner region. The reactive flux of particles to the reactive center can be written

$$j = j_0 p \tag{18.1.9}$$

where j_r is the total normal component of the reactive flux through the surface at radius b and j_0 is the normal component of the flux of particles reaching this surface *for the first time.* p is the probability that a particle, having reached the surface at b, eventually reacts instead of escaping to infinity. It may recross the surface many times before it finally reacts.

j_0 has already been evaluated in Section 14.2. It is just the flux into the sphere of radius b with absorbing boundary conditions. The requirement that the boundary condition should be absorbing is equivalent to the requirement that only particles reaching the surface for the first time are to be counted in j_0. Thus $j_0 = 4\pi D_0 b \rho_0$ where D_0 is the diffusion coefficient for the relative motion of reacting species and ρ_0 is the concentration of diffusing reacting particles (compare eqn (14.2.9)). So the object is to compute p by Brownian dynamics, by following the trajectories of particles that have reached the surface at b and seeing which of them eventually react.

Particles that reach b can cross back into the outer region before they react. We must decide how long to follow these particles before declaring that they have escaped and will never (or with negligible probability) reenter the inner region and react. This can be done in the following way. Let β_∞ be the probability that a particle, having reached b, will impact the reactive center at least once before escaping to infinity. Let ξ be the probability of reaction upon an impact and let Δ_∞ be the probability that a particle which collides with the active center but does not react collides again before escaping to infinity. The probability that the diffusing particle will react after one collision is then $\xi\beta_\infty$. The probability that it will react on its second collision is $\beta_\infty(1-\xi)\Delta_\infty\xi$, on its third, $\beta_\infty(1-\xi)^2\Delta_\infty^2\xi$, and so on. Adding these all up,

$$p = \frac{\beta_\infty\xi}{1-\Delta_\infty(1-\xi)} \tag{18.1.10}$$

ξ is a quantity that must be specified as part of the model, but β_∞ and Δ_∞ must still be computed. To determine β_∞ we must, in principle, follow trajectories for all times to see which ones will recollide. This is clearly not possible to do numerically, so we introduce a trick. We introduce a new sphere with a new radius, s, $s > b$; if a trajectory passes into the region exterior to s, we assume it will escape to infinity and terminate its calculation. The probability that a trajectory will collide at least once with the reactive center and also survive this truncation is no longer β_∞, but a new quantity, β. Let Ω be the probability that a particle starting at $r = s$ will eventually return to $r = b$. Then an argument similar to that leading to eqn (18.1.10) shows that

$$1-\beta_\infty = \frac{(1-\beta)(1-\Omega)}{1-(1-\beta)\Omega} \tag{18.1.11}$$

or

$$\beta_\infty = \frac{\beta}{1-(1-\beta)\Omega} \tag{18.1.12}$$

The quantity β only depends on the motion of the particle in the region $r < s$, a finite domain. Moreover, Ω only involves diffusion in the outer region, a region in which the diffusion may be treated analytically by the methods of Section 14.2.

It therefore only remains to compute Δ_∞. An argument similar to those already used shows that

$$\Delta_\infty = \Delta + \beta_\infty(1-\Delta) \tag{18.1.13}$$

where Δ is a new probability, the probability that having collided with the reactive site and leaving it, will recollide rather than escape to $r = s$. This quantity also only depends on the behavior of trajectories in the finite region $r < s$.

Putting all of these results together, we obtain for the rate constant $k = j/\rho_0$

$$k = k_0(b)\frac{\beta\xi}{1-(1-(1-\beta)\xi}\cdot\frac{1}{1-(1-\xi)[\Delta+(1-\Delta)(\frac{\beta}{1-(1-\beta)\Omega})]} \tag{18.1.14}$$

where $k_0(b)$ is the Smoluchowski rate for a reactive site of radius b. If intermolecular forces are present, as in the case studied in Section 14.3, the value of $k_0(b)$ must be modified accordingly.

Equation (18.1.14) looks rather complicated and it has four new quantities in it, ξ, β, Ω, and Δ. However, all of these objects are quantities that can be estimated directly by counting Brownian dynamics trajectories that satisfy the appropriate defining conditions. The complications arose because we want to compute only quantities depending on the properties of trajectories in a finite domain of space. It can be seen that determining what to compute may require considerable ingenuity and is just as necessary for a successful calculation as is the programming of the computation itself.

Northrup *et al.* (1984) have implemented the numerical program just described. They computed the reaction rates for several cases for which analytical results were known. The agreement between numerical and analytical results was excellent. Zhou (1998) has discussed alternative algorithms for diffusion limited rate calculations using Brownian dynamics.

The method has been applied to numerous reactions; we give reference to one as an example and a source of other references: (Zhou et al. 1997). Brownian dynamics has also been applied to phenomena other than diffusion limited reactions; compare (Fixman 1978), (He and Sheraga 1998) for application to the internal dynamics of proteins.

18.2 Brownian Paths as Fractals

The trajectories of a Brownian particle described by a Wiener process are very irregular. In fact, they are irregular on all scales and are differentiable almost nowhere. This is the origin of the $< \delta x^2 > \propto t$ behavior that we have noted again and again. Furthermore the trajectories are statistically self similar. This means that if distances are scaled by a factor $\lambda^{1/2}$ and time is scaled by a factor λ, all probabilities remain unchanged. Looking at a graph of a section of trajectory, we cannot tell at what magnification the graph is being viewed. Curves with this property are called *fractal curves.*[35]

There are two kinds of curves associated with a Brownian motion, and it is necessary to distinguish between them. The first kind is the graph of the position of a Brownian particle as a function of time. This kind of curve lives in an $N+1$ dimensional space, N space dimensions plus the time. Such a curve will be called a *Brownian function* (Mandelbrot 1982). The second kind of curve is a plot of the trajectory of the Brownian particle in its ambient N dimensional space; time varies along the curve, but one cannot tell how by looking at the curve. This curve is called a *Brownian trail* (Mandelbrot, loc. cit.).[36] It is the orbit of a Brownian particle.

Since fractal curves are irregular on all scales, it would be desirable to have a method to characterize the irregularity. Such a characterization is available in the concept of *fractal dimension.* We all have an intuitive feeling for what is meant

[35]The term *fractal* refers to objects more general than curves but we shall only be discussing the special case of curves in this chapter.

[36]The nomenclature is not uniform. For example, Addison (1997) uses the terms *trajectory* and *trace* instead of *function* and *trail.*

by the dimension of a geometric object. There are, however, a number of ways to make this intuitive idea precise. These ways are not all equivalent. Of course, all of them agree about the dimension of ordinary geometrical objects, e.g. smooth curves, smooth surfaces, solids bounded by smooth surfaces, etc. Otherwise we would not consider them as worthy of the name of dimension. But when applied to certain wild and irregular sets, the various definitions need not all give the same result. Thus non-standard objects have a plethora of dimensions, some of which we shall discuss below.

Mathematicians had studied the properties of some of these peculiar sets, for the most part regarding them as pathological examples. It was B. Mandelbrot in the 1970s who realized that these examples were not just bizarre inventions of the mathematician's imagination but were to be found in great profusion in nature. He coined the name *fractal* for such objects.

The simplest definition for the dimension of an object is the number of variables that must be specified in order to determine unambiguously the position of every point of the object. This definition presumes that the object is embedded in a space endowed with coordinates, a manifold. This dimension is called the *Euclidean dimension*, D_E of the object. For example, a curve in R^3 can be parameterized by three functions $x(t)$, $y(t)$, $z(t)$. Once these functions are given, only the value of the parameter t is needed to specify position in space along the curve. This concept coincides precisely with our intuitive idea of dimension. A more sophisticated definition of dimension is the *topological* dimension or *Brouwer* dimension, D_T, (Alexandrov, 1961). It is defined recursively by

(a) The empty set has dimension -1.
(b) The dimension of a set, X, is the smallest integer, D_T, such that every point of X has arbitrarily small neighborhoods whose boundaries have dimension less than D_T[37].

Thus a point, having no boundary, has topological dimension zero. The boundary of a line are its two end points which have dimension zero. Hence a line has dimension one. And so on. The mathematical advantage of this definition is that it is invariant under 1-1 continuous transformations. Note that no metric notions were necessary for its definition Thus D_T is a true topological invariant of a geometrical figure.

Both the Euclidean dimension and the topological dimension are, by their definitions, integers. Let us now go on to introduce several of the definitions of dimension, which need not be integer valued. These definitions are metric in nature and apply only to objects which are subsets of metric spaces. Perhaps the earliest of these is the Hausdorff–Besicovich dimension. This is defined as follows: cover the set in question with balls[38] of radius $\leq r$ and let

$$M(r) = \inf \sum_i r_i^D$$

[37]'Arbitrarily small' is here a topological rather than a metric concept. It means, given any point p, and any neighborhood, U, of p, there is an open set containing p and contained in U.

[38]A *ball* is a set of the form $\{x \mid |x - x_0| < a\}$ for some x_0 and for some a.

where the sum is over all of the sets of the covering and the inf (*infimum*, or greatest lower bound), is over all possible coverings with all $r_i < r$. Now take the limit of $M(r)$ as $r \to 0$. If D is large, this limit will be zero; if it is non-zero and finite for some D, it will vanish for any larger D. Similarly, for sufficiently small D, the limit will be infinity. There will be some particular value of D for which the limit exists. This value is D_H, the Hausdorff–Besicovich dimension of the set.

Why should we call this a dimension? Its results coincide with those of the intuitive notion of dimension for ordinary objects, lines, surfaces, solid bodies, etc. But for those sets irregular on all scales, D_H need not be the same as D_E or D_T. In fact, it need not be an integer. Mandelbrot's original definition of a fractal was a set whose Hausdorff dimension differed from its topological dimension. He soon retracted this definition for two reasons. First, there seemed to be no reason why the Hausdorff dimension was any more fundamental than some of the other generalized dimensions which have been suggested. Secondly, the Hausdorff dimension is essentially impossible to measure experimentally. One cannot pull objects with hypervolume of arbitrary dimension, D, off the shelf in the laboratory. The current definition of a fractal is, any object that has an interesting dimension different from its topological dimension. This is, of course, rather vague because 'interesting' is undefined, but this does not seem to cause any problem in practice.

A definition of dimension that is related to the Hausdorff dimension (but not equivalent to it) is the *box counting* dimension. To compute this dimension, divide the ambient space into non-overlapping boxes of side r and count the number, $N(r)$, of these boxes that contain points of the geometric object. Suppose it is found that $N(r) \sim r^{-D}$. Then $D = \lim[-\ln N/\ln r]$ where the limit is taken as $r \to 0$.

The reason this may be called a dimension is that, for ordinary objects the number of boxes pierced by a line varies as the length of the line divided by the side length of a box, r^{-1}. The number of boxes intersected by a surface varies as the area of the surface divided by the area of a box face, r^{-2}, etc. Notice that the Hausdorff dimension and the covering dimension will be the same if taking all of the covering sets to be the same size yields the infimum of $M(r)$. This will usually, but not always, be the case. In the rest of this chapter we shall understand the words *fractal dimension* to mean box counting dimension.

To put flesh upon these bones, we give two examples of fractal curves. The first is a much used example, the Koch curve. Take a line segment of length 1 and remove the middle third. Above the blank space on the line, erect a $\wedge$ consisting of two sides of a triangle of length $1/3$ (see Fig. 18.1). On each of the four line segments of length $1/3$, repeat the same construction; the straight line portions now, of course, have length $1/9$ and there are 16 of them. Repeat the process again; the straight line portions have length $1/27$ and there are 64 of them.

After n repetitions one has a very crinkly curve made up of 4^n straight line segments, each of length $(1/3)^n$. Thus $r = (1/3)^n$ and $N = 4^n$ so that

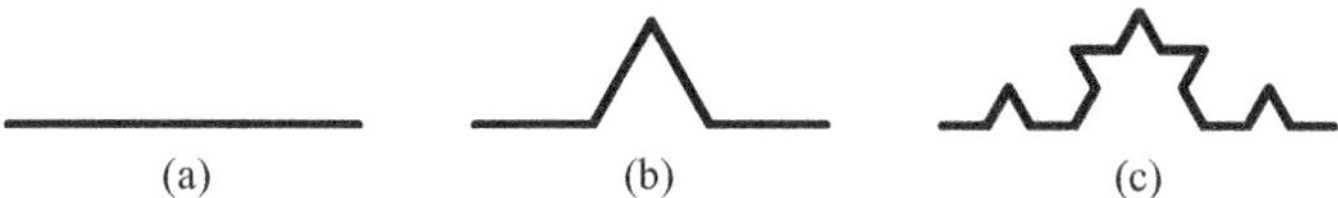

FIG. 18.1. The first three stages of the construction of the Koch curve.

$$D = -\frac{\ln N}{\ln r} = \frac{\ln 4}{\ln 3} = 1.2619\ldots \tag{18.2.1}$$

For finite n the curve just constructed is quite an ordinary curve, rather jagged, but of finite length and smooth except at finitely many points. In the limit as $n \to \infty$ however, it becomes infinitely jagged, has infinite length, and is non-differentiable. It behaves somehow as though it is 'fatter' than an ordinary curve and indeed, its box counting dimension is greater than that of an ordinary curve, though less than that of a plane area.

The Koch curve whose construction recipe has just been given, has a very regular rule for its construction; it is called a deterministic fractal. Many fractals are constructed using rules with a probabilistic element; they are called random fractals. We give one simple illustration by adding a random element to the construction of the Koch curve. The first step of the construction occurs as before. In the second step, instead of erecting the $\wedge$ shape always on the same side of the curve (say, the right side as we proceed from right to left along the curve), we erect it randomly, with probability $1/2$, either on the right side or on the left side. Three stages of a realization of this construction are shown in Fig. 18.2. It is obvious from the rules of construction that the box counting dimension of the probabilistic Koch curve is the same as that of the regular Koch curve. However, the curve does look more irregular to the eye.

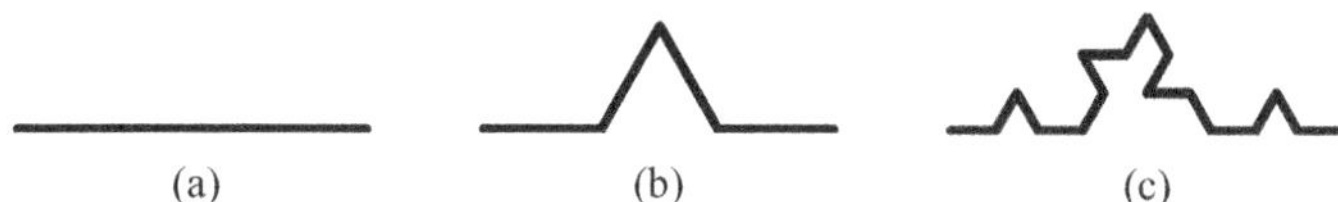

FIG. 18.2. The first three stages of the construction of a realization of the random Koch curve.

So far we have been considering mathematical objects for which the limit $r \to 0$ is meaningful. We should like to use these mathematical objects to model physical objects for which the limit may make no physical sense. For example, the Heisenberg uncertainty principle will certainly limit the length scale on which the geometric properties of a material object can be examined. This is an outside limit; in practice, deviations from an r^{-D} scaling are likely even before such length scales are attained. What is done is to measure or compute $N(r)$ over as wide a range of r as possible. Make a log–log plot of N versus r. If there is a range of several decades of r over which the log–log plot is linear with slope D, then we say that the object being studies is a fractal with fractal dimension D. For small r, presumably new physical processes become important, for example the fact that matter is atomic in nature. For very large distances the overall size of the object obviously provides a limit on the applicability of scaling.

What is the fractal dimension of a Brownian function? Of a Brownian trail? The question of lower length scale, discussed in the preceding paragraph, is not a problem here, since we shall be studying the Wiener process, not physical Brownian motion. After getting this far in this book, the reader should already be convinced that the Wiener process is a good representation of the physical Brownian motion over a wide range of length and time scales.

Since the Wiener process is a random process, a given Brownian curve is a particular realization of the process. Thus the number of boxes that cover one realization and the number that cover a different one (of the same time duration) may not be the same. However, different portions of the function are statistically the same if one rescales the time by a factor λ, and the space variable by the factor $\lambda^{\frac{1}{2}}$. Furthermore, we have seen that the difference between the maximum and minimum values of the realization in the time interval Δt are most likely to be of order of magnitude $(\Delta t)^{\frac{1}{2}}$. The covering of the part of the graph of the realization that lies over Δt then takes approximately $1/(\Delta t)^{1/2}$ squares.

It takes $T/\Delta t$ intervals to cover the time interval, T, of the portion of t being considered. Thus it takes $T(\Delta t)^{-3/2}$ boxes to cover the curve. Therefore, using the definition of box counting dimension (with $r = \Delta t$), we find $D = 3/2$ for the fractal dimension of a Brownian function.

The next question is the dimension of a Brownian trail. Starting from a given point on the trail, the portion of the trail traversed when the parameter changes by Δt lies within a box with side of order $(\Delta t)^{\frac{1}{2}}$ in length. Thus the number of boxes needed to cover the trail is of order $(\Delta t)^{-\frac{1}{2}}$ and the fractal dimension is 2. Thus a Brownian trail is rather plane like. The trail does not fill the plane, but it is clearly not an ordinary tame line. The fractal dimension of a simulated Brownian motion for a particle in a fluid has been calculated by Rappaport (1985); the result of the numerical experiment is $D = 2$ as expected.

These concepts can be generalized to a phenomenon called *fractional Brownian motion* (fBm). A fractional Brownian motion is a random process $\mathbf{X}(t)$:

(a) whose increments $\mathbf{X}(t) - \mathbf{X}(s)$ are Gaussian random variables with mean zero, and
(b) have variance $\mid t - s \mid^{2H}$, $0 < H < 1$.

H is called the *Hurst exponent*, after E. H. Hurst who introduced it in studying the statistics of the annual water level fluctuations of the Nile river. Arguments similar to those we have just given show that the fractal dimension of a fBm function is $2 - H$ while that of an fBm trail is $1/H$ except for motion in a plane. For a curve in the plane, the fractal dimension cannot exceed two so the dimension is $\min(1/H, 2)$ in that case.

Of course, ordinary, or regular, Brownian motion falls under the case $H = \frac{1}{2}$. Furthermore, for regular Brownian motion it is required that the increments be independent when the intervals do not overlap. There is no such requirement for fractional Brownian motion. A Gaussian process with independent increments is necessarily Markovian; fractional Brownian motion is not necessarily Markovian.

Fractional Brownian motion is useful in modeling certain physical phenomena. We have already mentioned hydrology; to this we might add landscapes and anomalous diffusion. On the other hand, there does not seem to be any dynamical theory giving rise to fBm.

An important use of fBm outside of the realm of physical science is computer graphics. Although that is beyond the scope of this book, it is worthwhile giving a brief outline of what is involved. Consider a stochastic process which is a function of two parameters, t_1 and t_2, $\mathbf{X}(t_1, t_2)$, and which is such that it is a fBm in each parameter separately. The variance of the increments of this process varies (by definition) as τ^H where

$$\tau = \sqrt((t_1 - t_1')^2 + (t_2 - t_2')^2)$$

The relation between the Hurst exponent, H, and the fractal dimension, D, is

$$D = D_E + 1 - H$$

where D_E is the Euclidean dimension of the object being considered. In the case now being discussed,the object is a surface, so $D_E = 2$.

Just as the graph of a Brownian function looks like the cross-section of a mountain range, the graph of a realization of such a two variable fBm, a surface lying over the (t_1, t_2) plane looks like the relief of a mountain range itself. For an ordinary Brownian motion, $H = 2$, so that $D = 5/2$. A surface so constructed (we shall not discuss algorithms for doing so) looks rather ragged. However, choosing $D = 2.1$ gives a surface that looks remarkably mountain like (Mandelbrot 1982; p. 264). Even more visually remarkable effects can be achieved by adding color to the algorithm. But these graphical applications would lead us too far afield and we shall not pursue them further.

18.3 Brownian Motion and Chaos

The motions of some physical systems are highly irreproducible in detail. There are two reasons for this. One reason is that there are random uncontrollable outside influences on the system ('noise'). The second is that there is an inherent sensitivity in the dynamics of the system; small deviations of the properties of the system give rise to large deviations in the eventual state of the system. This second circumstance is called *chaos* or, more descriptively, *deterministic chaos* to emphasize that randomness is not essential.

In this section we want to discuss an experiment that was alleged to demonstrates the chaotic character of Brownian motion. We should warn the reader that there is disagreement in the literature about the interpretation of the experimental results. However, the experiment is sufficiently interesting that we believe it worthwhile to present it. To do this, we shall first have to introduce the ideas and concepts on which the interpretation of the experiment is based.

No one has ever given an unambiguous definition which states necessary and sufficient conditions for chaos to be present. What is done instead is to list several characteristics which will enable chaotic behavior to be recognized when it is met, without pretending that the list is definitive. We shall illustrate this list on a simple one dimensional map, called the *binary shift map*. It is defined by

$$x_{n+1} = 2x_n \bmod 1 \tag{18.3.1}$$

This looks like a linear map, but it is not because of the *mod* 1 condition. It is not even a continuous map; $x = \frac{1}{2}$ is a point of discontinuity, but this will not affect our considerations. Let us express the numbers x in binary notation, as a decimal point followed by a string of zeros and ones. The effect of the transformation defined in eqn (18.3.1) is to erase the first binary digit to the right of the decimal point. That is the origin of the name of the map.

Now suppose we have two possible starting values x_0 and y_0, the first N binary digits of which are the same. In other words, the difference between x_0 and y_0 is less than $1/2^N$. After N iterations x_N and y_N will depend only on the binary digits of x_0 and y_0 from the $N+1$st onward. By hypothesis, there is no relation between these digits. Thus a small perturbation in the initial condition will give rise to a large difference in the eventual motions. Furthermore, if the initial point, say x_0 is only known to finite precision (perhaps the machine precision of a computer carrying out the iteration), say M binary digits, then, after M iterations the results of the computation will be meaningless, for they will depend only on the unknown digits, from the $M+1$st on. It is also true, though we shall not prove it here, that the iterates of a given point are dense in the unit interval with uniform density.

This brief discussion of chaos in discrete maps was for purposes of introduction. We are primarily interested here not in maps but in flows generated by differential equations. Again, we shall illustrate the points necessary for our purpose by exhibiting the simplest example, a set of three differential equations called the *Rössler* model (Rössler 1976). Three is the smallest number of dimensions for which deterministic chaos can exist in differential equation flows. The Rössler model has no physical interpretation, but is very useful for providing geometrical insight. The system of equations is

$$\begin{aligned}
\dot{x} &= -(x-y) \\
\dot{y} &= x + ay \\
\dot{z} &= b + xz - cz
\end{aligned} \tag{18.3.2}$$

a, b, and c are constants. Notice that there is only one, superficially innocuous, nonlinear term; it is this term which generates the chaos.

A graph illustrating a trajectory of these equations is shown in Fig.18.3. The

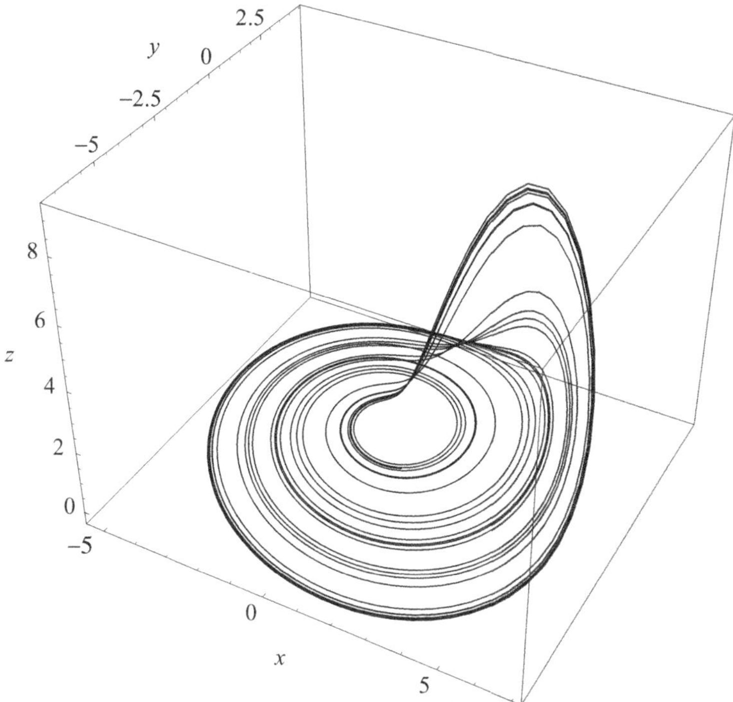

FIG. 18.3. The attractor of the system of eqn (18.3.2). The diagram was constructed by graphing a trajectory after it had moved close to the attractor. In order to show the structure of the attractor, the trajectory was terminated before it had become too dense.

graph is not itself a graph of a trajectory, but rather a graph of the *attractor*. The attractor is the set of points in configuration space towards which trajectories will tend if they start close enough. For practical purposes, therefore, it is the asymptotic trajectory. Typical examples of attractors are stable fixed points and stable periodic orbits. The attractor shown in Fig. 18.3 is neither of these. It is called a *strange attractor*. It is very intricately interwoven and is fractal in nature.

A given small volume of configuration space near the attractor is stretched, at least in some direction, as the time increases but this stretching cannot continue forever, for the attractor is confined to a finite volume. Eventually the attractor folds around and injects itself in an interleaved manner in a region where it has already been present. This combination of small scale stretching and large scale folding and reinjection is characteristic of chaotic systems.

It is important to have a way to characterize the rate of stretching quantitatively. This is furnished by quantities called *Lyapunov exponents*. We shall define these for a general autonomous system of differential equations[39]

$$\dot{\mathbf{x}} = \mathbf{F}(\mathbf{x}) \tag{18.3.3}$$

and suppose we have a solution, $\mathbf{x}(t)$, of this system. Consider a small perturbation of the initial point of this solution and see how it evolves in time. Since the initial perturbation, $\delta\mathbf{x}$, is small, we shall not attempt to resolve eqn (18.3.3) for $\mathbf{x} + \delta\mathbf{x}$. Rather, we shall linearize eqn (18.3.3) about the known solution, $\mathbf{x}(t)$ and solve the resulting linear equation

$$\delta\dot{\mathbf{x}} = (\frac{\partial \mathbf{F}(\mathbf{x}_0)}{\partial \mathbf{x}_0}) \cdot \delta\mathbf{x} \tag{18.3.4}$$

for $\delta\mathbf{x}(t)$.

A small sphere about a point (ξ_0, η_0, ζ_0) on the new trajectory will, in the course of time go over into an ellipsoid about a new point (ξ, η, ζ) on the perturbed trajectory. Let $\sigma_i(t)$ be the lengths of the principle semiaxes of the ellipsoid and σ_0 the radius of the original sphere. Then the Lyapunov exponents are defined by

$$\lambda_i = \lim_{t\to\infty} \lim_{\sigma_0\to 0} \frac{1}{t} \ln \frac{\sigma_i(t)}{\sigma_0} \tag{18.3.5}$$

The logarithm enters the definition because the rate of stretching is expected to be exponential in time and it is the exponent that is of interest. The limit $\sigma_0 \to 0$ occurs because, as time increases, the distortion of an initial sphere become more complicated than ellipsoidal unless the sphere becomes smaller. In other words, the Lyapunov exponents measure the stretching of the phase space near trajectories. The exponents may also depend on which trajectory is being considered and on which initial point is picked.

Although we have discussed Lyapunov exponents for three dimensional systems, there is no difficulty in extending the concept to any finite dimension, from one upwards.

The last concept we need to introduce is the *Kolmogorov–Sinai entropy.* This should not be confused with the thermodynamic entropy, as will be seen from its definition. It is a measure of how fast information is lost in the intricate meanderings of a dynamical system. To begin with, we divide the configuration space of the dynamical system up into cells, $\{C_\alpha\}$, where α belongs to some set of symbols used for indexing purposes. Also, divide the elapsed time up to the present into n intervals of length Δt. A coarse grained trajectory can then be specified by the sequence of symbols $\alpha_1, \alpha_2, \alpha_3, ..., \alpha_n$. This means that, at time Δt the system is in cell C_{α_1}, at time $2\Delta t$ it is in cell C_{α_2}, at time $3\Delta t$ in cell C_{α_3}, etc.

Let $I_\alpha(\mathbf{x})$ be the characteristic function (sometimes called indicator function) of the set C_α; i.e. it is the function that is unity when $\mathbf{x}$ belongs to C_α and is

[39] *Autonomous* means that the right- hand sides are independent of t.

zero otherwise. Let μ be an invariant measure associated with the flow. If the dynamical system is Hamiltonian, for example, the invariant measure is just $d\mathbf{x}d\mathbf{p}$, by Liouville's theorem. Now define

$$\mu(\alpha_1, \alpha_2, \ldots, \alpha_n) = \int d\mu I_{\alpha_1}(\mathbf{x}(\Delta t)) I_{\alpha_2}(\mathbf{x}(2\Delta t)) \ldots I_{\alpha_n}(\mathbf{x}(n\Delta t)) \qquad (18.3.6)$$

This is the n-point correlation function of the coarse grained trajectory with respect to the particular cell partition $\{C_\alpha\}$.

Now we are in a position to define the entropy of the trajectory with respect to the particular partition as

$$H(\{C_\alpha\}) = -\lim_{n\to\infty} \sum_{\alpha_1,\ldots,\alpha_n} \mu(\alpha_1, \ldots, \alpha_n) \ln \mu(\alpha_1, \ldots, \alpha_n) \qquad (18.3.7)$$

The $\mu \ln \mu$ terms in the sum are characteristic of entropy, either in thermodynamics or information theory. We have to sum over all possible states in order to obtain the average of the logarithm of the probability, which is the entropy. This definition depends on the partition of the state space into cells which we have chosen, and is, in this sense arbitrary and unnatural. In order to remove this dependence on the partition, we define the Kolmogorov–Sinai entropy as

$$H_{KS} = \sup_{\{C_\alpha\}} H(\{C_\alpha\}) \qquad (18.3.8)$$

where sup denotes *supremum*, or least upper bound.

The Kolmogorov–Sinai entropy can be shown to be zero for an ordered system, infinity for a completely random system, and a have a non zero value for a chaotic system. Its importance for our present considerations is that it provides a lower bound to the sum of the positive Lyapunov exponents of the system.

$$H_{KS} \leq \sum_{\lambda_i > 0} \lambda_i \qquad (18.3.9)$$

This non trivial fact is a theorem, which we shall not prove here (Pesin 1977).

If we could measure the Kolmogorov–Sinai entropy for an experimental system, and it turned out to be positive, this would mean that the system has at least one positive Lyapunov exponent. Hence there is exponentially fast growth of the distance between initially neighboring points in at least one direction. In other words, the system is chaotic. Unfortunately, H_{KS} is difficult to measure directly. However, Gaspard and coworkers (Gaspard *et al.* 1998) have measured a different kind of entropy, which they call the pattern entropy.

The experiment consisted of measuring the position of a Brownian particle of diameter 2.5 μm suspended in water in a small cell. The position was measured as a function of time for about 6.75 hr at intervals of 1/60 s. Thus a time series of

145,612 positions was generated. The configuration space was divided into cells of size ϵ and the pattern entropy was calculated from the time series. Unfortunately the definition and method of calculation of the pattern entropy is not clearly described in the paper. One must go to the cited literature (Grassberger and Procaccia 1983) to discover what is meant.

The point of this calculation is that the pattern entropy is a lower bound for the Kolmogorov– Sinai entropy. Recall that the Kolmogorov–Sinai entropy is a lower bound for the sum of the positive Lyapunov exponents. Consequently, if the pattern entropy is positive, then the system being studied must have at least one positive Lyapunov exponent. Consequently it is chaotic. Gaspard *et al.* assert that this demonstrates the chaotic nature of the underlying dynamics.

There have been several objections to this interpretation (Dettman et al. 1999; Grassberger and Schreiber 1999); rebuttal by Gaspard *et al.* (1999). The arguments are that sample systems that are known *not* to exhibit chaos in their basic microscopic dynamics can be shown to have the same kind of behavior in their entropy as was found by Gaspard *et al.*, 1/60 s is too long a time period to display anything about the underlying molecular dynamics, and that the metric used to compute the distance between trajectories (when exponentially fast separation is being discussed) may not be appropriate.

The situation has been clarified considerably by work of Dettman and Cohen (2000). These authors have done extensive numerical simulations on two-dimensional models in which one particle moves throughout an array of stationary particles. The time step in these calculations was microscopic, of the order of the time it takes for the moving particle to traverse a length of the size of the fixed particles. Some of these models were known on independent grounds to be chaotic, to have a positive Lyapunov exponent, while others were known to be non-chaotic, to have no positive Lyapunov exponent. Nevertheless, all of the models appeared to be chaotic according to the Grassberger–Procaccia test.

This is a practical rather than an in principle failure of the test. As a particle moves in the medium of fixed particles, it is exploring its surroundings. It has no prevision of the correlation of scatterer positions prior to collisions with the scatterers and so there is a kind of short time quasichaos. As time goes on, however, the local surroundings have been explored, and correlations between scatterers are determined. It would take an impractically long computing time to reach this regime. Yet it is important to reach this regime for computing the Kolmogorov-Sinai entropy because the definition, eqn (18.3.7), requires $n \to \infty$. Consequently the long time behavior of the system will ultimately dominate the entropy.

There are two major differences between the work of Gaspard *et al.* and that of Dettman and Cohen. First, the former is a laboratory experiment on an actual physical system, while tha latter is numerical simulations on idealized models. Second, the time interval between measurements in the experiment was macroscopic, 1/60s, while that of the simulations was microscopic. Even with data being taken on the microscopic time scale it was not possible to distinguish between chaotic and non chaotic dynamics because the required length of the time series needed for the validity of the algorithm being used was too great

It is extremely probable, even likely, that Brownian motion is chaotic on the microscopic time scale. But this has not yet been demonstrated, either theoretically or by the experiment we have just discussed. Further details of the experiment and discussion of its interpretation have been given by Briggs *et al.* (2001). There are still many open questions about how to distinguish chaotic systems in a practical way. Brownian motion may not be the best system in which to study these questions because of the latge ratio of the Brownian particle time scale to the microscopic time, the medium particle time scale.

Nevertheless, the question of the character of the microscopic dynamics of Brownian motion is a worthwhile question in its own right. We have here discussed one very interesting attempt to answer this question and have tried to expose the subtlety of the reasons the result was not defunitive. The problem is still open and a worthwhile topic for further research.

18.4 Concluding Remarks

In this chapter we have discussed some aspects of Brownian motion concerned with individual trajectories rather than average behavior. To do this we had to introduce some concepts about fractals and chaos that were not needed elsewhere in this book. Both of these areas are large fields of research, and we have given here only the minimum necessary information for understanding the relation to Brownian motion. The reader who wishes to investigate these fascinating fields further is invited to consult the following literature.

A useful introduction to the numerical treatment of stochastic differential equations can be found in the book of Honerkamp (1994). This book also contains other topics of interest for the subject of this one. For fractals, the standard reference is Mandelbrot's book (1982). Two books which try to bring together the topics of fractals and chaos are by Addison (1997) and by Peitgen *et al.* (1992). A semi-popular book with very eclectic applications is by Schroeder (1991). Finally, three books on chaos that are very close to statistical physics are Gaspard (1998), Dorfman (1999) and Beck and Schlögl (1993).

We should like to end this book by just mentioning some experiments by Libchaber *et al.* studying the behavior of Brownian particles in optical traps generated by intense laser beams (Faucheux *et al.*1995a; Faucheux *et al.* 1995b). Other experiments on the directed motion of Brownian particles are reviewed by Astumian (2000). So Brownian motion, understood for almost one hundred years, still has some surprises for us, and some interesting questions, both experimental and theoretical.

APPENDIX A

THE APPLICABILITY OF STOKES' LAW

In the transcription of the 1911–1912 lectures of H.A. Lorentz (1921) on kinetic problems, there is a section in which Lorentz considers the validity of Stokes' law for the description of Brownian motion. Since this appears to be very little known the present author's translation of the relevant section is presented here.[40]

§7. Investigation of the validity of Stokes' law for Brownian motion

One can now ask the question, what the resistance would be if the sphere had an arbitrarily changing motion. This is important for Brownian motion. One must, however, restrict oneself to low velocities, for in the calculation the terms $\mathbf{u}\partial\mathbf{u}/\partial x$ and so forth[41] were neglected. If one knew the velocity as a function of time for the duration of the experiment, then one could develop this in a Fourier series and apply the result (24).[42] Through this, the resistance will then be known. However, this is not well adapted to a general discussion.

The formula of Stokes may not be applied to Brownian motion; the motion is much too fast for that. In (24), the term $R(n\rho/2\mu)^{1/2}$ may be neglected with respect to 1 when the period, T, is large compared to $\pi\rho R^2/\mu$. Let $\pi\rho R^2/\mu = \theta$. Then, for very slow vibrations, where the period T is large with respect to θ, Stokes law may be applied, and this holds also for other motions in which the velocity changes but little in the time θ.

As an example of this, we may check whether Stokes' law may be applied to the decay of the motion of a sphere as a consequence of the friction in the fluid. If m is the mass and v the velocity of the sphere, and we apply Stokes' law, then

$$m\frac{dv}{dt} = -6\pi\mu Rv$$

so that

$$v = v_0 e^{-6\pi\mu Rt/m}$$

v drops to $1/e$ of v_0 after the time $m/6\pi\mu R$. If the density of the sphere is ρ_1 then this time is

[40]This material is presented with the permission of Koninklijke Brill NV, Leiden, The Netherlands

[41](Translator's note) This refers to the nonlinear terms in the Navier–Stokes equations for the fluid.

[42]Equation (24) of the original reads 'Thus the resistance is $-6\pi Rw_{bol}(1 + R\surd(n\rho/2\mu))$.' R is the sphere radius, μ the viscosity, ρ the fluid density, n the vibrational frequency, and $w_{bol} = w_0 \exp(int)$.

$$\tau = \frac{m}{6\pi\mu R} = \frac{2}{9}\frac{\rho_1 R^2}{\mu}$$

If ρ and ρ_1 are comparable with each other, then we see that θ and this time τ are of the same order of magnitude. Stokes' law may then not be applied to the decay of the motion.

For granules which undergo Brownian motion, θ becomes very small. If $R = 5 \times 10^{-5}$, $\mu = 18 \times 10^{-5}$ then $\theta = \pi \times 25 \times 10^{-5}/18 \times 10^{-5} \times \rho = 4 \times 10^{-5}\rho$ approximately.

Certainly the motion of suspended particles will change very strongly in this time so that the details of the Brownian motion cannot be calculated with Stokes' law. In many other cases, however, the motion will change little in the time θ and the law can be applied.[43]

[43] Since this translation was made, I have discovered a published English translation (Lorentz 1927), which is, however, not very accessible.

APPENDIX B

FUNCTIONAL CALCULUS

In this appendix we outline the calculus of functionals. A *functional* is a mapping from a set of functions to a set of numbers. It is a generalization of the idea of a function of several variables to the realm of a function of infinitely many variables, the values of the argument function. For example, $\sin(x^2)$ is not a functional. It assigns a number to every value of x, and hence is an ordinary function of x. On the other hand

$$I_1[f] = \int_0^1 f(x)dx \tag{B.1}$$

is a functional of f; it assigns a numerical value to the entire function, f. It is an analog of the finite sum Σf_i.

Functionals can simultaneously be functions of a numerical variable. One example of how this can occur is

$$I_2[f;t] = \int_0^1 g(x,t)f(x)dx \tag{B.2}$$

where g is a given function of two variables. I_2 is clearly an ordinary function of t for fixed f. For fixed t it a functional of f of the same type as I_1. Now let $f(x,t)$ be a function of two variables and consider

$$I_3[f(,t)] = \int_0^1 g(x)f(x,t)dx \tag{B.3}$$

where g is again some fixed function. This is another way that a functional can be an ordinary function at the same time. There are many other examples that could be invented of how this might occur.

Not all functionals are defined by integrals. The simplest example of one that is not is the Dirac delta 'function', $\delta(x)$. This is often described as the function that is zero everywhere except at the origin, where it is infinite in such a way that

$$\int_{-\infty}^{\infty} \delta(x)dx = 1 \tag{B.4}$$

Of course, there is no such function for any reasonable definition of the integral sign. The delta 'function' is really a functional, assigning to any function, its value at the origin. We may write

$$\delta[f] = f(0) \tag{B.5}$$

This is what is really meant by the usual notation

$$\int_{-\infty}^{\infty} f(x)\delta(x)dx = f(0) \tag{B.6}$$

There is a differential and integral calculus for functions of finitely many variables. Is there an analog for functionals? Function spaces are huge and, in general it is not possible to define measures on them in a way that would make integral calculus possible. But there are exceptional special cases. Chapter 6 was devoted to one of them. The Wiener integral is an integral of a functional over the space of continuous functions on an interval., albeit with a very special measure. The Feynman integral is also an integral of a functional over a function space. On the other hand, differentiation of functionals is possible. Suppose we take a functional $I[f]$, and ask how it changes when we make a small change in f, to $f + \delta f$. More particularly, we ask for the *linear* part of the change in I, which we denote by δI. This is the result of changes at all of the different values of x in the domain of f so we write it as

$$\delta I[f, \delta f] = \int \frac{\delta I[f]}{\delta f(y)} \delta f(y)dy \tag{B.7}$$

The quantity $\delta I[f]/\delta f(y)$ is called the *functional derivative* of I with respect to f. It is the coefficient of δf in the linear part of the change in I. Equation (B.7) is analogous to the formula for the differential of a function of several variables $\delta f = \Sigma(\partial f/\partial x_i)\delta x_i$. In general, the functional derivative $\delta I/\delta f(y)$ is both a functional of f and an ordinary function of y.

Let $\delta f(y)$ be of the special form $\epsilon h(y)$ where h differs from zero only in the interval $(y - \mu, y + \mu)$ and be such that $\int h(x)dx = 1$. Because the functional derivative is supposed not to depend on the increment δf, the integral on the right of eqn (B.7) may be approximated by

$$\delta I[f, \delta f(y)] = \epsilon \frac{\delta I}{\delta f(y)} \tag{B.8}$$

Thus

$$\frac{\delta I[f]}{\delta f(y)} = \lim_{\epsilon \to 0, \mu \to 0} \frac{\delta I[f, \epsilon h]}{\epsilon} \tag{B.9}$$

Since $h(y)$ approximates the delta functional when used as an integral kernel, eqn (B.9) is often written as

$$\frac{\delta I}{\delta f(y)} = \lim_{\epsilon \to 0} \frac{I[f + \epsilon\delta(x - y)] - I[f]}{\epsilon} \tag{B.10}$$

This formula leads to a formal expression that is often useful in evaluating functional derivatives expressed in the form of definite integrals

$$\frac{\delta f(x)}{\delta f(y)} = \delta(x - y) \tag{B.11}$$

This is as far as we need to go in the differential calculus of functionals for the purposes of this book. There is an extensive literature on the functional calculus that can be consulted by the reader who wishes to study the subject further (Volterra 1959; Lévy 1941).

APPENDIX C

AN OPERATOR IDENTITY

We derive the operator identity given in eqn (10.4.10). Consider an operator $A + B$. In the case of eqn (10.4.10), A can be taken as $i(1 - \wp)L$ and B as $i\wp L$. Define an operator $J(s)$ by

$$e^{(A+B)s} = e^{As} J(s) \tag{C.1}$$

Differentiate with respect to s, obtaining

$$(A + B)e^{(A+B)s} = e^{As}\dot{J} + Ae^{As}J(s) \tag{C.2}$$

Canceling and rearranging,

$$\dot{J} = e^{-As}Be^{As}J \tag{C.3}$$

From the definition, it is clear that $J(0) = 1$, so that integration of this equation yields

$$J(s) = 1 + \int_0^s e^{-Au}Be^{Au}J(u)du \tag{C.4}$$

Finally, multiplying by $\exp(-As)$ and using the definition of J, we get the final result

$$e^{(A+B)s} = e^{As} + \int_0^s e^{A(s-u)}Be^{(A+B)u}du \tag{C.5}$$

This is the operator identity that we used in Chapter 10.

Alternatively, if we had defined $K(s)$ by

$$e^{(A+B)s} = K(s)e^{As} \tag{C.6}$$

similar manipulations would have yielded the alternative operator identity

$$e^{(A+B)s} = e^{As} + \int_0^s e^{(A+B)u}Be^{A(s-u)}du \tag{C.7}$$

APPENDIX D

EULER ANGLES

In this appendix we discuss the Euler angles specifying the orientation of a rigid body in space. Erect an orthogonal coordinate system (x''', y''', z''') fixed in the body, and a coordinate system (x, y, z) fixed in space. The orientation of the body is specified by the angles relating these two sets of coordinates. A particularly useful set of three independent angles has been found by Euler. They are defined as follows.

Start with the body fixed axes (triple primed) coinciding with the space fixed axes (unprimed). We move the body to its final orientation in three steps. First, rotate the body, and consequently the body fixed axes, by an angle α about the z axis. Call the new position of the body fixed axes (x', y', z'). The second step is to rotate the body about the y' axis by an angle β. The new position of the body fixed axes are called $x'', y'', z'')$. Note that the y' and y'' axes are the same. Finally, rotate the body about the z'' axis by an angle γ. The new position, (x''', y''', z''') is the final position of the body fixed axes. Thus the overall effect is to express the total rotation in terms of the product (sequential rotations) of rotations by the three angles (α, β, γ). Figure D.1 shows the relative positions of the initial, intermediate, and final coordinate systems.

The rotations α, β, and γ are carried out about body fixed axes, although the body rotates and the body fixed axes change their positions with respect to the space fixed axes during the course of the overall rotation.

We now want to outline the transition from eqn (15.3.1) to eqn (15.3.2). The operators $\partial/\partial\xi_i$ are the generators of infinitesimal rotations about the body fixed (triple primed) axes. They transform like vectors. $\partial/\partial\alpha$, $\partial/\partial\beta$, and $\partial/\partial\gamma$ generate infinitesimal rotations about the z, y' and z''' axes respectively. They also transform like vectors. From Fig. D.1 it can be seen that

$$\partial/\partial\alpha = \cos\beta\partial/\partial\xi_3 - \sin\beta\cos\gamma\partial/\partial\xi_1 + \sin\beta\sin\gamma\partial/\partial\xi_2 \tag{D.1}$$

$$\partial/\partial\beta = \sin\gamma\partial/\partial\xi_1 + \cos\gamma\partial/\partial\xi_3$$

$$\partial/\partial\gamma = \partial/\partial\xi_3 \tag{D.2}$$

These equations can be solved for the $\partial/\partial\xi_i$, yielding

$$\begin{pmatrix} \partial/\partial\xi_1 \\ \partial/\partial\xi_2 \\ \partial/\partial\xi_3 \end{pmatrix} = \begin{pmatrix} -\frac{\cos\beta}{\sin\gamma} & \sin\gamma & \cot\beta\cos\gamma \\ \sin\gamma & \cos\gamma\sin\beta & -\sin\gamma\cos\beta \\ 0 & 0 & \sin\beta \end{pmatrix} \begin{pmatrix} \partial/\partial\alpha \\ \partial/\partial\beta \\ \partial/\partial\gamma \end{pmatrix} \tag{D.3}$$

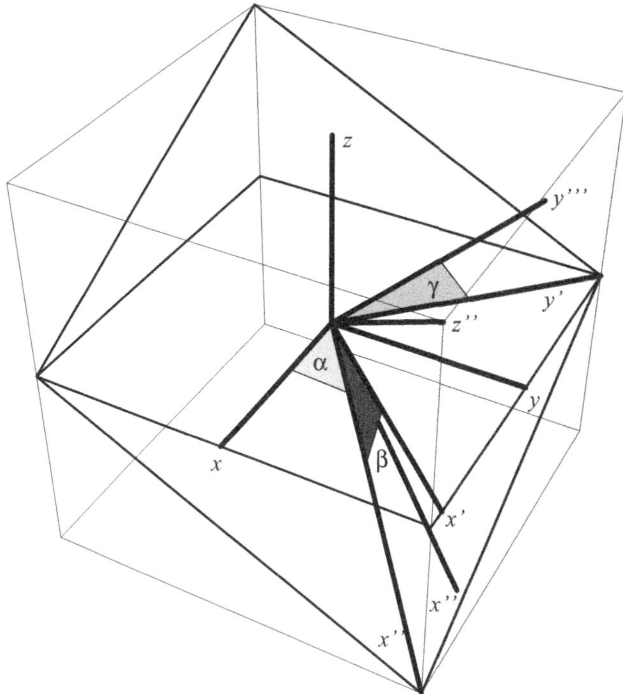

FIG. D.1. The Euler angles α, β, and γ specifying the position of the body-fixed axes (x''', y''', z''') with respect to the space-fixed axes (x, y, z).

It is now a routine though tedious matter to compute the right-hand side of eqn (15.3.1) in terms of the variables α, β and γ. The result is eqn (15.3.2) of the main text when $D_1 = D_2 \neq D_3$. The general case, when all of the Ds are unequal, can be handled by the same method.

APPENDIX E

THE OSEEN TENSOR

What is the fluid velocity at position $\mathbf{r}'$ caused by the application of a point force applied at position $\mathbf{r}$ in a fluid described by the linearized Navier–Stokes equations for an incompressible fluid? This is the question we seek to answer in this appendix. We study here only the time independent case.

The equations governing the fluid flow are

$$\begin{aligned} -\nabla p + \eta \nabla^2 \mathbf{v} + \mathbf{F}\delta(\mathbf{r}) &= 0 \\ \nabla \cdot \mathbf{v} &= 0 \end{aligned} \tag{E.1}$$

The notation is: p is the pressure, $\mathbf{v}$ is the fluid velocity, η the viscosity, and ρ the mass density of the fluid. $\mathbf{F}$ is the magnitude of the point force applied at $\mathbf{r}$. Thus, finding $\mathbf{v}$ is the same as finding the Green's function for the linearized time-independent Navier–Stokes equations. We proceed by taking the Fourier transform of eqns (E.1).

The notational convention of this appendix is to denote the Fourier transform of a function by the same letter as the original function but with the Fourier transform variable as a subscript. That is, the Fourier transform of $f(\mathbf{r})$ will be denoted by $f_{\mathbf{k}}$. Taking the Fourier transform of eqns (E.1), we obtain

$$\begin{aligned} -i\mathbf{k}p_{\mathbf{k}} - \eta k^2 \mathbf{v}_{\mathbf{k}} + \mathbf{F} &= 0 \\ i\mathbf{k} \cdot v_{\mathbf{k}} &= 0 \end{aligned} \tag{E.2}$$

Take the scalar product of the first equation of (E.2) with $\mathbf{k}$ and use the second equation to obtain

$$p_{\mathbf{k}} = -\frac{i\mathbf{k} \cdot \mathbf{F}}{k^2} \tag{E.3}$$

This can be used to eliminate $p_{\mathbf{k}}$ from eqn (E.2) with the result

$$\mathbf{v}_{\mathbf{k}} = \frac{1}{\eta k^2}[\mathbf{1} - \frac{\mathbf{kk}}{k^2}] \cdot \mathbf{F} \tag{E.4}$$

It only remains to transform this back into $\mathbf{r}$ space.

The inverse Fourier transform of k^{-2} is $1/4\pi R$. The inverse Fourier transform of $\mathbf{kk}/k^2$ is $(1/8\pi R)(\mathbf{1} - \mathbf{RR}/R^2)$. The Fourier transforms of these singular functions are actually distributions, or generalized functions. This is not too surprising since the equation we started with contains a distribution, the Dirac delta. Performing the inverse transform of eqn (E.4) using these two inverse transforms, we arrive at

$$\mathbf{v}(\mathbf{r}') = \mathbf{T}(\mathbf{r} - \mathbf{r}') \cdot \mathbf{F}$$
$$\mathbf{T}(\mathbf{R}) = \frac{1}{8\pi\eta R}[\mathbf{1} + \frac{\mathbf{RR}}{R^2}]$$

T is known as the *Oseen tensor.*

The same technique can be used to derive the Oseen tensor when the point force is changing in time (Chang and Mazo 1976). We have to replace zero on the right side of eqn (E.1) by $\partial \mathbf{v}/\partial t$ and take the Fourier transform in time also. We shall not need this generalization in this book.

APPENDIX F

MUTUAL DIFFUSION AND SELF-DIFFUSION

In this appendix we want to discuss the difference between mutual diffusion and self-diffusion from the point of view of the thermodynamics of irreversible processes. A fuller discussion of this topic can be found in Tyrrell and Harris (1984).

F.1 Mutual Diffusion

We first consider mutual diffusion in a two component mixture. Component 0 denotes the solute and component 1 denotes the solvent. Let $\mathbf{J}_i$ denote the diffusion flux of component i in the barycentric coordinate system. That is, the flux is measured relative to the local center of mass motion of the fluid. The basic description of diffusion is Fick's law

$$\mathbf{J}_i = \sum_j L_{ij} \nabla \mu_j \tag{F.1.1}$$

Because we are in the barycentric coordinate system, $\mathbf{J}_0 + \mathbf{J}_1 = 0$ and only one of the fluxes is independent. It then follows that $L_{00} = -L_{01}$. The relation between the flux $\mathbf{J}_0$ and the thermodynamic forces, the gradients of chemical potential, is then

$$\mathbf{J}_0 = -L_{00}(\nabla \mu_0 - \nabla \mu_1) \tag{F.1.2}$$

Using the Gibbs–Duhem relation $c_0 \nabla \mu_0 + c_1 \nabla \mu_1 = 0$, eqn (F.2) can be written

$$\begin{aligned} \mathbf{J}_0 &= -L_{00} \nabla \mu_0 \cdot (1 + \frac{c_0}{c_1}) = \frac{L_{00}}{x_1} \nabla \mu_0 \\ &= -\frac{L_{00}}{x_1} (\frac{\partial \mu_0}{\partial c_0})_{T,P} \nabla c_0 = -D_m \nabla c_0 \end{aligned} \tag{F.1.3}$$

where x_1 is the mole fraction of component 1. Consequently the relationship between the phenomenological coefficient L_{00} and the mutual diffusion coefficient, is

$$D_m = \frac{L_{00}}{x_1} (\frac{\partial \mu_0}{\partial c_0})_{T,P} \tag{F.1.4}$$

F.2 Self-Diffusion

Now we consider the same system except that now some of the solute molecules (component 0) are tagged somehow. The tagged mo;ecules are denoted as component α and the untagged solute molecules as component β. The fractions of tagged and untagged molecules are y_α and y_β respectively. $c_\alpha + c_\beta = c_0$. It is assumed that molecules of type α and of type β are identical except for tagging.

The fluxes of the solute species can be written as

$$\begin{aligned} -\mathbf{J}_\alpha &= l_{\alpha\alpha}(\nabla\mu_\alpha - \nabla\mu_1) + l_{\alpha\beta}(\nabla\mu_\beta - \nabla\mu_1) \\ -\mathbf{J}_\beta &= l_{\beta\alpha}(\nabla\mu_\alpha - \nabla\mu_1) + l_{\beta\beta}(\nabla\mu_\beta - \nabla\mu_1) \end{aligned} \tag{F.2.1}$$

In other words, the tagged system is treated as a three component mixture. The l coefficients are the phenomenological coefficients appropriate to the discussion of the three component system. The chemical potentials of the tagged species are

$$\begin{aligned} \mu_\alpha &= \mu_0 + RT\ln y_\alpha \\ \mu_\beta &= \mu_0 + RT\ln y_\beta \end{aligned} \tag{F.2.2}$$

Here, μ_0 is the chemical potential of component zero. It is not a standard state chemical potential; there is no implication that the solution is ideal.

In a self diffusion experiment, the concentrations of species 0 and 1 are uniform. Thus $\nabla\mu_0 = \nabla\mu_1 = 0$. Since $y_\alpha + y_\beta = 1$, $\nabla y_\alpha = -\nabla y_\beta$. Consequently

$$\begin{aligned} -\mathbf{J}_\beta &= RT(\frac{l_{\beta\beta}}{c_\beta} - \frac{l_{\beta\alpha}}{c_\alpha})\nabla c_\beta \\ -\mathbf{J}_\alpha &= RT(\frac{l_{\alpha\beta}}{c_\beta} - \frac{l_{\alpha\alpha}}{c_\alpha})\nabla c_\beta \end{aligned} \tag{F.2.3}$$

Writing $\mathbf{J}_\beta = -D_s\nabla c_\beta$ as the definition of the self-diffusion coefficient, D_s we find that

$$D_s = RT(\frac{l_{\beta\beta}}{c_\beta} - \frac{l_{\beta\alpha}}{c_\alpha}) \tag{F.2.4}$$

F.3 Relation between D_m and D_s

Now we return to the general situation where components 0 and 1 are not necessarily uniform in space. Since $\mathbf{J}_0 = \mathbf{J}_\alpha + \mathbf{J}_\beta$, we have

$$\begin{aligned} -\mathbf{J}_0 = {} & l_{\alpha\alpha}(\nabla\mu_0 + RT\ln y_\alpha) + l_{\alpha\beta}(\nabla\mu_0 + RT\ln y_\beta) \\ & + l_{\beta\alpha}(\nabla\mu_0 + RT\ln y_\alpha) + l_{\beta\beta}(\nabla\mu_0 + RT\ln y_\beta) \end{aligned}$$

$$= (l_{\alpha\alpha} + 2l_{\alpha\beta} + l_{\beta\beta})\nabla\mu_0 + RT(\frac{l_{\alpha\alpha}}{c_\alpha} - \frac{l_{\alpha\beta}}{c_\beta} + \frac{l_{\beta\alpha}}{c_\alpha} - \frac{l_{\beta\beta}}{c_\beta})c_0\nabla y_\alpha \tag{F.3.1}$$

where we have used the Onsager relation $l_{\alpha\beta} = l_{\beta\alpha}$. For this equation to be compatible with eqn (F.1.3) the coefficient of $\nabla\mu_0$ must be L_{00}/x_1 and the coefficient of ∇y_α must vanish. These conditions yield

$$\frac{L_{00}}{x_1} = l_{\alpha\alpha} + 2l_{\alpha\beta} + l_{\beta\beta}$$

$$0 = \frac{l_{\alpha\alpha}}{c_\alpha} + 2l_{\alpha\beta}(\frac{1}{c_\alpha} - \frac{1}{c_\beta}) - \frac{l_{\beta\beta}}{c_\beta} \tag{F.3.2}$$

where we have again used the Onsager relations. The second of eqns (F.3.2) can be used to eliminate $l_{\alpha\beta}$ from the first, and we find

$$D_m = \frac{l_{\alpha\alpha} - l_{\beta\beta}}{y_\alpha - y_\beta}(\frac{\partial\mu_0}{\partial c_0})_{T,P} \tag{F.3.3}$$

The equation for D_s becomes

$$D_s = (\frac{RT}{c_0})\frac{y_\alpha^2 l_{\beta\beta} - y_\beta^2 l_{\alpha\alpha}}{y_\alpha y_\beta(y_\alpha - y_\beta)} \tag{F.3.4}$$

To go any further it is necessary to know more about the y dependence of the l coefficients. Irreversible thermodynamics is silent on the subject of the dependence of phenomenological coefficients on thermodynamic state. It just assumes that the coefficients are independent of gradients so that the theory is linear. It is clear, however, that there is no a priori reason why eqns (F.3.3) and (F.3.4) should be the same. First of all, (F.3.3) contains the thermodynamic derivative $(\partial\mu_0/\partial c_0)_{T,P}$ as a factor while (F.3.4) does not. Secondly, the remaining factors are not necessarily the same. In the limit $c_0 \to 0$, however, the thermodynamic derivative does approach RT/c_0. Nevertheless, non thermodynamic arguments are needed to conclude that $D_m \to D_s$ as $c_0 \to 0$, that is, at infinite dilution of solute.

REFERENCES

Addison, P. S. 1997. *Fractals and chaos.* Inst. of Physics Publishers, Bristol.

Adelman, S. A. 1976. Fokker–Planck equations for simple non-Markovian systems. *J. Chem. Phys.* **64**, 124.

Alexandrof., P. 1961. *Elementary concepts of topology.* Dover, NY.

Altenberger, A.R. and Deutch, J. M. 1973. Light scattering from dilute macromolecular solutions. *J. Chem. Phys.* **59**, 894.

Aguirre, J. L. and Murphy, T. J. 1971. Diffusion equation for a pair of Brownian particles. *Phys. Fluids* **14**, 2050.

Aguirre, J. L. and Murphy, T. J. 1973. Brownian motion of N interacting particles II. Hydrodynamic evaluation of the diffusion tensor matrix. *J. Chem. Phys.* **59**, 1833.

Arnold, L. 1973. *Stochastische Differentialgleichungen, Theorie und Anwendungen.* R. Oldenbourg, Munich.

Astumian, R. D. 2000. Rachets, rectifiers, and demons: the constructive role of noise in free energy and signal transduction. In *Self-organized biological dynamics and non linear control.* Ed. J. Walleczek, Cambridge University Press, Cambridge.

Balazs, N. 1959. Statistical mechanics of strings. *Phys. Rev.* **113**, 1178.

Barnes, R. B. and Silverman, S. 1934. Brownian motion as a natural limit to all measuring processes. *Rev. Mod. Phys.* **6**, 162.

Bartlett, M. S. 1955. *An introduction to stochastic processes.* Cambridge University Press, Cambridge.

Batchelor, G. K. 1976. Brownian diffusion of particles with hydrodynamic interaction. *J. Fluid Mech.* **74**, 1.

Batchelor, G. K. 1977. The effect of Brownian motion on the stress of a suspension of spherical particles. *J. Fluid Mech.* **83**, 97.

Beck, C. and Schlögl, F. C. 1993. *Thermodynamics of chaotic systems, an introduction.* Cambridge University Press, Cambridge.

Ben-Naim, E., Redner, S., and ben-Avraham, D. 1992. Bimodal distribution in bimodal shear flows. *Phys. Rev.* **45** 7207.

Berg, H. C. 1983. *Random walks in biology.* Princeton University Press, Princeton.

Berichte der Bunsen Gesellschaft 1991. **95**, 225 *et seq.* (entire issue).

Berne, B. J. 1971. Time-dependent properties of condensed media. In *Physical chemistry, an advanced treatise.* v. VIII B. Ed. D. Henderson. Academic Press, NY.

Berne, B. 1977. Projection operator techniques in the theory of fluctuations. In *Modern theoretical chemistry* v. 6, *Statistical mechanics*, part B. Ed. B. Berne.

Plenum Press, NY.

Berne, B. J. and Pecora, R. 1979. *Dynamic light scattering with applications to chemistry, biology, and physics.* Wiley, NY.

Bernoulli, J. 1713. *Ars conjectandi.* Basel.

Bird, R. B., Hassager, O., Armstrong, R. C., and Curtiss, C. J. 1977. *Dynamics of polymeric liquids.* v. 2, *Kinetic theory.* Wiley, NY.

Bocquet, L. and Piasecki, J. 1997. Microscopic derivation of non-Markovian thermalization of of a Brownian particle. *J. Stat. Phys.* **87** , 1005.

Bocquet, L. 1998. From a stochatic to a microscopic approach to Brownian motion. *Acta Phys. Pol.* **B 29**, 1551.

Boguliubov, N. N. 1946. *Problems of a dynamical theory in statistical physics.* Gostechizdat (in Russian). English translation in *Studies in statistical mechanics*, v.1. Ed. J. de Boer and G. E. Uhlenbeck. North Holland, Amsterdam, 1962.

Böttcher, C. J. F. 1952. *Theory of electric polarization.* Elsevier, Amsterdam.

Brenner, H. and Edwards, D. A. 1993. *Macrotransport phenomena.* Butterworth, Boston, MA.

Briggs, M. E., Sengers, J. V., Francis, M. K., Gaspard, P., Gammon, R. W., Dorfman, J. R., Calabrese, R. V. 2001. Tracking a colloid particle for the measurement of dynamic entropies. *Physica* **A296**, 2001.

Brown, R. 1828. *A brief account of microscopical observations ,made in the months of June, July, and August, 1827 on the particles contained in the pollen of plants; and on the general existence of active molecules in organic and inorganic bodies.* Privately printed. Reprinted in *Edin. New Phil. J.* 1828, **5**, 358–371.

Brown, R. 1829. Additional remarks on active molecules. *Edin. J. Sci.* **1**, 314.

Browne, E. J. 1950. *Charles Darwin:Voyaging, Volume 1 of a biography.* Knopf, NY.

Brush, S. G. 1961. Functional integrals and statistical physics. *Rev. Mod. Phys.* **33**, 79.

Brush, S. 1968. A history of random processes I. Brownian movement from Brown to Perrin. *Arch. Hist. Exact Sci.* **5**, 1.

Burgers, J. M. 1938. *Second report on elasticity and plasticity.* Ch. III, Amsterdam Academy of Sciences. Nordeman, Amsterdam.

Cameron, R. H. 1960/61. A family of integrals serving to connect the Feynman and Wiener integrals. *J. Math. and Phys.* **39**, 126.

Cartier, P. 1997. (with C. De Witte-Morette, A. Wurm, and D. Collins). A rigorous mathematical foundation of functional integration. In *Functional integration, basics and applications.* Ed. C. Dewitt-Morette, P. Cartier, and A Folacci. Plenum Press, N Y.

Cerf, R. and Sheraga, H. A. 1952. Flow birefringence in solutions of macromolecules. *Chem. Rev.* **51**, 185.

Chandrasekhar, S. 1943. Stochastic processes in physics and astronomy. *Rev. Mod. Phys.* **15** ,1.

Chang, E. L. and Mazo, R. M. 1976. Line widths in the vibrational spectra; Polymers in dilute solution. *J. Chem.Phys.* **64**, 1389.

Chang, E. L., Mazo, R. M., and Hynes, J. T. 1974. On the Fokker–Planck equation for a linear chain. *Mol. Phys.* **28**, 1004.

Chapman, S. and Cowling, T. G. 1952. *The mathematical theory of non-uniform gases.* Cambridge University Press, Cambridge.

Chaudisaignes. 1908. Le mouvement brownien et la formule d'Einstein. *C. R. Acad. Sci. Paris.* **147**, 1044.

Cichocki, B. Felderhof, B. U., and Schmitz, R. 1988. Hydrodynamical interactions between two spherical particles. *Physico Chem. Hydro.* **10**, 383.

Cichocki, B. and Felderhof, B. U. 1988. Short time diffusion coefficient and high frequency viscosity of dilute solutions of spherical Brownian particles. *J. Chem. Phys.* **89**, 1049.

Cichocki, B. and Felderhof, B. U. 1989. Sedimentation and self diffusion in suspensions of spherical particles. *Physica.* **A154**, 213.

Cichocki, B., Felderhof, B. U., Hinsen, K., Wajnryb, E., and Bladziewicz, J. 1994. Friction and mobility of many spheres in Stokes flows. *J. Chem. Phys.* **100**, 3780.

Coffey, W. T., Kalmykov, Yu. P., and Waldron, J. T. 1996. *The Langevin equation with applications in physics, chemistry and electrical engineering.* World Scientific, Singapore.

Collins, F. C. and Kimball, G. 1949. Diffusion-controlled reaction rates. *J. Coll. Sci.* **4**, 425.

Courant, R. and Hilbert, D. 1937. *Methoden der Matematische Physik*, Band II. Springer-Verlag, Berlin.

Cukier, R. I. and Deutch, J. M. 1969. Microscopic theory of Brownian motion; the multiple time scale point of view. *Phys. Rev.* **197,** 240.

Czech, R. and Kehr, K. W. 1986. Depolarization of rotating spins by random walks on lattices. *Phys. Rev.* **B34** 261.

Darwin, F. ed. 1896. *Charles Darwin: his life told in an autobiographical chapter and a selected series of his collected letters.* D. Appleton, NY.

de Gennes, P.-G. 1979. *Scaling concepts in polymer physics.* Cornell University Press, Ithaca.

de Haas-Lorentz, G. L. 1913. *Die Brownsche Bewegung.* Vieweg.

Debye, P. 1942. Reaction rates in ionic solutions. *Trans. Electrochem. Soc.* **82,** 65.

Debye, P. 1929. *Polar molecules.* Chemical Catalog Co., NY, reprinted by Dover, NY.

Debye, P. and Bueche, A. M. 1948. Intrinsic viscosity, diffusion and sedimentation rate of polymers in solution. *J. Chem. Phys.* **16**, 573.

Dettman, C. P., Cohen, E. G. D. and van Beijeren, H. 1999. Microscopic chaos from Brownian motion? *Nature.* **401**, 875.

Dettman, C. P. and Cohen, E. G. D. 2000. Microscopic chaos and diffusion. *J. Stat. Phys.* **101**, 775.

Deutch, J. M. and Oppenheim, I. 1971, Molecular theory of Brownian motion for several particles. *J. Chem. Phys.* **54**, 3547.

Deutch, J. M. and Silbey, R. 1971. Exact generalized Langevin equation for a particle in a harmonic lattice. *Phys. Rev.* **A3**, 2049.

Dhont, J. G. H. 1996. *An introduction to the dynamics of colloids.* Elsevier, Amsterdam.

Doi, M. and Edwards, S. F. 1987. *The theory of polymer dynamics.* Oxford University Press, Oxford.

Doob, J. L. 1953. *Stochastic processes.* Wiley, NY.

Doob, J. L. 1942. The Brownian movement and stochastic equations. *Ann. Math.* **43**, 351.

Dorfman, J. R. 1999. *An introduction to chaos in non-equilibrium statistical mechanics.* Cambridge University Press, Cambridge.

Dvoretsky, A. and Erdos, P. 1950. Some problems on random walk in space. In *2nd Berkeley symposium on statistics and probability.* Ed. J. Neyman. California, Berkeley.

Dunlap, D. H., La Violette, R. A., and Parris, P. E. 1994. The spatial evolution of particles diffusing in the presence of randomly placed traps. *J. Chem. Phys.* **100**, 8293.

Edmonds, A. R., 1957. *Angular momentum in qunatum mechanics.* Princeton University Press, Princeton, NJ.

Einstein, A. 1905. Die von der Molekularkinetischen Theorie von Wärme geforderte Bewegung von in ruhenden Flüssigkeiten suspendierten Teilchen. *Ann. Phys.* **17**, 549.

Einstein, A. 1906a. Eine neue Bestimmung der Molekuldimensionen. *Ann. Phys.* **19**, 289.

Einstein, A. 1906b. Zur Theorie der Brownschen Bewegung. *Ann. Phys.* **19**, 371.

Einstein, A. 1907. Theoretische Bemerkungen über die Brownsche Bewegung. *Zeit. f. Electrochem.* **13**, 98.

Einstein, A. 1908. Elementaire Theorie der Brownsche Bewegung. *Zeit. f. Electrochem.* **14**, 235.

Einstein, A. 1910. Theorie der Opalescenz der homogenen Flüssigkeiten und Flüssigkeitmischen in der Nähe des kritisches Zustandes. *Ann. d. Phys.* **33**, 1275.

Einstein, A., 1911. Berichtigung zu meiner Arbeit: Eine neue Bestimmung der Molekuldimensionen. *Ann. Phys.* **34**, 591.

Einstein, A. 1917. Marian von Smoluchowski. *Naturwiss.* **50**, 107.

Eliot, G. 1989. *Middlemarch.* Penguin Books,NY.

Erdélyi, A. 1956. *Asymptotic expansions.* Dover, NY.

Erpenbeck, J. and Kirkwood, J. G. 1963. Erratum:Statistical mechanics of irteversible processes in polymer solutions [J. Chem. Phys. **29**, 909, (1958)]. *J. Chem. Phys.* **38**, 1023.

Ermak, D. L. and McCammon, J. A. 1978. Brownian dynamics with hydrodynamic interactions. *J. Chem. Phys.*. **69**, 1352.
Exner, F. 1900. Notiz zu Brown's Molekularbewegung. *Ann. d. Phys.* **2**, 843.
Faucheux, L. P., Bourdieu, L. S., Kaplan, P. D., and Libchaber, A. D. 1995a. Optical thermal rachet. *Phys. Rev. Lett.* **74**, 1504.
Faucheux, L. P., Stolovitsky, G, and Libchaber, A. D. 1995b. Periodic forcing of a Brownian particle. *Phys. Rev. E.* **51**, 5239.
Favro, L. D. 1960. Theory of the rotational Brownian motiuon of a free rigid body. *Phys. Rev.* **119**, 53.
Favro, L. D. 1965. Rotational Brownian motion. In *Fluctuation phenomena in solids.* ch. 3. Ed. R. E. Burgess. Academic Press, NY.
Felderhof, B. U. 1978. Diffusion of interacting Brownian particles. *J. Phys.* **A11**, 929.
Felderhof, B. U. 1988. Many-body hydrodynamic interactions in suspensions. *Physica* **A151**, 1.
Felderhof, B. U. and Vogel, J. 1992a. Wave number- and frequency-dependent diffusion coefficient of interacting Brownian particles. *Physica.* **A183**, 54.
Felderhof, B. U. and Vogel, J. 1992b. Long time collective diffusion coefficient of semi-dilute suspensions of spherical Brownian particles. *J. Chem. Phys.* **96**, 6978.
Feller, W. 1950. *An introduction to probability theory and its applications*, v1. Wiley, NY.
Feller, W. 1966. *An introduction to probability theory and its applications*, v2. Wiley, NY.
Feynman, R. P. 1950. Space-time approach to non-relativistic quantum mechanics. *Rev. Mod. Phys.* **20**, 367.
Feynman, R. P. and Hibbs, A. J. 1965. *Quantum mechanics and path integrals.* McGraw-Hill, NY.
Fixman, M. 1978. Simulation of protein dynamics. I General theory. *J. Chem. Phys.* **69**. 1527.
Fokker, A. D. 1914. Die mittlere Energie rotierenden electrische Dipole im Strahlungsfeld. *Ann. d. Phys.* **43**, 810.
Frisch, H. L. and Simha, R. 1956. The viscosity of colloidal suspensions and macromolecular solutions, in *Rheology, theory and applications*, v. 1. Ed. F. Eirich. Academic Press, NY.
Fürth, R. 1920. Die Brownsche Bewegung mit Berücksichtigung einer Persistenz der Bewegungsrichtung. Mit Anwendungen auf der Bewegung lebender Infusorien. *Zeits. f. Phys.* **2**, 244.
Gallagher, W. H. and Woodward, C. K. 1989. The concentration dependence of the diffusion coefficient of bovine pancreatic trypsin inhibitor: a dynamic light scattering study of a small protein. *Biopolymers.* **28**, 2001.
Gard, T. C. 1988. *Introduction to stochastic differential equations.* Marcel Dekker, NY.
Gardiner, C. W. 1983. *Handbook of stochastic processes.* Springer-Verlag, Berlin.

Gaspard, P., Briggs, M. E., Francis, M. K., Semgers, J. V., Gammon, R. W., Dorfman, J. R., and Calabrese, R. W. 1998. Experimental evidence for microscopic chaos. *Nature*. **394**, 865.

Gaspard, P., Briggs, M. E., Francis, M. K., Semgers, J. V., Gammon, R. W., Dorfman, J. R., and Calabrese, R. W. 1999. Microscopic chaos from Brownian motion? *Nature*, **401**, 876.

Gaspard, P. 1998. *Chaos, scattering, and statistical mechanics.* Cambridge University Press, Cambridge.

Gel'fand, I. M. and Yaglom, A. M. 1960. Integration in function space and its application to quantum physics. *J. Math. Phys.* **1**, 48.

Gerlach, W. and Lehrer, E. 1927. Über die Messung der Brownsche Bewegung mit Hilfe einer Drehwaage. *Naturwiss.* **15**, 15.

Ghikman, I. F. and Skorohod, A. V. 1975. *The theory of stochastic processes.* v.I. Springer-Verlag, New York.

Gillespie, D. T. 1992. *Markov processes, an introduction for physical scientists.* Academic Press, San Diego, CA.

Gnedenko, B.V. 1978. *The theory of probability.* Mir, Moscow.

Golay, M. J. E. 1949. The theoretical and practical sensitivity of the pneumatic infrared detector. *Rev. Sci. Inst.* **20**, 816.

Goldstein, H. 1951. *Classical mechanics.* Addison-Wesley, Cambridge, MA.

Gradshteyn, I. S. and Ryzhik, I. M. 1980. *Tables of integrals, series, and products.* Academic Press, NY.

Granger, C. W. J. and Morgenstern, O. 1970. *Predictability of stock market prices.* Heath, Lexington, MA.

Grassberger, P. and Procaccia, I., 1982. The long time properties of diffusion in a medium with static traps. *J. Chem. Phys.* **77**, 6281.

Grassberger, P. and Procaccia, I. 1983. Estimation of the Kolmogorov entropy from a chaotic signal. *Phys. Rev.* **A28**, 2591.

Grassberger, P. and Schreiber, T. 1999. Microscopic chaos from Brownian motion? *Nature.* **401**, 876.

Halmos, P. R. 1950. *Measure theory.* Van Nostrand, NY.

Hänggi, P, Talkner, P. and Borkovec, M. 1990. Reaction rate theory; fifty years after Kramers. *Rev. Mod. Phys.* **62**, 251.

Happel, J. and Brenner, H. 1973. *Low Reynolds number hydrodynamics.* Noordhoff, Leiden.

He, S. and Sheraga, H. 1998. Macromolecular conformational dynamics in torsion angle space. *J. Chem. Phys.* **108**, 721.

Helfand, E. 1979. Flexible *vs.* rigid constraints in statistical mechanics. *J. Chem. Phys.* **71**, 5000.

Hess, W. and Klein, R. 1983. Generalized hydrodynamics of systems of Brownian particles. *Adv. Phys.* **32**, 173.

Honerkamp, J. 1994. *Stochastic dynamical systems.* VCH, NY.

Hubbard, P. S. 1972. Rotational Brownian motion. *Phys. Rev.* **A6**, 2421.

Hughes, B. D. 1995. *Random walk and random environments. v.1 Random walks.* Oxford University Press, Oxford.

Hughes, B. D. 1996. *Random walks and random environments. v.2. Random environments.* Oxford University Press, Oxford.

Huygens, C. 1657. *De rationciniis in ludo alea.* The Hague.

Hynes, J. T. and Deutch, J. M. 1975. Nonequilibrium problems - projection operator techniques. In *Physical chemistry, an advanced treatise,* v. XIB. Ed. D. Henderson and H. Eyring. Academic Press, NY.

Ikeda, Y. 1956. Theory of irreversible processes of high polymer solutions-II. General discussion of diffusion constants and diffusion equation (in Japanese). *Kobayashi Rigaku Kenkusho Hokoku* (*Bull. Kobayashi Inst. of Phys. Research*). **6**, 44.

Ingen-Houtz, J. 1789. Remarks on the use of the microscope. In *Vermischte Schriften physisch-medizinschen Inhalts.* C. F. Wappler, Vienna.

Iranpour, R. and Chacon, P. 1988. *Basic stochastic processes, the Mark Kac lectures.* Macmillan, NY.

Ising, G., 1926. A natural limit for the sensibility of galvanometers. *Phil. Mag.* **1**, 827.

Ito, K. 1944. Stochastic integral. *Proc. Imp. Acad. Tokyo.* **20**, 519.

Kac, M. 1959. *Probability and related topics in the physical sciences.* Interscience, NY.

Kac, M. 1966. Wiener and integration in function space. *Bull. Am. Math. Soc.* **72** Part II, 52.

Kappler, E. 1931. Versuche zur Messung der Avogadro–Loschmidtchen Zahl aus der Brownsche Bewegung einer Drehwaage. *Ann. d. Phys.* **11**, 233.

Kappler, E. 1932. Experimentalle Untersuchungen über den zeitlichen Verlauf der Brownsche Bewegung einer Drehwaage. *Ann. d. Phys.* **15**, 1545.

Kehr, K. W. and Kuttner, R. 1982. A random walk on a random walk. *Physica* **A110**, 535.

Kerker, M. 1974. Brownian motion and molecular reality prior to 1900. *J. Chem. Ed.* **51**, 764.

Keyes, T. and Oppenheim, I. 1973. Bilinear hydrodynamics and the Stokes–Einstein law. *Phys. Rev.* **A8**, 937.

Kirkwood, J. G. 1949. The statistical mechanical theory of irreversible processes in solutions of flexible macromolecules; viscoelastic behavior. *Rec. Trav. Chim. Pays-Bas.* **68**, 649.

Kirkwood, J. G. 1953. The general theory of irreversible processes in solutions of macromolecules. *J. Poly. Sci.* **12**, 1.

Kirkwood, J. G. and Riseman, J. 1948. The intrinsic viscosities and diffusion coefficients of flexible macromoleculles in solution. *J. Chem. Phys.* **16**, 565.

Klein, R. 1997. Interacting Brownian particles: the dynamics of colloidal suspensions. In *The physics of complex fluids*, corso cxxiv, International School of Physics 'Enrico Fermi'. Ed. F. Malamace and H. E. Stanley. IOS Press, Amsterdam.

Klein, O. 1922. Zur Statistische Theorie der Suspensionen und Lösungen. *Arkiv f. Math., Astr., och Phys.* **16**, 1.

Kolmogorov, A. N. 1950. *Foundations of the theory of probability.* Chelsea, NY.

Kops-Werkhoven, M. M. and Fijnaut, H. M. 1981. Dynamic light scattering and sedimentation experiments on silica dispersions at finite concentrations. *J. Chem. Phys.* **74**, 1618.

Kramers, H. A. 1940. Brownian motion in a field of force and the diffusion model of chemical reactions. *Physica* **7** 284.

Kramers, H. A. 1946. The behavior of macromolecules in inhomogeneous flow. *J. Chem. Phys.* **14**, 415.

Kuhn, W. 1934. Über die Gestalt fadenformigen Moleküle in Lösungen. *Koll. Zeit* **68** 2.

Kuhn, W. and Kuhn, H. 1945. Abhangigkeit der Viskosität von Stromungsgefälle bei hochverdünnten Suspensionen und Lösungen. *Helv. Chim. Acta.* **28**, 97.

Landau, L. D. and Lifshitz, E. M. 1959. *Fluid mechanics.* Trans. by J. B. Sykes and W. H. Reid. Pergamon Press, NY.

Landau, L. D. and Lifshitz, E. M. 1986. *Theory of elasticity.* 3rd ed. Trans. by J. B. Sykes and W. H. Reid. Pergamon, Oxford.

Lawson, J. L. and Uhlenbeck, G. E. 1950. *Threshold signals.* McGraw-Hill, NY.

Langevin, P. 1908. Sur la theorie de mouvement brownien. *C. R. Acad. Sci. Paris.* **146**, 530.

Lax, M. 1966. Classical noise IV: Langevin methods. *Rev. Mod. Phys.* **38**, 541.

Lebowitz, J. and Rubin, E. 1963, Dynamical study of Brownian motion. *Phys. Rev.* **131** , 2381.

Lebowitz, J. and Résibois, P. 1965. Microscopic theory of Brownian motion in an oscillating fluid; connection with macroscopic theory. *Phys. Rev.* **139 A** , 1101.

Lévy, P. 1941. *Problèmes concrètes d'analyse functionelle.*2nd ed. Gauthier-Villars, Paris.

Lorentz, H. A. 1921. *Lessen over theoretische naturkunde*, v. V, *Kinetische problemen (1911–1912).* Ed. E. D. Bruins and J. Reudler. E. J. Brill, Leiden.

Lorentz, H. A. 1927. *Lectures on theoretical physics.* Trans. by Silberstein, L. and Trivelli, A. P. H. Macmillan, London.

Mabberly, D. J. 1985. *Jupiter Botanicus, Robert Brown of the British Museum.* J. Cramer, Braunschweig.

MacDonald, D. K. C. 1948. Spontaneous fluctuations. *Rep. Prog. Phys.* **12** ,56.

Mandelbrot, B. 1982. *The fractal geometry of nature.* W. H. Freeman, San Francisco.

Margenau, H. and Murphy, G. M. 1943. *The mathematics of physics and chemistry.* Van Nostrand, NY.

Markov, A. A. 1907. Investigation of a remarkable case of dependent trials (in Russian). *Izv. Akad. Nauk St. Petersburg.* **6**,61.

Marcus, A. H., Lin, B. and Rice, S. A. 1996. Self-diffusion in dilute quasi-two-dimensional hard sphere suspensions. Evanescent wave light scattering and video microscopy studies. *Phys. Rev.* **E53**, 1765.

Martin, Ph. A. 1996. *Une initiation á l'integrale functionelle en physique quantique et statistique.* Presses Polytech. et Univ. Romand, Lausanne.

Masoliver, J. and Weiss, G. H. 1992. First passage times for a generalized telegrapher's equation. *Physica* **A183**, 537.

Matheron, G. and de Marsily, G. 1980. Is transport in porous media always diffusive? A counterexample. *Water Resources Res.* **16**, 901.

Mazo, R. M. 1969a. On the theory of Brownian motion: I. Interaction between Brownian particles *J. Stat. Phys.* **1** , 89.

Mazo, R. M. 1969b. On the theory of Brownian motion II. Nonuniform systems. *J. Stat. Phys.* **1**, 101.

Mazo, R. M. 1969c. On the theory of Brownian motion III. Two body distribution function. *J. Stat. Phys.* **1**, 559.

Mazo, R. M. 1971. Theory of Brownian motion IV. a hydrodynamic model for the friction factor. *J. Chem. Phys.* **54**, 3912.

Mazo, R. M. 1973. Theory of Brownian motion VI. asymptotic solution of a retarded Fokker–Planck equation. *J. Chem. Phys.* **58**, 108.

Mazo, R. M. 1975. On the theory of Brownian motion VIII. A Brownian particle in a dilute gas with inelastic collisions. *J. Stat. Phys.* **12**, 427.

Mazo, R. M. 1978. Aspects of the theory of Brownian motion. In *Lecture Notes in Physics, No. 84, Stochastic processes in nonequilibrium systems.* Ed. L. Garrido, P. Seglar, and P. J. Shepard . Springer, Berlin.

Mazo, R. M. and Van den Broeck, C. 1986. Spectrum of an oscillator making a random walk in frequency. *Phys. Rev.* **A34**, 2364.

Mazur, P. and Openheim, I. 1970. Molecular theory of Brownian motion. *Physica* **50**, 241.

McCombrie, C. W. 1953. Fluctuation theory in physical measurements. *Rep. Prog. Phys.* **16,** 266.

Michaels, I. A. and Openheim, I. 1975. Long-time tails and Brownian motion. *Physica.* **81A**, 221.

Montroll, E. M. 1952. Markov chains, Wiener integrals, and quantum theory. *Comm. Pure App. Math.* **5**, 415.

Montroll, E. W. and Schlesinger, M. F. 1984, On the wonderful world of random walks. In *Studies in statistical mechanics, v.11, Non-equilibrium phenomena from stochastics to hydrodynamics.* Ed. J. L Lebowitz and E. W. Montroll. Elsevier, NY.

Montroll, E. W. and Weiss, G. H. 1965. Random walks on lattices II. *J. Math. Phys.* **6**, 167.

Moran, P. A. P. 1948. The statistical distribution of the length of a rubber molecule. *Proc. Camb. Phil. Soc.* **44**, 342.

Mori, H. 1965a. Transport, collective motion, and Brownian motion. *Prog. Theo. Phys.* **33**, 423.

Mori, H. 1965b. A continued fraction representation of the time correlation functions *Prog. Theo. Phys.* **34**, 399.

Morse, P. M. and Feshbach, H. 1952. *Methods of theoretical physics.* v. 1, McGraw-Hill, NY.

Murphy, T. J. and Aguirre, J. L. 1972. Brownian motion of N interacting particles I. Extension of the Einstein relation to the N-particle case. *J. Chem. Phys.* **57**, 2098.

Nelson, E. 1964. Feynman integrals and the Schrödinger equation. *J. Math. Phys.* **5**, 332.

Northrup, S. H., Allison, S. A. and McCammon, J. A. 1984. Brownian dynamics simulation of diffusion-influenced bimolecular reactions. *J. Chem. Phys.* **80** , 1517.

Noyes, R. M. 1961. Effects of diffusion rates on chemical kinetics. In *Progress in reaction kinetics.* v. 1. Pergamon Press, Oxford.

Nyquist, H. 1928. Thermal agitation of electrical charges in conductors. *Phys. Rev.* **32**, 110.

Oppenheim, I., Shuler, K., and Weiss, G. 1977. *Stochastic processes in chemical physics: the master equation.* MIT Press, Cambridge.

Ornstein, L. S. 1920. The Brownian movement. *Proc. Amst. Acad. Sci.* **21**, 96.

Osborne, M. P. M. 1959. Brownian motion in the stock market. *Op. Res.* **7**, 145.

Ostwald, W. 1909. *Grundriss der algemeinen Chemie.*4. Auflage. Verlag von Wilhelm Engelmann, Leipzig.

Paley, R. E. A. C. and Wiener, N. 1934. *Fourier transforms in the complex domain.* American MathMathematical Society, NY.

Pawula, R. F. 1967. Approximation of the linear Boltzmann equation by the Fokker–Planck equation. *Phys. Rev.* **162**, 186.

Pearson, K. 1905. The problem of the random walk. *Nature.* **72**, 294.

Perrin, J. 1908a. La loi de Stokes et le mouvement brownien. *C. R. Acad. Sci. Paris.* **147**, 475.

Perrin, J. 1908b. L'origine de mouvement brownien. *C. R. Acad. Sci. Paris.* **147**, 530.

Perrin, J. 1909. Mouvement brownien et realité moleculair. *Ann. de Chim. et de Phys.* **18**, 1.

Perrin, F. 1926a. Vie moyenne de l'état actif des molecules fluorescentes. *C. R. Acad. Sci. Paris.* **182**, 219.

Perrin, F., 1926b, Polarization de fluorescence et vie moyenne des molecules excitées. *J. de Phys. et la Rad.* **7**, 930.

Perrin, F. 1934. Mouvement Brownien d'une ellipsoide (I) Dispersion dielectrique pour les molecules ellipsoidales. *J. de Phys. et la Rad.* **5**, 497.

Pesin, Ya. B. 1977. Characteristic Lyapunov exponents and smooth ergodic theory. *Russian Math. Surveys.* **32**:4, 55.

Peitgen, H-O, Jürgens, H., and Saupe, D. 1992. *Chaos and fractals, new frontiers of science.* Springer, NY.

Planck, M. 1917. Über einen Satz der Statistischen Dynamik und seine Erweiterung in der Quantentheorie. *Sitz. Konig. Preuss. Akad. Wiss.* **1**, 324.

Poincaré, H. 1908. *Thermodynamique.* 2nd ed. Gauthier, Paris.

Rappaport, D. C. 1985. The fractal nature of molecular trajectories in fluids. *J. Stat. Phys.* **40**, 751.

Rayleigh 1905. The problem of the random walk. *Nature* **72**, 318.

Résibois, P. and Davis, H. T. 1964. Transport equation of a Brownian particle in an external field. *Physica* **30**, 1077.

Rice, S. A. 1985. Diffusion-limited reactions. In *Comprehensive chemical kinetics.* v. 25. Ed. C. H. Bamford, C. R. H. Tipper, and R. G. Compton. Elsevier, Amsterdam.

Risken, H. 1989. *The Fokker–Planck equation: methods of solution and applications.* 2nd ed. Springer-Verlag, Berlin.

Rössler, O. 1976. An equation for continuous chaos. *Phys. Lett.* **57A**, 397.

Rouse, P. E. Jr. 1953. A theory of the linear viscoelastic properties of coiling polymers. *J. Chem. Phys.* **21**, 1272.

Rozanov, Yu. A. 1987. *Introduction to random processes.* Springer-Verlag, Berlin.

Rubin, R. J. 1960. Statistical dynamics of simple cubic lattices. Model for the study of Brownian motion. *J. Math. Phys.* **1**, 309.

Rubin, R. J. 1961. Statistical dynamics of simple cubic lattices. Model for the study of Brownian motion II. *J. Math. Phys.* **2**, 373.

Rubin, R. J. 1968. Momentum autocorrelation function for a heavy particle in a finite crystal. **90**, 3061.

Samoletov, A. A. 1999. A remark on the Kramers problem. *J. Stat. Phys.* **96**, 1351.

Schroeder, M. 1991. *Fractals, chaos, and power laws. Minutes from an infinite paradise.* W. H. Freeman, NY.

Schofield, J., Marcus, A. H., and Rice, S. A. 1996. Dynamics of quasi two-dimensional colloidal systems. *J. Phys. Chem.* **100,** 18950.

Schuss, Z. 1980. *Theory and applications of stochastic differential equations.* Wiley, NY.

Shulman, L. S. 1981. *Techniques and applications of path integration.* Wiley, NY.

Sirovich, L. 1971. *Techniques of asymptotic analysis.* Springer, NY.

Smoluchowski, M. 1906a. Kinetische Theorie der Brownsche Bewegung und der Suspensionen. *Ann. D. Phys.* **21**, 756. Polish and French versions were published earlier that year in the Bulletin of the Krakow Academy of Sciences.

Smoluchowski, M. 1906b. Sur le chemin moyenne parcouru par les molécules des gaz et sur son rapport la theorie de la diffusion. *Bull. Internat. Acad. Krakovie.* 202.

Smoluchowski. M. 1908. Molekularkinetische Theorie der Opalescenz von Gasses im Kritische Zustand, sowie einige verwandte Erscheinungen. *Ann. d. Phys.* **25**, 205. Polish and French versions were published earlier in the Bulletin of the Krakow Academy of Sciences.

Smoluchowski, M. 1913. Einige Beispiele Brownsche Molekularbewegung unter Einfluss äusserer Kräfte. *Bull. Int. de l'Acad. de Cracovie, Classe de Sci. math-nat.* **A** 418–434.

Smoluchowski, M. 1915. Über Brownsche Molekularbewegung unter Einwirkung äusser Kräfte und deren Zusammenhang mit der veralgemeinerten Diffusionsgleichung. *Ann. d. Phys.* **48**, 1103.

Smoluchowski, M. 1916. Drei Vorträge über Diffusion, Brownsche Molekularbewegung und Koagulation der Kolloidteilchens. *Phys. Zeit.* **17** 557, 587.

Smoluchowski, M. 1917. Versuch einen mathematischen Theorie der Koagulationkinetik kolloider Losüngen. *Zeit. f. Phys. Chem.* **92**, 129.

Sneddon, I. 1957. *Partial differential equations.* McGraw-Hill, New York.

Sommerfeld, A. 1917. Zum Andenken an Marian von Smoluchowski. *Physik. Zeit.* **22**, 533.

Sommerfeld, A. 1952. *Electrodynamics.* Trans. by E. G. Ramberg. Academic Press, NY.

Spitzer, F. 1964. *Principles of random walk.* Van Nostrand, Princeton.

Stratonovich, R. 1966. A new representation for stochastic integrals and equations. *SIAM J. Control.* **4**, 632.

Svedberg, Th. 1906. Über die Eigenbewegung der Teilchen in colloidalen Lösungen. *Zeit. für Elektrochemie.* **12**, 853, 909.

Taylor, G. I. 1953. Dispersion of soluble matter in a fluid flowing slowly through a tube. *Proc. Roy. Soc. Lond.* **A219**, 186.

Taylor, G. I. 1954. Conditions under which dispersion of a solute can be used to measure molecular diffusion. *Proc. Roy. Soc. Lond.* **A225**, 473.

Tchen, C. M. 1952. Random flight with multiple partial correlation. *J. Chem. Phys.* **20**, 214.

Teske, A. 1977. *Marian Smoluchowski, Leben und Werk.* Ossolineum, Wroclaw. This is a German translation of a book originally published in 1955 in Polish.

Titchmarch, E. C. 1939. *The theory of functions.* 2nd ed. Oxford University Press, Oxford.

Titulaer, U. M. 1978. A systematic solution procedure for the Fokker–Planck equation of a Brownian particle in the high friction case. *Physica* **91A**, 321.

Tokuyama, M. and Oppenheim, I. 1978. Statistical mechanical theory of Brownian motion-translational motion in an equilibrium fluid. *Physica* **94 A**, 501.

Turq, P., Lantelme, F. and Friedman, H. 1977. Brownian dynamics: its application to ionic solutions. *J. Chem. Phys.* **66**, 3039.

Tyrrell, H. N. V. and Harris, K. R. 1984. *Diffusion in liquids, a theoretical and experimental study.* Butterworths, London.

Uhlenbeck, G. E. and Goudsmit, S. 1929. A problem in Brownian motion. *Phys. Rev.* **34**, 145.

Uhlenbeck, G. E. and Ornstein, L. S. 1930. On the theory of Brownian motion. *Phys. Rev.* **36**, 823.

Ulam, S. 1957. Marian Smoluchowski and the theory of probability in physics. *Am. J. Phys.* **25**, 475.

Van den Broeck, C. 1988. Drunks, drift, and dispersion, Aggregation thesis, Faculty of Science, Vrije Universiteit Brussel.

Van der Pas, P. W. 1971. The discovery of Brownian motion. *Scien. Historiae* **13**, 17.

Van Kampen, N. G. 1965. Fluctuations in non-linear systems, in *Fluctuation phenomena in solids*, Ed. R. E. Burgess. Academic Press, NY.

Van Kampen, N. G. 1981a. *Stochastic processes in physics and chemistry.* North Holland, Amsterdam

Van Kampen, N. G. 1981b. Itô versus Stratonovich. *J. Stat. Phys.* **24**, 175.

Van Kampen, N. G. 1985. Elimination of fast variables. *Phys. Rep.* **124**, 69.

Van Kampen, N. G. 1986. Brownian motion on a manifold. *J. Stat. Phys.* **44,** 1.

Van Kampen, N. G. and Oppenheim, I. 1986. Brownian motion as a problem of eliminating fast variables. *Physica.* **138A**, 231.

.Volterra, V. 1959. *Theory of functionals and of integral and integrodifferential equations.* Dover, NY.

Wajnryb, E. and Dahler, J. 1997. The Newtonian viscosity of a moderately dense suspension. *Adv. Chem. Phys.* v. 102. Ed. I. Prigogine and S. A. Rice. Wiley, NY.

Wang, M. C. and Uhlenbeck, G. E. 1945. On the theory of Brownian motion II. *Rev. Mod. Phys.* **17**, 323.

Wang Chang, C. S. and Uhlenbeck, G. E. 1970. In *Studies in statistical mechanics.* v. 5. Ed. de Boer, J. and Uhlenbeck, G. E.. North Holland, Amsterdam. (reprinted from an unpublished University of Michagan report, 1956).

Wax, N. 1954. *Selected papers on noise and stochastic processes.* Dover, NY.

Weber, G. 1966. Polarization of the fluorescence of solutions. In *Fluorescence and phosphorescence analysis.* Ed. D. M. Hercules. Interscience, NY.

Weiss, G. H. 1994. *Aspects and applications of the random walk.* North Holland, Amsterdam.

Weiss, G. H. and Rubin, R. J. 1983. Random walks, theory and selected applications. In *Adv. Chem Phys.*, v. LI, Ed. S. Rice and I. Prigogine. Wiley, NY.

Westhaver, J. W. 1947. Concentration of potassium by countercurrent electromigration: some theoretical aspects of the operation. *J. Res. Nat. Bur. Stand.* **38**, 169.

Wilemski, G. 1976. On the derivation of Smoluchowski equations with corrections in the classical theory of Brownian motion. *J. Stat. Phys.* **14**, 153.

Wilson, E. B. Jr. 1952. *An introduction to scientific research.* McGraw Hill, NY.

Wolf, D. 1996. Stochastic differential equations. In *Advances in computer simulation.* Ed. J. Kertesz and I. Kondor. Springer-Verlag, Berlin.

Yamakawa, H. 1971. *Modern theory of polymer solutions.* Harper and Row, NY.

Zhou, H-K 1998. Comparison of three Brownian dynamics algorithms for calculating rate constants of diffusion-influenced reactions. *J. Chem. Phys.* **108,**

8139.

Zhou, H-X , Wong, K-Y, and Vajayakumar 1997. Design of fast enzymes by optimizing interaction potential in active site. *Proc. Nat. Acad. Sci. U.S.A.* **94**, 12372.

Zimm, B. H. 1956. Dynamics of polymer molecules in dilute solution: viscoelasticity, flow birefringence and dielectric loss. *J. Chem. Phys.* **24** 269.

Zsigmondy, R. 1917. Über Koagulation und Teilchenattraktion. *Zeit. für Phys. Chem.* **92**, 600.

Zwanzig, R. W. 1960. Ensemble method in the theory of irreversibility. *J. Chem. Phys.* **33**, 1388.

Zwanzig, R. W. 1961. Statistical mechanics of irreversibility. In *Lectures on theoretical physics* Ed. W. E. Britten and B. W. Downs. Interscience, NY.

Zwanzig, R. W., Kieffer, J. and Weiss, G. 1968. Validity of the Kirkwood-Riseman theory. *Proc. Nat. Acad. Sci. U.S.A.* **60**, 381.

INDEX

The manufacturer's authorised representative in the EU for product safety is Oxford University Press España S.A. of El Parque Empresarial San Fernando de Henares, Avenida de Castilla, 2 - 28830 Madrid (www.oup.es/en or product.safety@oup.com). OUP España S.A. also acts as importer into Spain of products made by the manufacturer.
Printed and bound by CPI Group (UK) Ltd, Croydon, CR0 4YY
06/07/2026
02157632-0007